GIS Processing of Geocoded Satellite Data

Springer
London
Berlin
Heidelberg
New York
Barcelona
Hong Kong
Milan
Paris
Santa Clara
Singapore
Tokyo

Jonathan Williams

GIS Processing of Geocoded Satellite Data

Springer

Published in association with
Praxis Publishing
Chichester, UK

PRAXIS

Jonathan Williams
GSS Consultant
Hutchison 3G Ltd
Maidenhead
Berkshire
UK

SPRINGER–PRAXIS BOOKS IN GEOPHYSICAL SCIENCES
SUBJECT *ADVISORY EDITOR*: Dr Philippe Blondel, C.Geol., F.G.S., Ph.D., M.Sc.,
Senior Scientist, Department of Physics, University of Bath, Bath, UK

ISBN 1-85233-368-5 Springer-Verlag Berlin Heidelberg New York

British Library Cataloguing-in-Publication Data
GIS processing of geocoded satellite data. –
(Springer-Praxis books in geophysical sciences)
1. Cartography – Remote sensing. 2. Geographic information systems
I. Title
526

ISBN 1-85233-368-5

Library of Congress Cataloging-in-Publication Data
GIS processing of geocoded satellite data / Jonathan Williams.
p. cm. – (Springer-Praxis books in geophysical sciences)
Includes bibliographical references (p.).
ISBN 1-85233-368-5 (alk. paper)
1. Cartography–Remote sensing. 2. Geographic information systems. I. Title. II. Series.

GA102.4.R44 W56 2001
526–dc21 2001041179

Printed by MPG Books Ltd, Bodmin, Cornwall, UK

Copy editing by Alex Whyte
Cover design: Jim Wilkie
Typesetting: BookEns Ltd, Royston, Herts., UK

Printed on acid-free paper supplied by Precision Publishing Papers Ltd, UK

This book is dedicated to my parents, my wife Judith and our two boys Mark and Adam

Contents

List of illustrations

Colour section appears between pages 206 and *207*

Front cover: Vegetation image of India (copyright CNES 1998, SPOT Image Distribution)
Back cover: SRTM coverage globes (courtesy NASA/JPL)

List of tables

Author profile

Jonathan Williams has been associated with research and applications development involving satellite data since 1978. Since that time he has worked on a number of important GIS projects including the UK ERS Processing and Archive Facility, the Integrated Geographic Information System (IGIS), the European ASTRON and Galileo programmes and as a GIS consultant to a variety of clients in government, utilities, environment and transport sectors. His previous work for Wiley-Praxis, *Geographic Information from Space*, was published in 1995. Jonathan is currently a consultant with the mobile multimedia company, Hutchison 3G, specialising in global spatial services.

Acknowledgements

I would like to thank my former colleagues in Logica, members of the British Association of Remote Sensing Companies (BARSC) and my many associates throughout the GIS community for their support and encouragement over the past few years. In particular I would like to extend my thanks to Damian Baker, Jim Barton, Stuart Brand, Peter Burnett, Phil Capp, Martin Cole, Jeremy Colls, Andy Coote, Bernard Denore, Lionel Elliott, Duncan Ferns, John Fuller, Mike Grimmett, Ray Harris, Neil Hubbard, Dan Isaac, Mike Jackson, Nikki Jones, Paul Kamoun, Michalis Ketselides, 'Rinco' de Koeijer, Roman Krawec, Richard Lenton, Jesus Meilan, Martin Milnes, Stuart Mills, David Morten, Mark Norman, Pat Norris, Dave Parker, Graham Peake, Francesco Pignatelli, Nigel Press, Lois Proud, Richard Proud, Clare Pryor, Roger Robinson, Mike Romaniuk, Richard Saull, Nic Snape, Mike Steven, Zof Stott, Matthew Stuttard, Martin Symonds, Neil Tubman, Richard Turner, Tony Underwood, Nick Veck, Mike Wooding and Sheena Wing.

I am also grateful for the fantastically creative environment being provided by Hutchison 3G and am pleased to confirm that the opinions expressed in the book are mine alone and should not be taken as representing an official Hutchison viewpoint. My apologies to any deserving soul who I may have forgotten, also to the many contributors of illustrations for this book; sadly far too many to thank you all individually.

The author would also like to pay tribute to the fine teas of Messrs Brooke Bond, the fine ales of the Hog's Back Brewery, and the even finer music of Mr Neil Young which have collectively helped to sustain me through many long evenings during the production of this manuscript. Also, my heartfelt thanks to my family for having the good sense to take avoiding action when Daddy starts shouting at the computer.

Enjoy!

1

Geocoding basics

In the six years since the publication of *Geographic Information from Space* (Williams 1995) there have been a number of major changes in the world of Earth Observation which have led to the need for a new title. These have included:

- true integration of GIS and image-processing systems;
- development of a more commercial market for EO applications;
- the impact of the internet, mobile telecommunications and other technological advances;
- the development of GPS into a mass-market technology;
- better information extraction techniques for satellite data;
- a plethora of new satellites and sensors including dedicated environmental programmes and the development of very high resolution (VHR) imagery which can discriminate objects as small as 1 metre.

This introductory chapter begins by introducing image-geocoding methods, and describing mathematical models for geocoding and particular techniques needed for Synthetic Aperture Radar (SAR). This is followed by the Global Positioning System (GPS), image enhancement techniques, a description of geocoded image maps and mosaics and, finally, a description of commercially available image products and processing levels.

1.1 IMAGE GEOCODING METHODS

1.1.1 Why geocoding is necessary

In an ideal world images taken from an orbiting satellite would be acquired directly in a usable map projection (UTM for instance) ready for immediate integration with other spatially referenced data. In reality, however, satellite images are subject to a number of effects which introduce distortion into the image. For example:

- the satellite platform experiences changes in orbit and attitude;
- the mechanics of the sensor do not perform to a linear specification;

Table 1.1 Categorisation of geometric errors

Category	Description
Orbit	Changes in satellite orbit Changes in satellite attitude High-frequency attitude variation (jitter)
Sensor	Sensor mechanics non-linearities Detector misalignment Band-to-band misregistrations Variations in swath width and timing
Ground	Earth curvature Earth rotation Local topographic effects

- the Earth is curved and rotates beneath the satellite while an image is formed;
- variations in terrain relief cause perspective shifts.

Consequently the image will not be acquired in a geometry that corresponds to a standard map projection, which means that the terrestrial location of any point or feature on an image is not known accurately. Additional problems include the inability to directly relate the image to other spatially referenced data (for example printed map sheets) and the fact that images taken of the same area at different times will not be co-registered.

Sources of geometric error in satellite images can be divided into three main categories: orbit related, sensor related and ground related. Examples are given in Table 1.1.

Another possible approach to categorisation of errors (Friedmann *et al.* 1983) is to divide them into low-, medium- and high-frequency errors (where frequency is defined relative to time taken to image a scene) as follows:

- **low** – where effects are much less rapid than scene-imaging time, such as those caused by variations in the satellite orbit and the Earth geoid;
- **medium** – where effects are of the same order of magnitude as the scene imaging time, for example attitude variations;
- **high** – where effects are much more rapid than scene-imaging time for example sensor scan rate or local Earth terrain variations.

To account accurately for all the possible geometric effects requires a precise mathematical model for each component contributing to the distortion; in practice, however, it may not be possible to formulate an exact model of terrain, for example, or to obtain all the parameters needed (for example, detailed satellite ephemeris or sensor engineering data). The use of such models is described in Section 1.2.

1.1.2 Bulk geocoding

For many applications the approximate transformation afforded by bulk geocoding may prove sufficient. Initially, the Landsat Thematic Mapper (TM) will be used to illustrate the basic concepts of image geometry, as this is widely used throughout the Earth Observation community. The Earth Resources Technology Satellite (ERTS) Program launched the first of a series of satellites (ERTS-1) in 1972. Part of the National Aeronautics and Space Administration's (NASA) Earth Resources Survey Program, the ERTS Program and the ERTS satellites were later renamed Landsat to better represent the civil satellite program's prime emphasis on remote sensing of land resources. Landsats 1, 2, and 3 carried the Multi-spectral Scanner (MSS) sensor and experimental Return Beam Vidicon (RBV) cameras. The Landsat 4 satellite carried the MSS and Thematic Mapper (TM) sensors as does the still currently flying Landsat 5 satellite. The sixth satellite in the Landsat series was unsuccessfully launched and did not achieve orbit. The Landsat 7 satellite carries the Enhanced Thematic Mapper Plus (ETM+) sensor. Sensor enhancements include the addition of the panchromatic band and improved resolution for the thermal band. A technical specification for Landsat can be found in Section 1.7.1.

Later in the chapter other sources of satellite images such as SPOT and NOAA-AVHRR are also discussed. Bulk geocoding uses relatively simple models of the satellite orbit and sensor geometry in order to determine the terrestrial location of each pixel. The key components of the orbit model is the satellite position at a given moment in time and its attitude (the direction in which the platform is pointing). The position of the sensor relative to the platform will be known, together with its orientation. These, along with a simple model of the Earth's surface (such as a sphere or an ellipsoid) provide the basis of the mathematical model.

One fundamental category of distortion to which all images are subject is that caused by the shape of the Earth. For bulk geocoding purposes the Earth can be regarded as a sphere or an ellipsoid (more accurate models for the Earth's geometry are described in Section 1.2). Another terrestrial parameter that must be accounted for is the Earth rotation effect. This is caused by the fact that during the time taken to acquire an image the Earth will have rotated relative to the satellite ground track. By considering the relative velocity of the satellite vector to the Earth's rotation vector, the magnitude of the effect can be calculated and a simple correction can be derived from this to slip the image every few lines, which leads to the familiar parallelogram appearance of Landsat images.

The parameters used as input to the bulk geocoding model are the satellite ephemeris data and the satellite/sensor engineering data, such as angle of intermediate field of view (IFOV) and sensor geometry. Applying the transform will result in an image in a nominal map projection (although inaccuracies, for example, in orbit timings will persist through the model into the map projection).

The bulk image processing for Landsat TM consists of three steps:

1. Ingest the raw image data.
2. Apply the bulk geometric correction algorithm.
3. Generate user products. (Clark 1990)

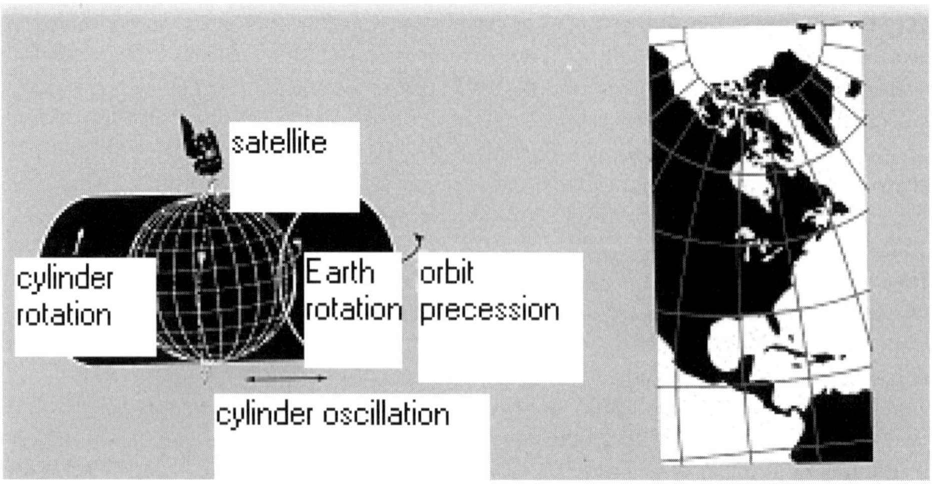

Fig. 1.1 Space Oblique Mercator projection (courtesy USGS).

The default map projection used for geocoded products is the satellite-centred Space Oblique Mercator (SOM) with standard cartographic projections such as Universal Transverse Mercator (UTM) or Polar Stereographic available on request. SOM was invented by Colvocoresses of the United States Geological Survey in 1974 and is particularly suited for the geometric processing of satellite images (Brooks *et al.* 1981). The SOM, shown in Fig. 1.1, is a conformal mapping along the ground track of the satellite and uses the along-track and across-track directions as axes. The geocoded image is resampled from the original image values using a cubic convolution method (see Section 1.2.2 for further details).

Bulk geocoding will usually result in typical errors of the order of 1 km or less for Landsat – recent improvements produce an accuracy of 300 m. It should be noted that most bulk geocoded images are intrinsically very geometrically accurate, a typical length distortion being 0.15 per cent (Westin 1990), and if the relative error offset (due to miscalculation of the origin) is subtracted, much smaller absolute errors may be obtained. This can, for example, be performed by using a single ground control point (see Section 1.1.3) or overlaying coastline data (see Section 1.2.4). Such an approach is necessary if the data is to be overlaid with other images or incorporated with digital maps or other spatially-referenced data in a GIS (see Chapter 4 for a fuller description).

1.1.3 Ground Control Point methods

A more accurate geocoded product can be obtained by identifying a network of Ground Control Points (GCPs). A GCP (Davison 1986) is a feature which can be readily identified on both the image to be transformed and the map being used to establish its geographic location. Suitable GCPs will contrast strongly with their background and should not be too large in extent that it precludes accurate determination of their location. Ideal GCPs include such linear features as motorway junctions, railway crossings and runways. Coastal features such as islands and headlands are suitable as long as there is no noticeable effect due to tidal variation. Similarly, lakes and reservoirs can often be used as long as changes in water level do not effect the apparent location of the GCP. Features subject to seasonal or annual variation should be avoided – for example, woodland boundaries are often clearly identifiable on an image but may not still be in the location shown when the reference map was last updated.

The geographic location of GCPs can be established in various ways depending on the accuracy required and the equipment available for ground surveys. The position in image coordinates is usually determined using a Geographic Information System (GIS). The imaging capabilities of a GIS are very much system dependent but will usually include the abilities to pan and zoom images and perform various image enhancement operations (see Section 1.5 for examples). After the operator has read in the image and displayed it on the GIS, the cursor is used to select a sub-image containing a GCP. The sub-image is then displayed on the GIS and, using the cursor, a coarse location for the GCP is selected. At this stage it is convenient to magnify the area around the coarse GCP location and use the cursor to select a precise location for the GCP. This is repeated until a satisfactory distribution of GCPs has been achieved. The simplest method of determining the map location is to measure by eye (for example, using a roamer marked with a map grid) from a standard map. More accurate results, however, may be obtained by using a digitising tablet. This is an electronic device that uses a mouse or puck equipped with crosswires, which is moved across the map, attached to the surface of the tablet. The puck is placed above the GCP to be measured and the exact location determined, typically by using a fine wire grid embedded in the tablet or by means of electrical wave phase. Alternative methods are to use a digitised version of the map on the display (this is very prevalent nowadays as the boundaries between image-processing systems and GIS become blurred) and the GPS, which is described in Section 1.4.

1.1.4 Calculating the transformation matrix

The simplest geometric model used is where the GCPs form a representative sample of the entire image; and a transformation matrix determined from them is therefore applicable to the entire image. The usual model assumed is a two-dimensional polynomial. After the set of GCPs has been determined for an image, two lists of control point locations will have been determined – the first describes their locations in map coordinates, and the second in image coordinates. The transformation matrix describes the equation that relates one set to the other. For example, using a linear polynomial:

$$\mathbf{X} = M\mathbf{x} + \mathbf{c} \qquad (1.1)$$

where $\mathbf{X}$ = vector location in image coordinates
$\mathbf{x}$ = vector location in map coordinates
$M = 2 \times 2$ transformation matrix to be determined
$\mathbf{c}$ = constant offset vector

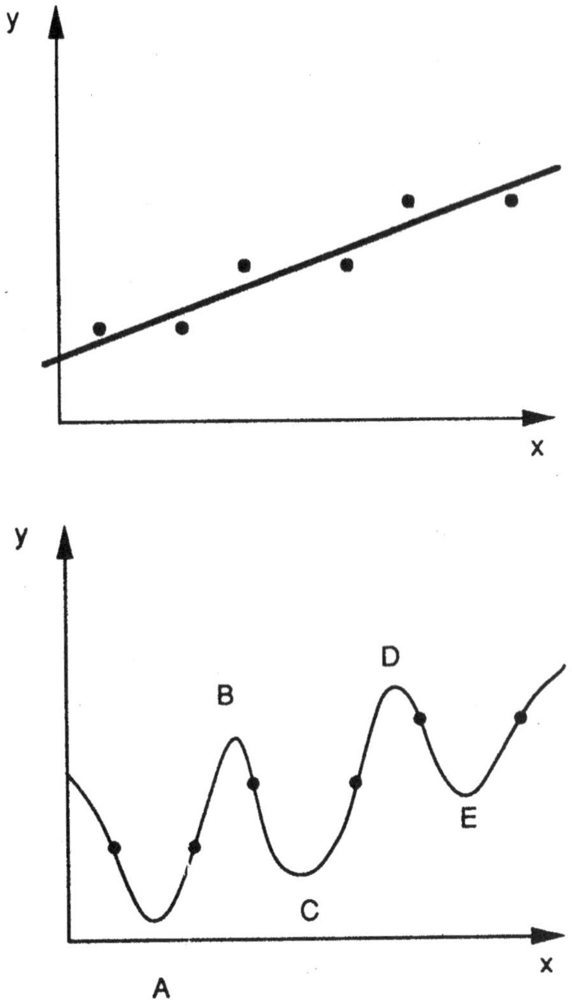

Fig. 1.2 Effect of polynomial distortion between control points: (a) a linear function provides the best-fit straight line between control points but does not pass through any of them; (b) a higher-order polynomial passes through all the control points exactly but does not provide a good fit at intermediate points (marked A to E).

Since there may be typically 50 pairs of GCPs, the equation for M is overdetermined – that is, more information is provided than is needed to solve the equation. As it is not therefore possible to solve the equation exactly, a least-squares technique (Ford and Zanelli 1985) is used to determine the best-fit transformation. The best-fit transformation is defined to be that which minimises the sum of squares of the residual GCP vectors. The residual vector is defined to be the difference between a GCP's true image position and that given by the transformation, that is:

$$\mathbf{R} = \mathbf{X} - (M\mathbf{x} + \mathbf{c}) \tag{1.2}$$

The resultant matrix, M, is then stored for use in the actual transformation. It is possible to use quadratic, cubic or even higher-order polynomials as the basis for the model but care must be taken. Although such polynomials may seem better because they fit the control point set, increased instability can lead to the introduction of artificial errors at intermediate points and thus lead to unwanted image distortions. This effect can be seen in one dimension in Fig. 1.2.

Another approach to calculating the transformation matrix (Derenyi and Pollock 1990) is to display a roughly transformed image on the GIS together with a digital raster or vector map in the desired projection. This can then be used to incrementally refine the transformation matrix in areas where the fit appears to be poor (for example, where distribution of GCPs is sparse or local terrain-height variations have induced distortion).

1.1.5 Image georeferencing

There are two generally used approaches to the transformation of satellite date (Emery *et al.* 1989):

- transform a map, or set of map points, to fit the image
- transform the image to the map projection.

The first method, known as georeferencing (or direct image referencing), is shown in Fig. 1.3.

The advantage of this method is that the data to be transformed (such as digitised maps or grids) is usually stored in vector format, which is much more compact than the raster format used to store image data. This means that the volume of transformed data and the amount of processing required are reduced, although with modern hardware this is not such an important consideration as previously. The main advantage of this approach is that there is no loss of information due to resampling (see Section 1.2.2). The disadvantage is that the image cannot be directly compared to other sources of spatially referenced data or used for multi-temporal work, which means that the method is generally only useful for single images rather than multi-image sets.

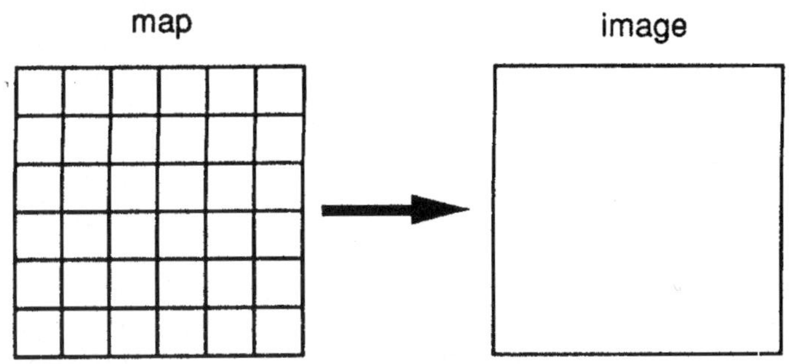

Fig. 1.3 Direct image referencing: the map grid is transformed to fit a particular image.

Since the advent of processing power that enables processing to be done in near real time, georeferencing is less widely used. One example of the applications of georeferencing was a project to locate the position of two Scottish islands (Benny 1985). The islands, Sule Skerry and Stack Skerry, lie 50 km north of the Scottish mainland and to the west of the larger Orkney islands. The Admiralty surveyed the islands in the nineteenth century with a positional error of a few hundred metres. The locations were more accurately triangulated by the Ordnance Survey in 1897, but it was not considered worth while to incorporate the more accurate positions in the Admiralty charts. The investigation was carried out in 1983 as part of a project that involved the determination of GCPs for the whole of Scotland. Using the above methods to calculate the transformation on the GCP set enabled the islands' actual positions to be georeferenced to Landsat MSS data. This showed substantial errors with their recorded map position and subsequent investigation revealed close agreement with the unincorporated 1897 triangulation data.

1.1.6 Image geocoding

The second and more common method, known as geocoding, involves transforming the image to the map projection. The transformation matrix used with geocoding is the inverse of that used for georeferencing, since we are now transforming the image to fit the map (see Fig. 1.4).

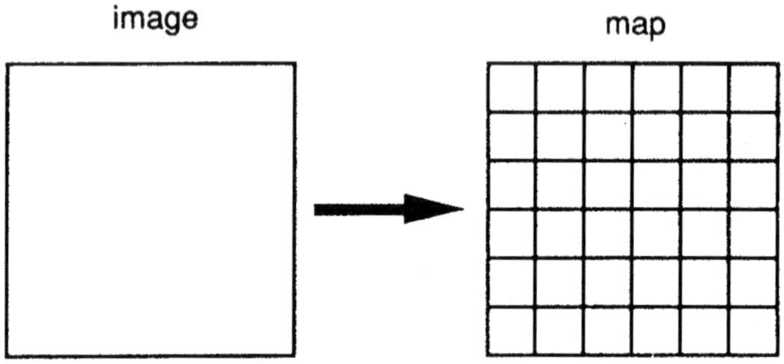

Fig. 1.4 Image geocoding: the image is transformed to fit the map grid.

It is usual to produce a transformed image on a regular grid as this makes it easier to calculate map locations and combine with other spatially referenced datasets. The grid spacing for a particular data source is normally about the same size (or slightly smaller) than the input pixel size. Typical grid spacing for various data sources are shown in Table 1.2.

The extent of the transformed image can be specified in various ways (see Fig. 1.5), for example:

- entire input image (in this case the output will usually be padded with blank values to produce a rectangular image)
- the largest rectangle on the map that fits inside input image
- specifying a sub-image in map coordinates.

Table 1.2 Typical grid spacings for image geocoding

Satellite	Sensor	Grid spacing (m)
NOAA	AVHRR	1,000
Landsat	MSS	100
		50
	TM	30
		25
SPOT	XS	20
	PA	12.5
		10
IRS	PAN	5

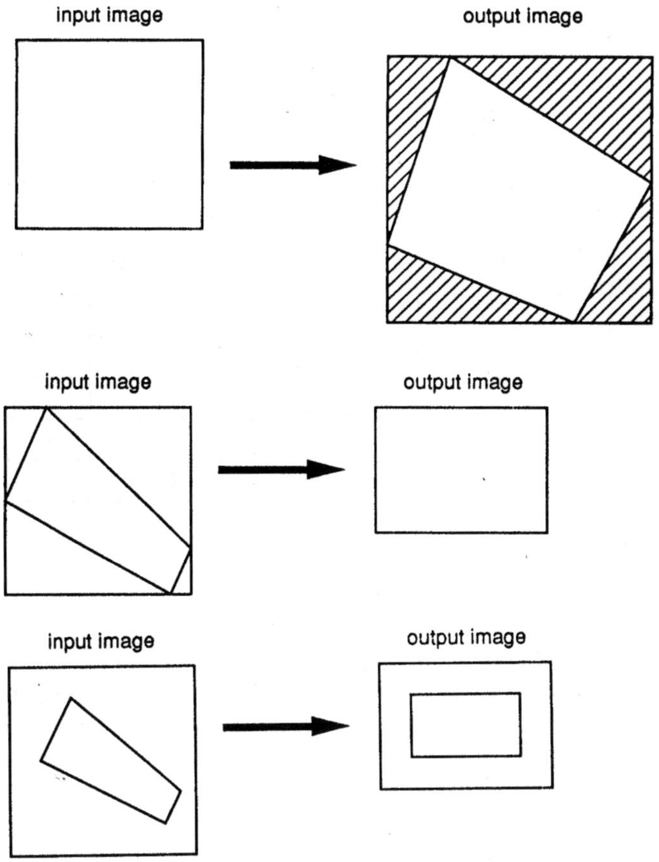

Fig. 1.5 Methods for specifying the extent of the output image: (a) the entire input image is selected and the output is padded with blanks; (b) the largest map rectangle that entirely fits the input image is selected; (c) a rectangular sub-image is specified in map coordinates.

The actual transformation algorithm proceeds as follows: first a grid in the required map projection is set up at a suitable (regular) spacing. For each line in the output grid, and for each point in that line, the inverse transformation is applied to that point to calculate its position in the original image. (If the transformed point is outside the original image then the output pixel is set to blank value.) If the point lies within the boundary of the original image then the nearest pixel in the original image is calculated and the output pixel set to the value of that pixel. The resampling step of calculating the nearest input pixel to the transformed image point is necessary because the transformation will (in general) not correspond to an exact integer location. This method of resampling is known as nearest neighbour; it is simple to implement but not particularly accurate. Expressed mathematically:

$$O(x,y) = I(X',Y') \tag{1.3}$$

where O = value in output (geocoded) image
I = value in input image
(x, y) = location in map coordinates
$(X,Y) = f(x,y)$ [f is a transformation function]
$X' = \text{int}(X + 0.5)$
$Y' = \text{int}(Y + 0.5)$

More sophisticated resampling techniques are described in Section 1.2.2.

1.1.7 Image registration

An alternative transformation method is image registration: this is usually of value when the images are compared directly (for example multi-temporal analysis of changes) or when a reference map is unavailable. The advantage of such methods is that they require little image-specific information (such as satellite orbital information) and only one image needs to be resampled, thus preserving radiometric fidelity. The registration algorithm uses tie-points that can be identified and located in both images. As their absolute map reference does not have to be known, thus this method can be used when local maps are unobtainable or unreliable. One image is selected as the base image, and the second image is registered to this. The actual registration process is carried out using a GIS. For the purposes of image registration the display is often divided into two 'linked' windows . One shows a sub-image of the base image, the other the corresponding region of the second image.

The image registration process commences by an operator selecting an area that is likely to contain a suitable control point. A tie-point that can be identified in both image windows is selected. Facilities to assist selection of tie-points include contrast stretching and edge enhancement (see Section 1.5 for details), as well as the ability to progressively zoom-in to control points to enable the operator to place a cursor accurately on the tie-point location. After a few points have been selected in this manner a fairly accurate matrix can be calculated. This allows the program to predict the location in the second image, given the cursor position in the base image to a few pixels, and greatly helps the operator select the next control point. When sufficient control points have been selected the transformation matrix can be calculated in the normal way. Transformation and resampling proceed using the method described in Section 1.1.6.

1.1.8 A GCP database

Since the same GCPs can be utilised for any images covering similar areas it is possible to eliminate much of the tedious effort involved by setting up a GIS database of GCP locations in map coordinates. Table 1.3 shows an extract from such a database used by the UK National Remote Sensing Centre, showing a selection of GCPs from Kenya. The database identifies for each GCP: the name of the country; the map sheet number; the GCP sequence number on that map; easting, northing and elevation in metres; and finally a brief descriptive text.

Table 1.3 Extract from a GCP database (based on Davison 1986)

Country	Map	GCP	Grid reference		Height	Description
			East	North	(m)	
EAK	1742	1	383092	9774976	780	Bend in river nr Yatta Plateau
EAK	1742	2	385101	9760470	750	Bend in river Athi nr Milembwa
EAK	1742	3	378209	9764706	800	Bend in river Kikku nr Kamisitu
EAK	1742	4	368383	9765682	850	River junction Nr Kiteet
EAK	1742	6	384292	9767628	700	Bend in river nr Yatta
EAK	1744	1	367354	9749532	900	Bridge on Kiboko Ranch
EAK	1744	2	479067	9749842	1,200	Top point on Kilema Range
EAK	1744	4	378545	9728230	1,100	Umani
EAK	1744	5	386676	9733640	890	Bend in road through plantation
EAK	1744	6	387490	9738302	894	Bend in road Kalungu
EAK	1751	2	407824	9771752	650	Bend in river nr Ikutha
EAK	1751	4	396627	9755238	725	Bend in river nr Athi new bridge

Typical database operations include addition, deletion and modification of GCP details (for example, entering height value for a previously defined point). Another useful check is to determine points which are very close to a GCP being entered and which may simply be that point subject to a certain measurement error. For example, a warning may be issued when a candidate GCP is within 100 metres of a GCP that has been previously entered on the database. The approximate locations stored in an image header, or calculated by bulk transformation, provide an initial estimate of the area covered by a particular image. Allowing a margin of error of a few kilometres, depending on the images used, produces a geographical search area in which the GCPs for the image can be expected to lie. The database query facilities can be used to produce a candidate set of GCPs. This can be used to display the rough locations on the GIS, which can then be fine-tuned interactively by the operator.

1.1.9 Automatic selection of GCPs

If GCPs have to be located for each new image, even the use of a GIS database of known points to speed the operation can still entail a very labour-intensive task. It can, however, be greatly improved by the use of automatic relocation techniques, which involves setting up a library of GCP 'chips' extracted from actual images. The following describes a typical method of using GCPs that has proved successful for use with Landsat TM imagery (Benny 1983). The standard satellite information is used to produce an initial transformation matrix; however, as with most bulk geocoding techniques, the result may well be intrinsically very accurate geometrically but still require location to the correct origin. Typically a single GCP will suffice to complete the initial matrix. Based on this the GCP library is searched for all GCP chips that can reasonably be expected to lie on the image. The set retrieved is sorted in order of their distance from the first point. The initial matrix is used to provide the best guess for the location for the second GCP and a spiral search commenced at this point until the real best-fit location is found. This process is continued for all subsequent GCPs with, at each stage, the true locations fed back into fine-tuning the transformation matrix. To safeguard against spurious matches two constraints may be placed on the search criteria for each GCP:

- the minimum correlation value that is regarded as a successful match
- the maximum distance from the expected location that a successful match can be accepted (this value can be adaptive – decreasing in value as the matrix iteratively becomes more accurate).

The algorithm for GCP chip location consists of the following steps:

1. Using the GIS cursor, locate a central GCP on the image.
2. Retrieve the chip corresponding to this GCP.
3. For each point (x, y) in a search area around this initial location, calculate the correlation between chip and image at (x, y).
4. Set the GCP location to that of the correlation maximum in the search area.
5. Use the location of the initial GCP to define the origin of transformation.
6. Calculate an approximate image transformation using bulk geocoding with known parameters (such as spacecraft height and heading).
7. Retrieve the candidate set of GCPs that should appear on the image and sort into ascending distance from the origin.

Repeat steps 8 to 11 for each GCP in the set:

8. Using the current approximate image transformation estimate the location of the GCP in the image.
9. For each point (x, y) in the search area calculate the correlation between chip and image at (x, y).
10. If maximum correlation is less than threshold, mark GCP as 'not found'.
11. Otherwise set the GCP location to that of the correlation maximum in the search area and recalculate the transformation matrix.

For cloud-free summer images, typically over 95 per cent of the GCPs for a Landsat TM image can be relocated by this method. For more complex images, such as SPOT, simple correlation techniques such as used for Landsat can be unsuitable for chip matching. Landsat images, with their fixed nadir viewing, ensure that chips from different passes have the same geometry. Off-nadir imagery from SPOT has a variable viewing angle that necessitates the comparison of chips with different geometries – for example, using image-matching techniques (see Section 3.2). It is also possible to synthesise image chips from data of a smaller pixel size, for example, making use of Landsat MSS to produce AVHRR chips, or aerial photography to produce Landsat chips. As the need for multi-sensor registration becomes more widespread, greater use of geographic models of image features is likely. These could, for example, consist of digital models of a GCP's location, shape, edges, texture, background and seasonal variation. The model for a particular GCP can then be used to simulate the appearance of an image chip for a particular sensor geometry.

1.1.10 Validation techniques

The validation of geocoded images is an essential part of the geocoding process. It is vital to be able to have confidence in the accuracy and repeatability of any particular geocoded product, and we return to this topic several times throughout this book. There are many possible definitions and approaches to validation but three complementary approaches can be identified:

- *Blunder detection* – Trapping of gross errors or physically impossible results, thus ensuring that all products will always be correct to a 'first order'.
- *Process validation* – Using the model to produce quality control parameters; for example, generation of quality products and reports as an intrinsic part of the geocoding process. This is often used, for example, as a preliminary check on the accuracy of a transformation matrix before resampling is commenced.
- *Product validation* – Quality control of the final output product against expected accuracy standards (for example, remeasuring selected GCPs against their known map location).

One example of blunder detection described previously is GCP chip location where the search is abandoned if the search area extends too far from the estimated GCP location or if the correlation maximum is below a certain threshold. Process validation is often derived from statistical analysis of the set of GCP locations and the transformation derived from them; however, single metrics (such as RMS error) may not be particularly useful. A better approach (Harrison *et al.* 1989) is to produce an entire 'error band' showing the accuracy of the transformation at all locations in the image. Because the transformation equation is overdetermined the resultant matrix will not fit the GCPs exactly. One measure of the accuracy of a transformation is to examine the residual error at each GCP. This is determined by applying the transformation to each GCP and comparing it with its known position, using Equation (1.2) to determine the residual error. A simple way to examine the residuals is to display them as vectors overlaid on the original image. More precise mathematical analysis is also possible. To provide an unbiased result certain of the original set of GCPs may be assigned as check points: these are not used to calculate the matrix but are used as part of the residual plot, to provide a better idea of the effectiveness of the geocoding process some distance from the control point network.

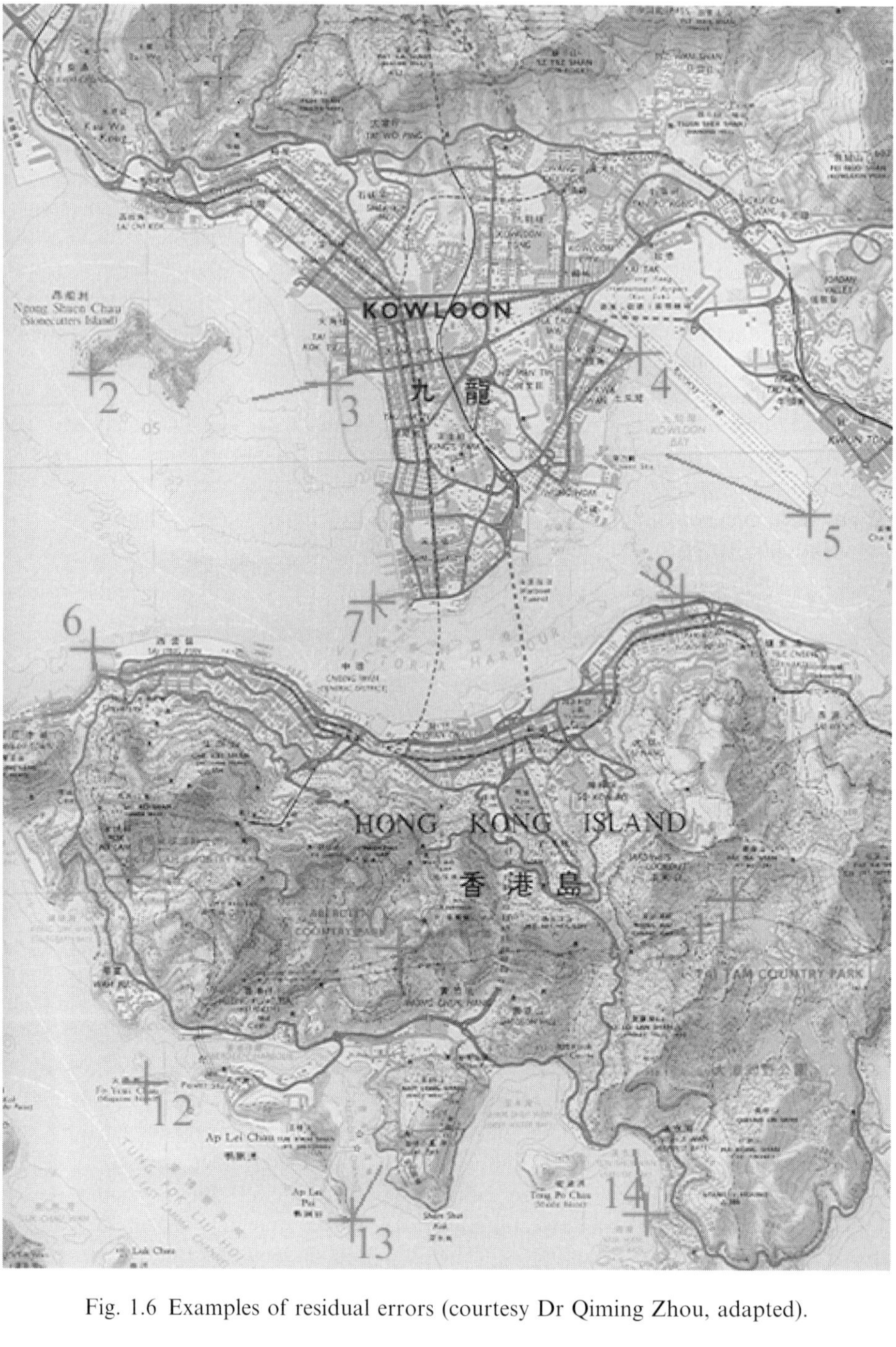

Fig. 1.6 Examples of residual errors (courtesy Dr Qiming Zhou, adapted).

Other approaches to process validation include analysis of the GCP distribution, transformation function analysis and production of an error budget which ascribes errors to each part of the model used (for example, error due to a Digital Elevation Model (DEM), or error due to digitising GCPs from maps). An interesting automatic approach to validation of the GCP selection was developed at Nottingham University (Heard *et al.* 1992). The geometric rectification system (GERES) is a prototype expert system which can advise on poor spatial distribution of GCPs, advise the operator of areas that are particularly lacking in coverage and point out new candidate GCPs in these areas. Figure 1.6 shows some graphical examples of residual errors.

Finally, product validation consists in assessing the overall geometric accuracy of the final product. This can be performed by measuring the location of points on the image against their true location on the map (essentially this is repeating the GCP selection exercise). Other methods include comparison of the image with a geocoded product of known accuracy, either by superimposition on a GIS or by identification of tie points (the transformation between two images geocoded into the same map space should be identical to the unit matrix). Another example of product validation is a method used for quality assessment of geocoded Meteosat images (Adamson *et al.* 1988) before dissemination to the meteorological community. Meteosat is a geosynchronous satellite, located over the equator at a longitude of 0°, that provides 2.5 km pixel size visible images and 5 km pixel size infrared images. In theory a geostationary satellite should produce images that need no rectification (apart from transforming to a convenient map projection) but in practice variations in orbit and attitude, particularly after a satellite manoeuvre, introduce distortions that require removal by geocoding. The validation process automatically calculates the displacement from image features to a library of landmarks (typically 10×10 pixels in size) which are mainly water boundary features such as peninsulas or inland lakes. The best landmarks are isolated, symmetrical islands such as Mallorca. A correlation technique is used to find the best-fit location of landmarks in the actual image and, hence, to validate the geocoding accuracy.

1.2 THE MODEL-BASED APPROACH TO GEOCODING

1.2.1 Overview

To produce a highly accurate geocoded image it is essential to adopt a model-based approach to geocoding. This allows the setting up and manipulation of exact mathematical equations that describe every aspect of the image geometry. It not only allows a controlled and methodical means of producing a geometrically accurate product but also allows errors to be tracked to particular parts of the model that can subsequently be refined. The sequence of steps performed in a typical geocoding operation are as follows:

1. Set up the components of the mathematical model and calculate the transformation between image (pixel) coordinates and the desired map projection by introducing parametric data to the mathematical model.

2. Resample the original image data onto the output grid (using the calculated transformation) to produce a geocoded image.
3. Validate the model – that is, determine that the transformation that has been calculated is physically reasonable.

As the sources of image distortion can be divided into orbit-related, sensor-related and ground-related errors, so the components of a typical geocoding model can be divided into three sub-models:

- orbit model
- sensor model
- ground model.

Orbit model
The purpose of the orbit model is to be able to locate the instantaneous position of the satellite sensor (at a time that a particular pixel was imaged) with respect to a geocentric coordinate system. It is also important to be able to determine which way the satellite is pointing (its attitude). Thus the orbit model can be conveniently broken down into two component models:

- *orbit position model* – the (x, y, z) position of a reference point on the satellite structure as a function of time
- *orbit attitude model* – the location of the three local orbital reference axes (pitch, roll and yaw), also as a function of time, coupled with the calculation of an attitude reference vector of a sensor with respect to these axes.

The conventional definition of the three orbital axes (derived originally from aircraft terminology) is as follows:

- *yaw* – a vector from the Earth's centre to the instantaneous position of the satellite's centre
- *roll* – a vector in the orbit plane (perpendicular to the yaw vector) in the direction of the satellite's motion
- *pitch* – a vector (also in the orbit plane) perpendicular to the yaw and roll vectors to complete a right-handed orthogonal system.

The attitude reference system is another right-handed orthogonal set, this time fixed in the satellite (with origin the satellite centre) and used to describe the three components of the satellite's rotation about its instantaneous centre. A specific example of derivation of an orbit model can be found in Section 1.2.3.

Sensor model
The sensor model can be one of the most complex sub-models to produce, involving the geometry, mechanics and optics of the sensor itself. It is necessary to consider each satellite sensor individually and to model it accordingly – for example (O'Brien and Prata 1990), the ERS-1 Along-Track Scanning Radiometer (ATSR) has a novel conical scanning system that views a point on the Earth's surface once on a forward

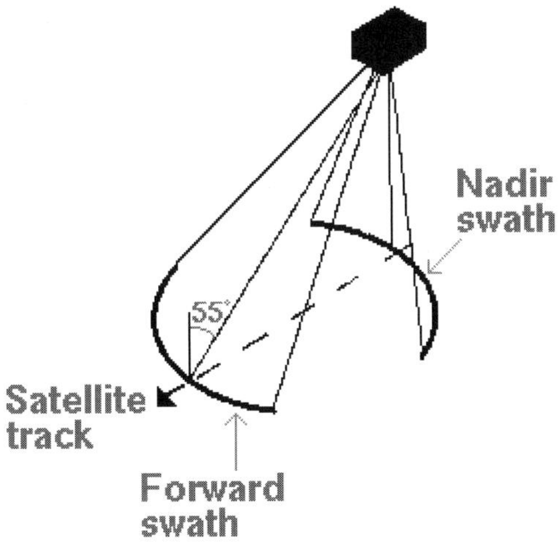

Fig. 1.7 ATSR scanning mechanism (Courtesy Rutherford Appleton Laboratory, www.atsr.rl.ac.uk).

scan and once on a backward scan. Thus the ATSR sensor model must describe the rotation scan mirror which traces a circle of 450 km radius on the Earth's surface in 0.15 second. Figure 1.7 shows the ATSR scanning mechanism.

With a scanning system such as Landsat (MSS or TM) the model will concentrate on the mechanics of the sensor – the rate of scan, for example. The aim is to be able to calculate the instantaneous look direction of each detector within the attitude reference. With push-broom sensors such as SPOT there is no corresponding mechanics as the motion of the sensor is provided by the motion of the satellite itself. When modelling a sensor it is generally not possible to completely separate the radiometric properties of an image from its geometric ones, thus there will always be some 'cross-talk' from the geometric aspects of a sensor model to its radiometric ones. One of the most important aspects of any sensor model is its 'resolution' – and although this term is often used interchangeably with 'pixel size' or 'instantaneous field of view' (IFOV), it is important to distinguish between them.

The intermediate field of view is determined by the sensor geometry and mechanics and describes the portion of the Earth's surface that can be seen at any one instant. The IFOV may not be a constant value, owing, for example, to changes in platform height and orientation. With scanning sensors the IFOV is commonly determined by a fixed angle of view, but this can also lead to variable sizes of IFOV – for example, with AVHRR images the very large swathe width, coupled with the effect of the Earth's curvature means that extreme off-nadir IFOV is significantly greater than central IFOV.

The pixel size is the terrestrial area represented by a single pixel in the satellite image, which may not necessarily be the same as the IFOV as many images will be deliberately resampled to have a constant, uniform pixel size. Finally, the definition of resolution is a more qualitative concept, the basic definition being based on the smallest size target that a sensor is capable of detecting. For terrestrial targets, however, their detectability is very much dependent on their geometry, texture and contrast with their surroundings (for example, bright objects that are smaller than pixel size may have sufficient contrast to significantly alter the pixel value above the surrounding background, and hence become detectable).

It is also important to consider the true spatial extent of the information 'encoded' into a pixel – in reality, sensors will not provide a simple block response exactly covering a square pixel area on the ground. Sensor response is actually characterised by a two-dimensional point spread function (PSF) centred on a pixel but extending beyond it. For Landsat (Wilson 1988) the PSF can be represented as the product of two one-dimensional line spread functions (LSFs).

$$\mathrm{PSF}(x, y) = \mathrm{LSF}(x) * \mathrm{LSF}(y) \tag{1.4}$$

where $\mathrm{PSF}(x, y)$ = two-dimensional point spread function
$\mathrm{LSF}(x)$ = one-dimensional across-track line spread function
$\mathrm{LSF}(y)$ = one-dimensional along-track line spread function.

A diagram of the across-track LSF(x) can be seen in Fig. 1.8; the function is largely determined by the combined effect of the sensor optics, electronics and detectors.

The along-track function LSF(y) is similar but the effect of electronics is negligible. The PSF will vary from sensor to sensor and a knowledge of this function will contribute greatly to accurate modelling of aspects such as radiometric calibration and design of resampling kernels.

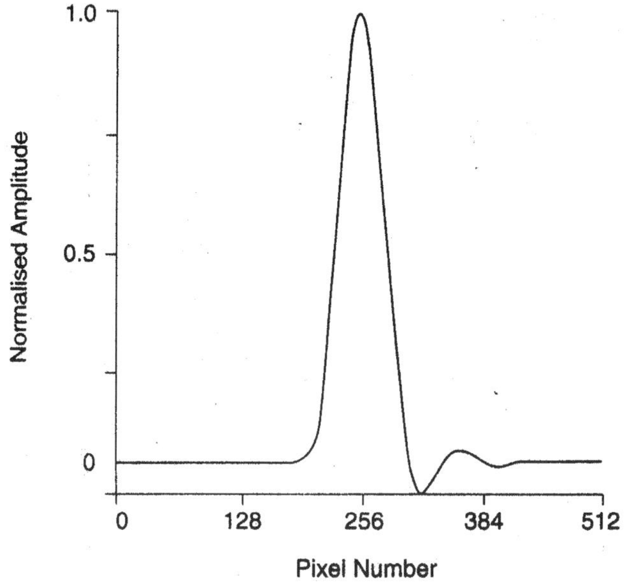

Fig. 1.8 Across-track line spread function for the Landsat 4 primary focal plane detectors (from Wilson 1988).

Ground model

The ground model consists of two components:

- *Earth model* – a description of the overall shape of the Earth
- *terrain model* – a more detailed model of local scene topography.

The equation for a spherical Earth model is given by:

$$x^2 + y^2 + z^2 = R_e^2 \tag{1.5}$$

where (x, y, z) = location of a point on the Earth's surface
R_e = mean equatorial radius (6378.138 km).

A more accurate model (Curlander *et al.* 1987) of the Earth as an oblate ellipsoid is given by the equation:

$$(x/R_e)^2 + (y/R_e)^2 + (z/R_p)^2 = 1 \tag{1.6}$$

where (x, y, z) = location of a point on Earth's surface
R_e = mean equatorial radius (6378.138 km) + average scene elevation
R_p = $(1 - 1/f)R_e$
f = flattening factor (f = 298.255).

To produce a geocoded image fully corrected for terrain effects (an orthoimage) needs a detailed model of the image geometry and the local terrain. This can only really be provided by a detailed Digital Elevation Model (DEM) of the Earth's surface. Ideally this should be of a similar grid-spacing to the image pixel size (for example, 30 metres for Landsat TM). Provision of DEMs is not very good for most of the Earth's surface. Thus the most suitable approach may well be to derive the DEM required from the satellite image itself as part of a stereo-pair or using other approaches such as SAR interferometry (these techniques are described in more detail in Chapter 3). This is one example of the synergy between satellite images and geospatial 'world models' – a topic we return to under the heading of 'Geographic Information Systems' in Chapter 4.

Although not strictly part of a ground model, the use of other terrestrial objects such as GCPs and coastlines is an important part of the geocoding process, having two main applications:

- To validate the model, for example by measuring the accuracy of the transformation at a set of independent check points (see Section 1.1.10).
- To provide a low-frequency offset to a model which typically will account for all the high-frequency geometric aspects but may need 'nudging' into position to define its map-origin correctly.

1.2.2 Resampling

Once the geocoding model has been set up, and the transformation calculated, the final step is to resample the values from the original image to points on a regular output grid. The basic requirement is to be able to transform a particular point in (x,

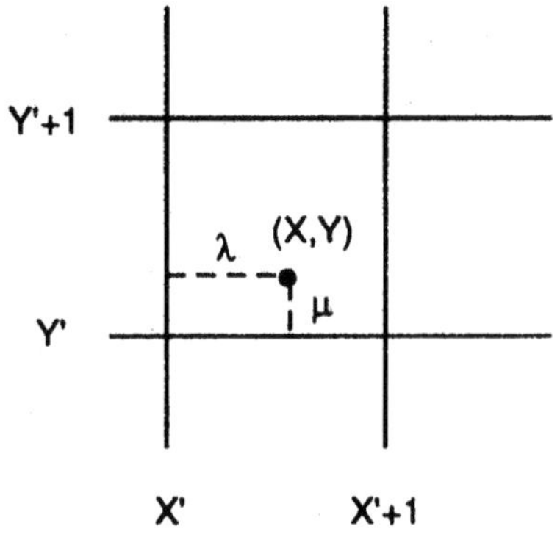

Fig. 1.9 The geometrical basis for bilinear interpolation.

y) map coordinates back into the original image coordinates and estimate what the corresponding image value should be at that point. The simplest method, and the one generally used for most bulk geocoded products, is to use the value of the nearest pixel in the original image – otherwise known as nearest neighbour interpolation. A more accurate method is bilinear interpolation, whereby a weighted average of the four surrounding pixels is used (see Fig. 1.9). The algorithm used is defined as follows:

$$O(x, y) = (1-\lambda)(1-\mu) * I(X', Y') + (1-\lambda)\mu * I(X', Y'+1) + \lambda(1-\mu) * I(X'+1, Y') + \lambda\mu * I(X'+1, Y'+1) \quad (1.7)$$

where O = value in output (geocoded) image
I = value in input image
(x, y) = location in map coordinates
(X, Y) = $f(x, y)$ [f is a transformation function]
X' = int(X)
Y' = int(Y)
λ = $X - X'$
μ = $Y - Y'$

Another commonly used algorithm is cubic convolution, which uses a cubic spline to approximate the theoretically ideal interpolator known as the 'sinc' function:

$$\text{sinc}(x) = \frac{\sin(\pi x)}{\pi x} \quad (1.8)$$

As this function has an infinite domain it cannot be computed exactly. Instead (Shlien 1979) it is represented by a piecewise cubic function. One such function (Moik 1980) which is continuous in value and slope is given by:

$$\begin{aligned} f(x) &= 1 - 2X^2 + X^3 && 0 \leqslant X < 1 \\ &= 4 - 8X + 5X^2 + X^3 && 1 \leqslant X < 2 \\ &= 0 && X \geqslant 2 \end{aligned} \quad (1.9)$$

where $X = \mathrm{abs}(x)$

1.2.3 SPOT geocoding model

One of the most important missions for cartographic applications is the French SPOT mission which commenced with the launch of SPOT-1 from Kourou in French Guiana in February 1986 and has been in continuous operation ever since.

SPOT sensor model

The SPOT sensor consists of two High Resolution Visible (HRV) instruments capable of resolution to 10 metres. SPOT replaces the mechanical scanning mirror systems of earlier satellites (typified by Landsat) with the newer technology of pushbroom sensors. The pushbroom sensor comprises linear arrays consisting of 1,728 charge-coupled device (CCD) detectors. The scanning of each line of a SPOT image is achieved by the measurement of the current generated by each CCD detector – four arrays are used to detect each spectral band. Another difference with SPOT is that the sensors are movable, by means of mirrors, in 0.6° steps to a maximum of 27° off-nadir. This allows the sensor to be pointed at a particular target, increasing the potential revisit frequency and providing the capability of acquiring stereo-pairs (see Section 3.1.1). One aspect of this mechanism is that unprocessed SPOT images can vary in size, from 60 × 60 km at nadir, to 60 × 80 km when fully tilted.

The SPOT camera model can be compared with that of an aerial photograph. The single most important factor in the applicability of an aerial photograph is that it is acquired almost instantaneously – that is, the camera platform does not move a significant distance in the time it takes to form a photographic image. This leads to a simple optical geometry known as single-point perspective where the light rays from every point on the surface to the corresponding point on the photograph all pass through a single point (the perspective point). A SPOT image, on the other hand, does not have this single-point perspective as it is effectively made up of a collection of one-dimensional images acquired at 1.5 ms intervals. Each of these images is taken by an array of detectors lying in a plane approximately defined as a vertical plane perpendicular to the image track. Note that the image plane is independent of the viewing angle as the geometry of the sensor mirror is defined as a rotation about an axis perpendicular to this plane, thus rotating the mirror simply causes the imaging geometry to change within the plane – it does not alter the geometry of the plane itself. As described previously, however, the satellite orbit and attitude does change significantly as the entire image is acquired – the envelope of the image geometry being defined by the continuous variation in the visible planes of each image line. Thus the image geometry can be regarded as being defined by a perspective locus given by the instantaneous perspective centre of each line.

For an image with single-point perspective six parameters are needed to determine its exterior orientation – namely, its three-dimensional position (x, y, z) and the three angles that describe its attitude (ω, ρ, κ). These parameters only need to be determined once per photograph. In order to be able to process 'dynamic imagery' it must be possible to model the continual variation of the instantaneous values of these parameters. For this purpose a good orbit model is essential.

SPOT orbit model

The SPOT image header contains detailed ephemeris data describing the motion of the satellite. The most important in terms of the SPOT geocoding model are:

- satellite position
- three-dimensional velocity vector
- attitude changes of sensor
- time when centre of scanline was imaged.

The attitude control system (Westin 1990) is designed to keep the satellite within 0.15° of a nominal reference. The SPOT orbit model can be derived by standard methods of orbital mechanics using the ephemeris data available for a particular orbit. The position–velocity data supplied with a SPOT image give the instantaneous position and velocity of the spacecraft at one-minute intervals (the attitude drift rates are recorded in the satellite header eight times every second and describe how the angular velocity of the satellite about its three axes varies with time). While not sufficient to provide instantaneous information for each line in an image (an entire SPOT image is acquired in 9 seconds) the orbit and attitude of the spacecraft is generally stable enough for accurate interpolation between the one-minute data. This leads to two possible models to determine the relationship between the local orbital reference and the attitude reference:

1. Assume that the relationship between the two is constant during the acquisition of a particular scene.
2. Estimate instantaneous relationship using satellite attitude drift rates.

The orbit of SPOT (as any other satellite) is determined by the laws of dynamics. An orbit can be described by six parameters (Westin 1990):

a = half-major axis
e = eccentricity
i = inclination
Ω = right ascension of the ascending node
ω = argument of the perigee
M = mean anomaly

For satellite images it is not necessary to model all these parameters accurately; for example, during the acquisition of a single image it is sufficient to model the orbit as a plane circular one. The primary parameters are derived from the satellite ephemeris data downlinked with the image. It is also necessary to model the satellite attitude, which is given approximately by the equations:

$$\begin{aligned} &\text{roll} && \omega(t) = \omega_0 + \Delta\omega(t) \\ &\text{pitch} && \rho(t) = \rho_0 + \Delta\rho(t) \\ &\text{yaw} && \kappa(t) = \kappa_0 + \Delta\kappa(t) \end{aligned} \tag{1.10}$$

where $(\omega_0, \rho_0, \kappa_0)$ = initial attitude
$\Delta\omega(t)$ = change in roll angle at time t (similarly for pitch and yaw)

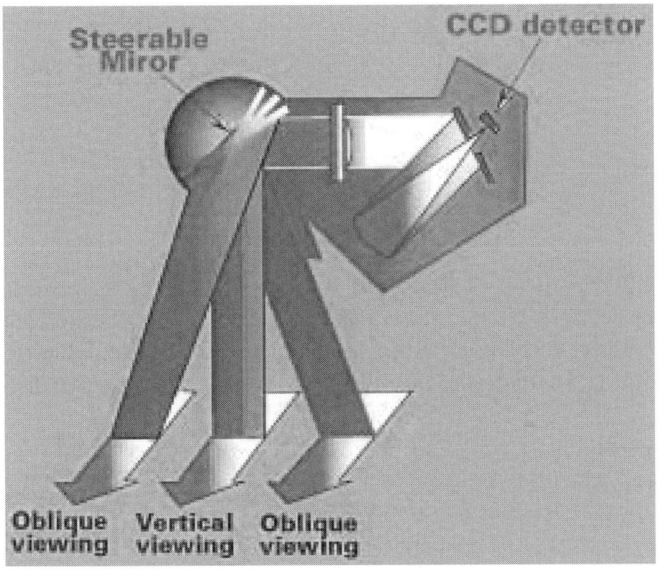

Fig. 1.10 SPOT imaging geometry (Courtesy SPOT Image).

It can be shown that none of the above approximations results in an error of greater than 0.2 metre, which is more than acceptable for geocoding purposes. The SPOT imaging geometry can be seen in Fig. 1.10.

Ground model

Another effect that must be accounted for in the precision geocoding of satellite images is the effect of the Earth's terrain on the image geometry. As can be seen in Fig. 1.11, because an optical scanner (or a pushbroom sensor) essentially measures

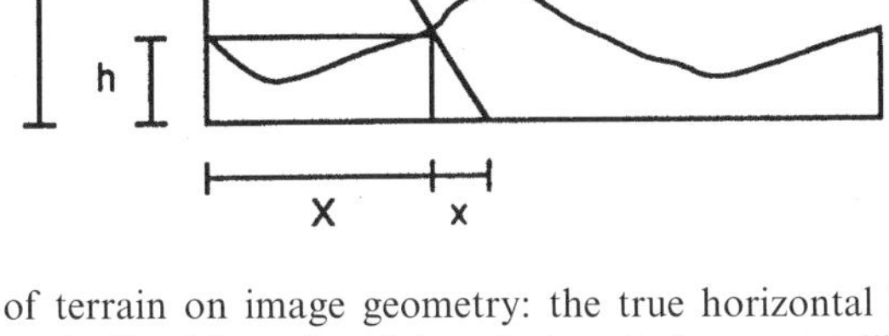

Fig. 1.11 The effect of terrain on image geometry: the true horizontal location of a point at height h above the Earth's surface (viewed at angle from a satellite orbiting at height H) is X, however it actually appears displaced by an additional distance x.

Table 1.4 Effect of relief on pixel displacement (adapted from Binnie and Colvocoresses 1987)

Elevation (m)	Horizontal distance from nadir (km)		
	10	50	100
100	1.6	7.9	15.8
200	3.2	15.8	32.0
500	7.9	39.5	79.0
1,000	15.8	79.0	158.0

Note: Values of displacements are in metres, for example an object 200 metres high, 50 km from nadir causes a displacement of 15.8 metres.

angles, an object extending above the surface of the Earth will cause an erroneous position to be recorded. An image that is fully corrected for all these geometric effects is called an orthoimage.

Note that as this effect depends on the across-track distance from the satellite nadir, it will be zero for points imaged directly at nadir and be most pronounced for targets imaged at large off-nadir angles such as produced by the steerable SPOT sensor. This effect is such that a 200-metre high object causes a relief displacement of nearly one pixel at the end of a Landsat TM scanline (see Table 1.4), the effect being even more pronounced for SPOT images with their off-nadir viewing angle.

With high resolution off-nadir sensors it is possible to obtain images of the same area with a vast range of viewing angles and a corresponding plethora of Sun–target–sensor geometries. One immediate consequence is that of terrain relief – given that the satellite altitude remains constant the apparent displacement is a combined function of the elevation of the point being imaged and its distance from the satellite nadir. This has two effects:

- The displacement is independent of pixel size – thus a relief displacement of 30 metres will produce an apparent shift of one pixel on a Landsat TM image but three on a SPOT PA image.
- Off-nadir images are far more effected by relief displacement.

Depending on the accuracy required, the relief effect can be modelled by a flat Earth model (usually sufficient for narrow swathe widths) or a curved Earth (normally used for the more off-nadir images). To produce geocoded images fully corrected for the relief effect (orthoimages) the relief correction must be applied to every point in the output image grid (by using values from a DEM), prior to resampling. As this is a computationally expensive exercise it may be beneficial to perform some interpolation between points on a 'super-grid', particularly where relief does not vary greatly over large horizontal distances, or in nadir areas of an image where the effect would not be expected to be pronounced.

Because of the stability of the SPOT geocoding model it can be applied to several adjacent scenes using a technique known as *pass processing*. This involves taking the

GCP adjustment applied to a single scene and applying it to a number (typically six to ten) of images in a single pass (Swann *et al.* 1988). This is useful for ocean areas or featureless interiors such as Australia or the Antarctic where good ground control is difficult and expensive to establish. Using this method, average requirements of less than one GCP per image can be attained.

1.2.4 AVHRR geocoding models

Bulk AVHRR geocoding

The AVHRR is one of the primary instruments carried aboard the NOAA series of polar-orbiting satellites and provides daily coverage of the entire globe at a resolution of 1 km. Although originally designed as a meteorological satellite, AVHRR has found a vast number of other applications, due to its synoptic coverage and rapid revisit capability (see, for example, Section 1.6.3 or 7.2.3).

In May 1998, a new series of operational environmental satellites began with the launch of NOAA-K. NOAA-K, L and M will be the successors to the NOAA satellite series ended with NOAA-14. These new satellites carry a series of instruments which have been modified and improved from previous NOAA series satellites still in orbit and operational. The Advanced Very High Resolution Radiometer (AVHRR/2) has been modified. The new instrument, AVHRR/3, adds a sixth channel in the near-IR, at 1.6 μm. This will be referred to as channel 3A and will operate during the daylight part of the orbit. Channel 3B corresponds to the previous channel 3 on the AVHRR/2 instrument, and will operate during the night portion of the orbit. A technical specification of the AVHRR sensor can be found in Section 6.2.1.

The TIROS-N satellites carrying the sensor are in a Sun-synchronous orbit at an altitude of 833–870 km. The repeat cycle of 12 hours is significantly different from that of Landsat or SPOT. The satellite does not have sufficient onboard recording capacity to store an entire global coverage of 1 km data known as Local Area Coverage (LAC), so for fully global cover it is resampled on board (Kidwell 1986). The resampling is performed by averaging the first four pixels out of every five and leaving a gap of 3 km across scanlines; this is often regarded in practice as being equivalent to producing imagery of 4 km resolution. As with high-resolution sensors, geocoding of AVHRR is an essential prerequisite to accurate information extraction. There are, however, major differences in the geometry of the images being considered, for example:

- The area being imaged means that there is significant distortion induced by the curvature of the Earth.
- The large pixel size means that it is not necessary to transform the images to the same absolute geometrical accuracy.
- Large-scale images are not susceptible to the same local relief effects that Landsat and SPOT images suffer from.

The Earth curvature effect is caused by the large swath width and the fact that the AVHRR pixels each represent a fixed angle rather than a fixed size on the ground.

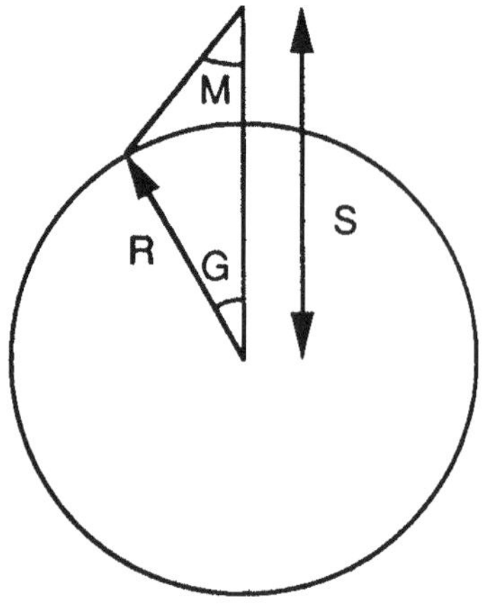

Fig. 1.12 Derivation of the geocentric angle for the AVHRR sensor (adapted from Brush 1988).

Consequently, pixels near the edge of an image are much larger than those at the centre. The relationship between the geocentric angle and the scanning radiometer angle (Brush 1988) can be seen in Fig. 1.12.

Using the trigonometric sine rule we obtain the formula:

$$[\sin(G+M)]/S = [\sin(M)]/R \qquad (1.11)$$

where G = geocentric angle from nadir
M = scanning radiometer scan angle
S = satellite distance from geocentre
R = Earth's radius.

From this it is possible to convert between the two angles and to resample to a constant geocentric angle, hence a constant pixel size.

Precision model

More accurate corrections for image distortion can be obtained by considering a simplified model of the satellite orbit (Emery *et al.* 1989). This models the satellite position and sensor-pointing direction based on the readily available satellite ephemeris data. The model used must also be able to incorporate the effect of Earth rotation and sensor mirror dynamics. The scanner optics consist of a telescope fixed relative to spacecraft and a mirror at 45° to the telescope. The rotation of the mirror directs radiation reflected from the Earth's surface (or from clouds) into the telescope optics. The mirror rotates at a constant rate (six times per second) relying on the satellite's forward motion to allow a scanned image to be built up. The simple scanner optics make it relatively straightforward to formulate a mathematical sensor model. Because of large pixel size it is rarely necessary to compensate for small-scale terrain distortions, thus a model of the Earth's surface as an ellipsoid (or even a sphere) will generally suffice. The remaining geometric distortions can then be eliminated by a standard template-matching technique using GCPs. For example, timing inaccuracies will often lead to a transformation which is intrinsically correct but displaced along the track. As a rule of thumb, 6 to 10 GCPs distributed evenly in the north–south direction will normally suffice to perform the necessary along-track nudge, and even a single GCP (Ho and Asem 1986) can be sufficient to produce an RMS error of better than two pixels (approximately 2 km). Alternatively, because of the large area involved, it may be preferable to allow an operator to navigate the image interactively using a digitised coastline.

It is also important to be able to transform large area images into a suitable map projection; for example, AVHRR products available from NOAA are provided in three standard projections namely Polar Stereographic, Plate Carrée and Mercator. Each of these has certain disadvantages for large area applications – for example, the Mercator introduces significant distortion in polar regions and the Polar Stereographic splits the Earth into two distinct hemispheres. Another disadvantage of these particular projections is that they are not of 'equal area' in that the true area represented by an individual pixel is not a constant but is a function of latitude. For this reason it may be desirable to transform AVHRR data into an equal-area radial projection such as Peters' projection (Lloyd and d'Souza 1987) shown in Fig. 1.13.

Automatic methods

As with high-resolution satellite imagery it is beneficial to automate the geocoding of AVHRR images by using pattern-matching techniques to relocate GCPs (Sun and Takagi 1987). The basic method is to produce a library of GCP chips which can be used to determine the location of control points automatically; for example, personnel at Dundee University have built a library of 337 AVHRR ground control points of the UK (Cracknell and Paithoonwattanakij 1989). This is particularly valuable as multiple AVHRR images of the same area are acquired frequently (sometimes several a day) and it is imperative to geocode them in near real time. The searching algorithm is seeded by providing approximate locations of three GCPs, and because the greatest distortion in AVHRR imagery is in the extreme off-nadir pixels it is advantageous to provide GCPs in the form of a triangle with vertices at top line (centre), bottom line (extreme left), and bottom line (extreme right). Alternatively the AVHRR orbit model can be used to determine approximate seed point locations automatically. A typical flowline for automatic geocoding of AVHRR imagery can be seen in Fig. 1.14.

More accurate chip relocation can be attained by using high-resolution chips matched to the low-resolution AVHRR images, for example, Cracknell and Paithoonwattanakij (1989) have shown that Landsat MSS data can be used to achieve registration accuracy of better than 0.25 pixel (approximately 250 metres). Sub-pixel geocoding is necessary for a number of applications – for example, derivation of ocean current vectors. With conventional geocoding accuracy (of the order of a pixel at nadir) the current vectors can be measured to an accuracy of ± 1.8 cm/s, if a geocoding accuracy of 0.2 km is available then they may be measured to an accuracy of ± 0.5 cm/s.

Fig. 1.13 World Map: Peters' Projection (copyright Akademische Verlagsanstalt, www.petersmap.com).

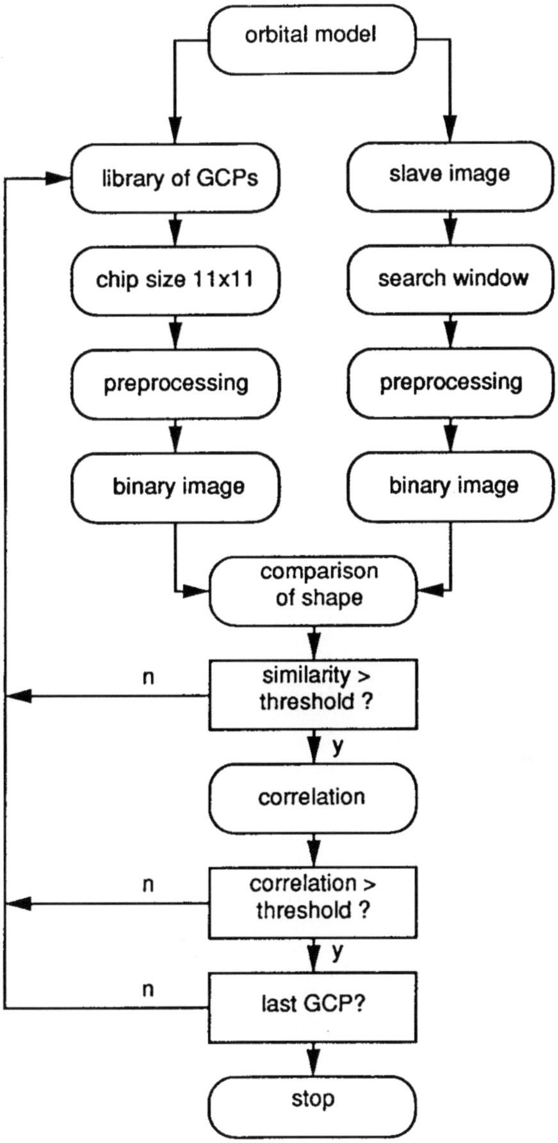

Fig. 1.14 Operational flowlines for AVHRR geocoding using image chips (adapted from Cracknell and Paithoonwattanakij 1989).

1.3 SAR IMAGE GEOCODING

All the major satellite sensors that have been described so far (such as Landsat, SPOT and AVHRR) have certain characteristics in common. The most important of

Table 1.5 Major spaceborne SAR missions

Mission	Country	Launch
Seasat	USA	1978
SIR-A	USA	1981
SIR-B	USA	1984
ERS-1	Europe	1991
JERS-1	Japan	1992
SIR-C	USA	1994
ERS-2	Europe	1995
Radarsat	Canada	1996
SRTM	USA	2000

these is that they are passive sensors relying for their energy source on either reflected sunlight or, in the case of thermal sensors, electromagnetic radiation from warm bodies. One consequence of this is that the sensors can only operate in relatively narrow spectral windows due to the atmospheric attenuation of incoming and reflected radiation at most spectral frequencies. Another, more serious, limitation is the inability of the sensors to operate in cloudy or hazy conditions. This can seriously impede operations in temperate regions such as northern Europe or in tropical regions when the cloudy seasons are often of the most interest. This has led to an increased utilisation of active microwave sensors which generate their own pulse of energy and then record the signal obtained by its reflection from the Earth's surface.

One of the most important spaceborne microwave sensors is the Synthetic Aperture Radar (SAR), examples of which were flown on: the US Seasat mission in 1978; two NASA space shuttle missions (as the Shuttle Imaging Radar, SIR-A and SIR-B); as the primary instrument on the European Space Agency ERS-1 and ERS-2 missions; the Japanese counterpart (JERS-1); the Canadian Radarsat mission (described in more detail in Section 1.7.5); and the Shuttle Radar Topographic Mission (SRTM) to produce detailed Digital Elevation Models (DEMs) of the Earth's surface (see Chapter 3). The major spaceborne SAR missions are summarised in Table 1.5.

Although images produced from SAR appear similar to those obtained from optical/infrared imagery they have several significant differences. For a start, because SAR sensors are not reliant on the Sun's energy for illumination their operation is far less affected by the time of year, the season of acquisition and the prevalent weather. The imaging mechanism is also fundamentally different – an optical sensor forms an image whose features are determined by the angles subtended at the sensor by objects on the Earth's surface, whereas the basic operation of a radar system is to calculate the time an electromagnetic wave takes to leave a transmitting antenna and return to a receiving antenna. Because the speed of the electromagnetic waves is a constant (the speed of light), there is a direct relationship between the time taken for the signal to return and the distance of the target being detected. The two main factors that determine the nature of the reflected radar beam are:

- the parameters of the imaging system (such as incidence angle, resolution, wavelength and polarisation)
- the properties of the target being imaged (such as the surface roughness, the dielectric constant, moisture content of the target, and local topography).

A radar that can be used to form an image of the Earth's surface is known as an imaging radar sensor. Non-imaging sensors, such as the scatterometer and the radar altimeter, are also widely used for Earth observation. Imaging radars are classified into two types:

- *Real Aperture Radars* (more usually flown on aircraft)
- *Synthetic Aperture Radars* (flown on both satellite and aircraft platforms).

The basic concept behind a SAR is to make use of the satellite motion to synthesise a larger antenna from the position of the platform at a sequence of points. The return from a point on the Earth's surface is detected at a number of different positions and the signals are corrected for phase and combined by a method known as coherent addition. The image format as generated by a SAR system can be seen in Fig. 1.15.

SAR data is amenable to precise geolocation (Schreier 1991) because time measurements in SAR are more precise than angular measurements in optical imagery. Given a sufficiently accurate orbital model it should therefore be possible to produce a precision geocoded product.

1.3.1 Geometric aspects of SAR

The geometry of SAR images has several differences from that of optical imagery. Two of the geometric aspects of SAR which are not found in optical images are

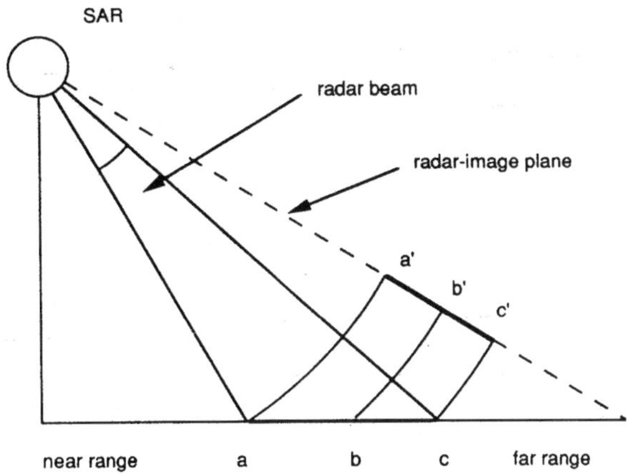

Fig. 1.15 Basic SAR image format on a level surface: the relationship between ground range images and slant range images can be seen by comparing ground points (marked *a*, *b*, *c*) with their counterparts in the radar image plane *a'*, *b'*, *c'* (adapted from Keydel 1992).

shadowing and *layover*. In an optical image the side of a raised feature such as a hill opposite to the illumination source will not be imaged if it is in an area of shadow. Radar images will also have the same areas of shadow; however, as long as there are no other surface targets at the same distance from the sensor the shadowing effect will cause the appearance of darker 'missing' features in the image. The layover effect is caused by objects at the same distance being imaged together. This is an example of the more general problem of ambiguous imagery whereby two distinct objects from different surface features are mapped onto the same resolution cell (pixel). The counterpart in optical images occurs when objects at the same angular position (that is on the same optical ray) but at different distances are imaged simultaneously – in this case the ambiguity is resolved because only the front object is actually imaged. These effects are illustrated in Fig. 1.16. In both types of imagery the effects will be most apparent in areas of high relief, particularly where the local terrain slope is similar or greater than the incidence angle of the sensor.

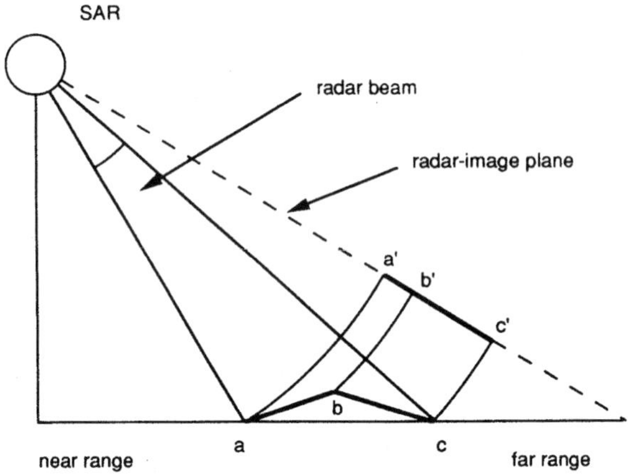

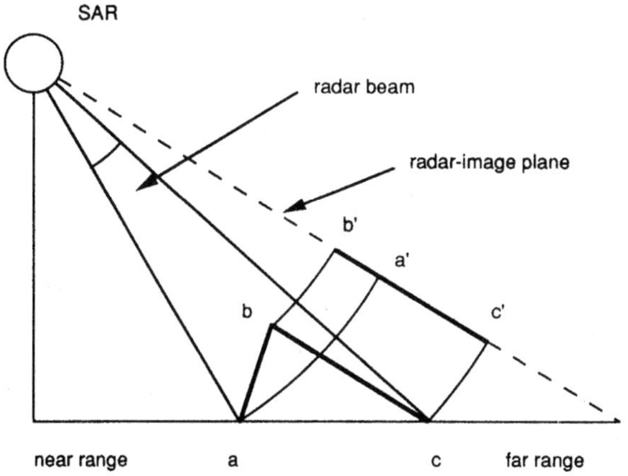

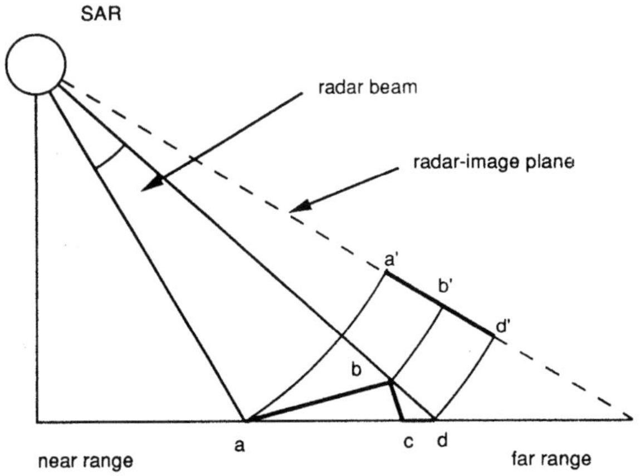

Fig. 1.16 Geometric effects in SAR images. (a) *Foreshortening* – slope *ab* is inclined towards the radar and slope *bc* is inclined away from the radar (back slope). The distance corresponding to *ab* in the image plane is much shorter than that corresponding to slope *bc*. (b) *Layover* – the angle of the slope is greater than the look-angle, the slant range for point *a* is now greater than that for point *b* and their relative positions (*a′b′*) have been reversed in the image plane. (c) *Shadow* – the slope *bc* causes a radar shadow in the image plane. (Adapted from Keydel 1992.)

In a similar manner to optical images, SAR images also experience a relief effect due to local topography. On a slope facing towards the sensor, neighbouring pixels will be roughly the same distance from the sensor resulting in a foreshortening effect. In the extreme case when the terrain slope is greater than that of the incident signal, layover will occur. On slopes facing away from the sensor the features will be portrayed more accurately, although in the limiting case there may be a shadowing effect. These effects not only produce geometric distortion but also disturb and destroy the backscatter information – geometric and radiometric relations for SAR are even more intricately intertwined than with optical images.

1.3.2 SAR geometry equations

For both optical and SAR imagery the basic purpose of geocoding is to associate an accurate terrestrial location with every point on an image; however, with optical images it is the spacecraft position and angular measurements that are the governing factors, and for SAR it is radar time delays and shifts in signal frequency. The basic task of SAR geocoding is to relate the time and range information obtainable from the radar to the corresponding location on the Earth's surface. The three fundamental equations which determine the SAR geometry (Curlander *et al.* 1987) are:

- the Doppler shift equation
- the range equation
- the Earth model equation.

By solving all three equations simultaneously the geodetic location of the required point (x, y, z) can be found.

The Doppler phenomenon describes the way that, within the radar footprint, signals reflected from targets moving towards the sensor are shifted to higher frequencies and those moving away from the target are shifted to lower frequencies. The digital processor filters the signal to focus the energy towards one specific point in the azimuth line. Most processors utilise the point of no frequency shift (zero Doppler) as reference in the azimuth focusing. For a point target the Doppler shift can be estimated from the formula:

$$f_D = (2/\lambda R)\,(\mathbf{V}_s - \mathbf{V}_t)\cdot(\mathbf{R}_s - \mathbf{R}_t) \tag{1.12}$$

where f_D = the Doppler value used in the azimuth phase reference function
λ = radar wavelength
R = sensor-target slant range
$\mathbf{V}_s$ = sensor velocity vector
$\mathbf{V}_t$ = target velocity
$\mathbf{R}_s$ = sensor position vector
$\mathbf{R}_t$ = target position vector

Note that to use this equation it is necessary to have imagery in slant range format – this is the format generated by most processors (Schreier 1991). In slant range imagery one axis represents the distance from the satellite position to the target, while the orthogonal axis represents the Doppler shift.

In general, targets such as fields and roads will be stationary relative to the Earth's surface, thus

$$\mathbf{V}_t = \boldsymbol{\omega}_E \times \mathbf{R}_t \tag{1.13}$$

where $\mathbf{V}_t$ = target velocity
$\boldsymbol{\omega}_E$ = Earth's rotational velocity (7.292×10^{-5} rad/s)
$\mathbf{R}_t$ = target position vector

From simple geometry the slant range from a sensor to the target is given by

$$R(i, j) = [(\mathbf{R}_s - \mathbf{R}_t)\,.\,(\mathbf{R}_s - \mathbf{R}_t)]^{1/2} \tag{1.14}$$

where $R(i, j)$ = slant range at pixel (i, j)
$\mathbf{R}_s$ = sensor position vector
$\mathbf{R}_t$ = target position vector

Finally, as with optical images a suitable basic model for the Earth can be that of a sphere or an ellipsoid. The SAR geometry model is shown in Fig. 1.17.

1.3.3 Bulk geocoding process

In simple terms the geocoding process will consist of the following phases:

- calculate components of model
- calculate transformation and apply to points on output grid
- resample using original data.

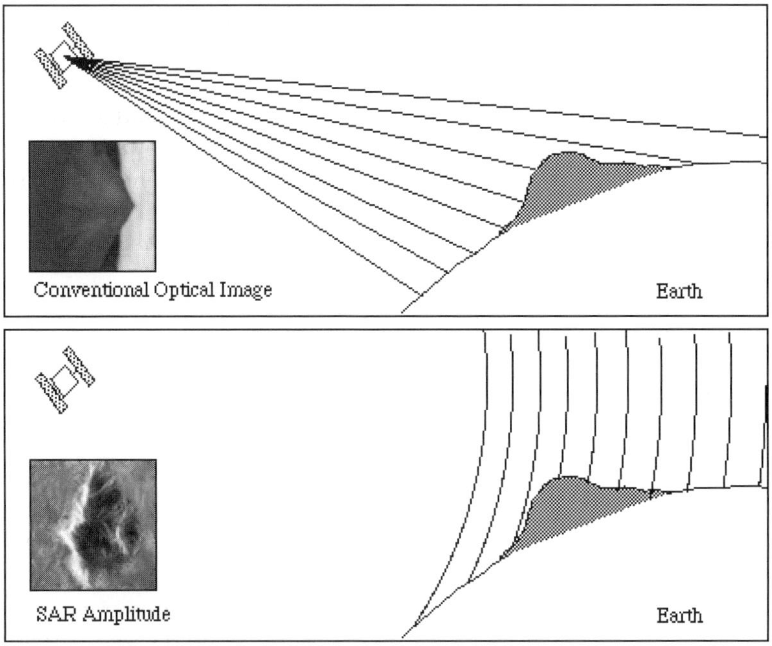

Fig. 1.17 SAR geometry model (Courtesy Alaska SAR Facility).

The initial estimate of the imaging model requires a spacecraft orbit model which is based on the transmitted satellite ephemeris. Following this, various algorithms can be used to determine the geographic location of a pixel (x, y). As with the geocoding procedure described in Section 1.1, the basic requirement (Coll and Pettigrew 1988) is to be able to determine a transformation between a regular grid in the desired output space and the uncorrected SAR image. In practice this reverse transformation is not simple to evaluate directly so the approach usually adopted is to determine the corresponding forward transform (from the uncorrected imagery to the map space) and then to invert it numerically. This process is best performed iteratively, for example (Kwok and Curlander 1987):

1. Determine nominal target location by intersecting range vector with the geoid.
2. Calculate the target Doppler vector.
3. Calculate the iso-Doppler lines (points of the Earth's surface with constant Doppler frequency, f_D).
4. Maintaining correct sensor–target range, rotate the range vector to intersect the appropriate iso-Doppler line given by f_D.
5. Iterate this process until f_D converges to a predetermined value.

As with optical imagery, the reverse transformation is used to map back from the desired output grid to the raw input data. In order to preserve information it is highly desirable to only use one resampling step, which is another reason for using slant range data rather than data that has already been resampled into ground range. Interpolation kernels are generally similar to those used for optical imagery (for example, nearest neighbour, bilinear interpolation and cubic convolution). More sophisticated, higher-order kernels such as the discrete Fourier transform (DFT) have also been proposed as suitable for SAR geocoding (Laycock 1990). The DFT is another approximation to the sinc interpolator and, as such, suffers from a general problem associated with truncated interpolators, i.e. it is possible to generate negative image values. By setting these values to zero the overall image statistics become biased; however, it appears that this problem is less severe when using multi-look SAR images than it is with optical images.

1.3.4 Orthoradar generation

Because of the nature of SAR geometry, even in areas of moderate changes in relief, the use of a DEM for geocoding becomes essential and such images are known as 'orthoradar corrected'. Various alternative approaches, e.g. using DEMs to perform range displacement correction, are possible. For example, Fig. 1.18 shows the data-processing flow for a three-pass correction scheme. The functions performed in each pass are as follows:

- *First pass*
  - rectify slant range image and map in the azimuth direction.
- *Second pass*
  - generate distortion map to account for relief displacements
  - combine with resampled locations from first pass
  - resample first pass image in range.
- *Third pass*
  - complete resampling to geocoded map format.

For operational SAR geocoding it is advantageous to maintain databases of DEMs and GCPs, for example the SAR geocoder (Schreier 1991) installed at the DLR processing and archiving facility for ERS-1 uses three major databases:

- a map library database storing the features of 8,000 topographic maps of areas in Europe
- a GCP database, including chips extracted from SAR images
- a DEM database where the data is stored in a common geographic format but with applications modules that permit the DEMs to be extracted in any recognised European map projection and at a suitable grid spacing.

DEMs can also be used (Schreier *et al.* 1990) in an alternative method of precision geocoding to generate a synthetic image. The appearance of a SAR image is dictated to a far greater extent than optical imagery by the terrain, making it possible to simulate the appearance of a SAR image based on a DEM and known radar parameters such as incidence angle. This can then be used for precise geometric

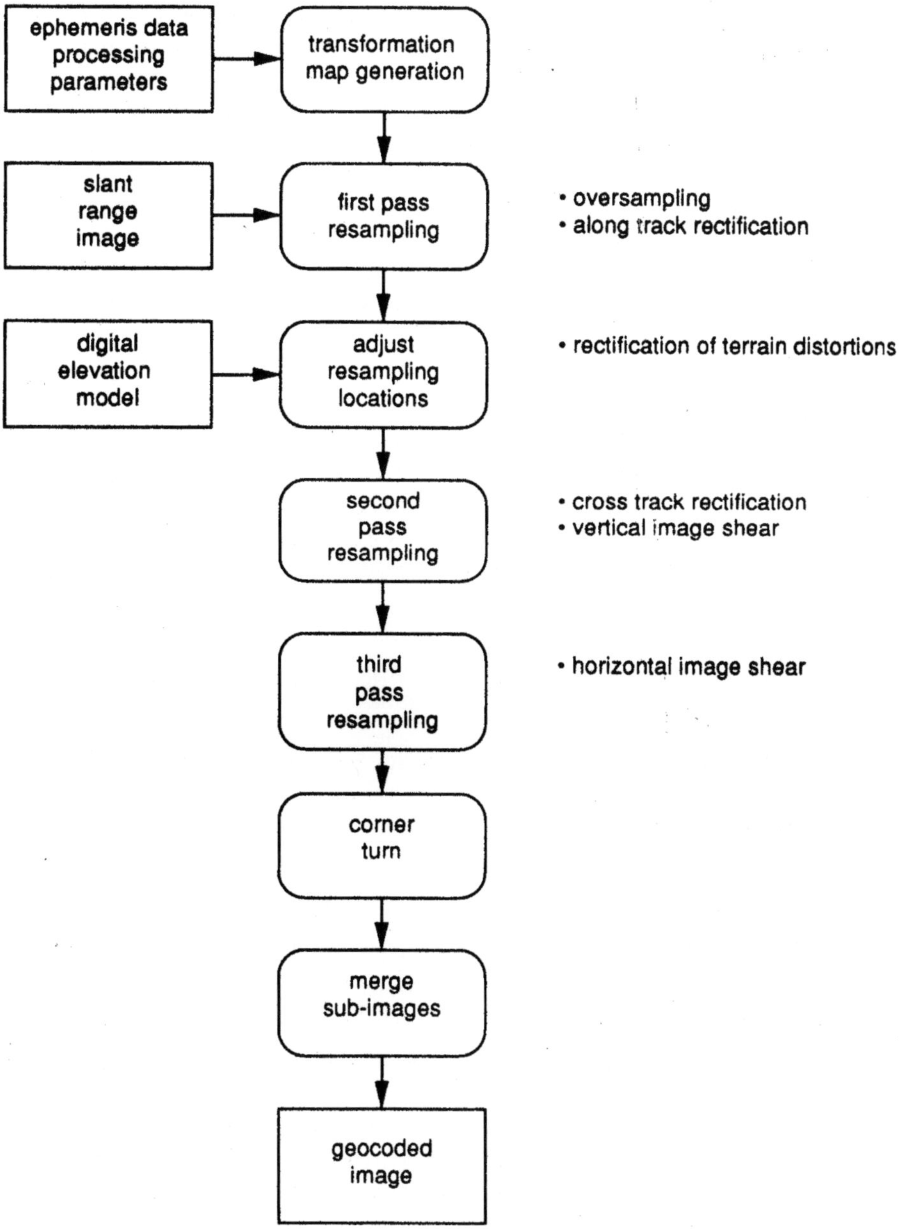

Fig. 1.18 Dataflow for SAR geocoding using a three-pass resampling method (adapted from Kwok and Curlander 1987).

control by registering an image to the synthetic image. This can usually be achieved by simple correlation techniques and has the advantage of producing increased geometrical fidelity, since the dataset being used for ground control is the same as that used for terrain correction. Another advantage is that it is possible to locate areas of difficult topography in advance and to pre-calculate simulated sub-images for these areas. Another use of simulated images (Tilley and Bonwit 1989) is to reduce layover distortion in images. The effect of layover is to concentrate the radar returns from slopes that face towards the radar receiver. The simulated image will show where this effect has taken place and a suitable redistribution of energy 'up slope' can be performed. One final use of DEMs in SAR geocoding is to 'pre-screen' hilly areas where terrain correction will be necessary.

Use of ground control points

Even with the use of relatively accurate DEMs there is still likely to be residual low-frequency errors which are most easily removed, as in the case of optical images, by the use of GCPs. The main sources of these residual errors (Schreier *et al.* 1990) are:

- remaining uncertainties in pixel location (estimated at 20 metres in range and 110 metres in azimuth in the case of ERS-1)
- shifts and scalings relative to the common ellipsoid (possibly up to 300 metres)
- errors in DEM generation (for example, the relatively accurate DEMs available for Germany are estimated to have errors of approximately 30 metres in plan and 7 metres in elevation).

The desirable attributes for GCPs for use in SAR geocoding (Weiler *et al.* 1986) are similar to the optical case:

- They are readily detectable and distinct from the surrounding area.
- They are well distributed throughout the image.
- They can be locatable on suitable maps (or their position accurately known, for example, by GPS survey).
- They should remain stable over time.

Using GCPs as part of a precision geocoding model, it is possible to achieve sub-pixel geocoding accuracy even in areas of rough terrain. For example, geocoded Seasat SAR imagery of Canada with a nominal resolution of 25 metres (Coll and Pettigrew 1988) achieved an RMS accuracy of 12.9 metres (after errors due to map measurement had been subtracted) on an image of moderate terrain relief and a similar result on an area with noticeable terrain relief.

The automatic selection of GCPs from SAR images is more problematical than with optical images because of their radiometric characteristics – primarily speckle – although this can be reduced to a certain extent by the use of non-linear filters such as the Lee sigma filter (Lee 1981). Even following such an operation, correlation methods to relocate control points may not prove successful. One alternative approach (Weiler *et al.* 1986) is the use of an 'interest operator' which identifies points in an image which are clearly distinguishable from the background. The algorithm used is as follows:

1. For each block of pixels in image calculate a threshold
2. For each 4×4 window in the block calculate the variance of the pixel values in window.
3. If variance > threshold, then set output pixel to 1, else set output pixel to 0.

The appropriate threshold is calculated by setting up a grid of regularly spaced 4×4 windows. For each window the variance of the pixel values in the window is calculated. The variances are then sorted into descending order and a weighted mean is calculated using the vector $\{0, 15, 10, 6, 3, 1, 0, 0, 0, 0, 0, 0, 0, 0, 0, 0\}$.

Alternatively, if several contiguous images from the same orbit are required for geocoding, a pass-processing scheme (Coll and Pettigrew 1988) analogous to that for optical images described in Section 1.2.3 may be employed. Finally, another technique which can be used for SAR is simulation-based geocoding (Guindon and Adair 1992). The simulated SAR image is generated using input DEMs and a model of the sensor viewing geometry and SAR instrument characteristics. Tie-points are identified in both the simulated and real images and used as the basis of the transformation function.

1.3.5 Validation

In the case of SAR geocoding, the 'impossible' points that blunder detection needs to avoid are those where there is no information to reconstruct the geocoded image (shadow areas) or where there is insufficient information to reconstruct the original signal accurately (layover areas). These areas are then omitted from the final geocoded product. It is often desirable to derive this information as a parallel process of the SAR geocoding; for example, the SAR geocoder (Schreier 1991) installed at the DLR processing and archiving facility produces an accompanying 'layover, shadow and incidence angle mask' which indicates the incidence angle and areas affected by layover and shadow. Blunder detection can also be applied to filter out 'bad' GCPs in the standard manner – for example, the following rules (Raggam 1990) can be used to remove blunders in GCP selection:

1. Eliminate GCPs with residuals greater than three times the RMS error.
2. Only use the best known GCPs.
3. Use a predefined residual limit and constrain the distribution of GCPs to be homogeneous over the image area.

The validation of the geocoding process itself (Dowman and Upton 1990) includes statistical and radiometric measures of quality and evaluation of error equations predicted from sensor and terrain parameters. Another set of validation parameters is available in the form of residual errors if GCPs are used. Product validation includes comparison of the output product with reference maps and comparison with synthetic images which use slope value derived from DEMs to predict the expected backscatter. A summary of validation functions can be seen in Table 1.6.

Table 1.6 Validation functions for SAR geocoding (based on Dowman and Upton 1990)

Validation function	Quality metric
Compare layover and shadow with geocoded product histograms	Shadow/layover masks
Evaluate error equations	Predicted errors from geocoding model
Calculate statistical quality measures	Radiometric quality
Compare geocoded product with maps/DEMs profile statistics	Residual vectors

1.4 GLOBAL POSITIONING SYSTEM

The principal navigation and positioning technologies used today include inertial navigation systems, radio-navigation systems, beacons, satellite wide-area systems and local radio links, cellular triangulation, vehicle wheel sensors and heading sensors, and dead-reckoning systems. All of these technologies can potentially contribute to the use of georeferenced data but the best synergy with EO data is provided by the rapidly developing application of GPS. The United States Global Positioning System (GPS) is a satellite-based high-accuracy positioning system which can be accessed from anywhere on the Earth's surface (Mack 1992). The complete system consists of 21 satellites (plus three back-up platforms) which can be used to locate the terrestrial position of a GPS receiver. GPS is increasingly being used in conjunction with GIS and satellite images to provide a rapid and accurate source of mapping information.

This section describes the basic operation of GPS to provide terrestrial location information. Later chapters describe applications of GPS, integration with GIS and the proposed European Galileo system. The GPS satellite configuration is shown in Fig. 1.19.

Originally designed as a US Department of Defense system the accuracy was deliberately degraded in selective availability (SA) mode for civilian use. In this mode the RMS accuracy attainable with a single receiver is in the order of 50 m. Selective Availability was removed in May 2000, and with differential GPS much higher accuracies can now be achieved. To obtain a positional fix (Sharun 1993) signals from at least three different satellites must be obtainable, in practice this means that they must be at least 15 degrees above the horizon. Signals from three satellites enable the horizontal position of the GPS receiver to be determined, and if four satellites are available then a full three-dimensional fix can be obtained.

In simple terms the time taken to receive the signal from the satellite provided a measure of its distance from the receiver. For accurate assessment of position, more complex algorithms are needed to account for timing inaccuracies in the receiver and corrections for water vapour in the troposphere and charged particles in the

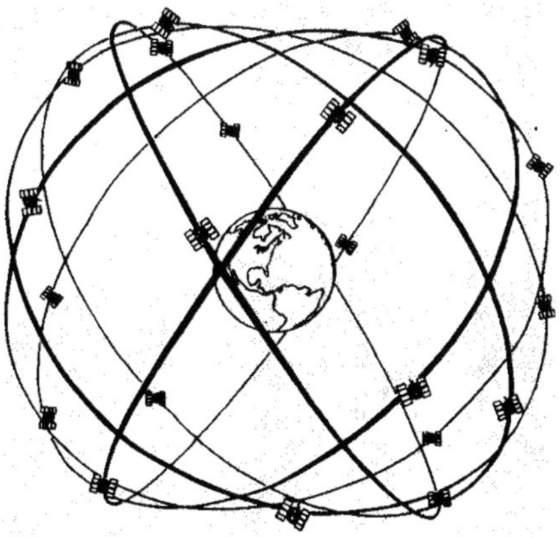

Fig. 1.19 GPS satellite configuration (from Larijani 1998).

ionosphere. Another major source of error is 'multipath' transmission, where the signal bounces off nearby buildings or other objects and can reduce the accuracy of the clear signal received direct from the satellite.

More accurate results can be attained using differential GPS (DGPS), whereby GPS stations at fixed, accurately known locations broadcast the range errors from each satellite in view. By correlating these correction messages with the standard GPS signal an accuracy of between 2 and 10 metres can be achieved, depending on the distance from the base stations. This enables features suitable for use as GCPs to be located to sub-pixel accuracy for most commonly used types of satellite images.

The successful use of GPS depends on line-of-sight visibility of the satellite, so it may not be effective in environments such as forest canopy or 'urban canyons'. In these cases systems that make use of GPS in conjunction with the Russian equivalent Glonass may give improved results.

1.5 IMAGE ENHANCEMENTS

There are a variety of methods that can be used to enhance the visual interpretability of the image. Among the techniques commonly used in image map production are:

- *removal of instrument errors* – replacement of missing pixels or lines caused by defects in data acquisition and removal of the effects of sensor imbalance
- *contrast stretch* – to improve the dynamic range of an image to make it easier to interpret visually
- *linear filter operations* – the use of digital filters to emphasise the linear features often associated with important cultural objects (such as roads and rivers) or to reduce the effects of noise in an image
- *non-linear filter operations* – often used to reduce the speckle present in SAR images

- *multi-band operations* – to choose a suitable colour combination of image bands, or to perform various enhancing transformations on multi-spectral data, including the merging of data from different image sources to maximise the particular information content of each.

1.5.1 Removal of instrument errors

Data dropout

A common problem with satellite images is data dropout which may occur during image transmission from the satellite to a groundstation or at a later processing stage (for example, transcription onto magnetic tape). This may manifest itself as missing individual pixels, groups of pixels or even entire lines. The first stage in repairing these defects is to determine where the data is missing. This can be performed interactively using a GIS or automatically using image statistics (for example, searching for groups of zero values in otherwise bright areas). Line repair can then be performed either by substituting neighbouring pixel values or by interpolation from neighbouring pixels – for example, replacing the missing pixel with the average of the pixel values above and below it, or with the average value of the three pixels above and three pixels below it. An alternative method of line repair is band substitution, where values are substituted from a similar band (for example, Landsat TM2 for TM1) either using unmodified values, or applying a contrast stretch based on the global statistics of the two bands. Alternatively values may be interpolated between two or more bands.

Striping

Another common instrument error is striping which occurs as follows: most satellite sensor systems use a bank of adjacent sensors for each spectral band – for example 6 (Landsat MSS) and 16 (Landsat TM). For pushbroom sensors a much larger number of detectors are used to record an entire line at a time. Miscalibration between the detectors can lead to a striped effect which may appear as horizontal stripes (as in the case of Landsat), vertical (SPOT) or occasionally a combination of both. The simplest solution is to calculate the image histogram corresponding to pixels acquired with each sensor and to adjust the histograms by applying a simple linear contrast stretch (see Section 1.5.2), this normalises the sensor output to that of a nominal median sensor. It is essential to perform this operation prior to the actual geocoding, which will (among other effects) rotate the stripes from the original grid pattern.

1.5.2 Contrast stretch

For many images only a small sub-range of the available dynamic range (0 to 255 for eight-bit data) are actually present in a particular image. When viewed an image will often appear dark and lacking in contrast and the human eye will not be able to discriminate the detail present in the image. The technique of modifying the image's dynamic range to suit the response of the human eye is known as contrast stretching.

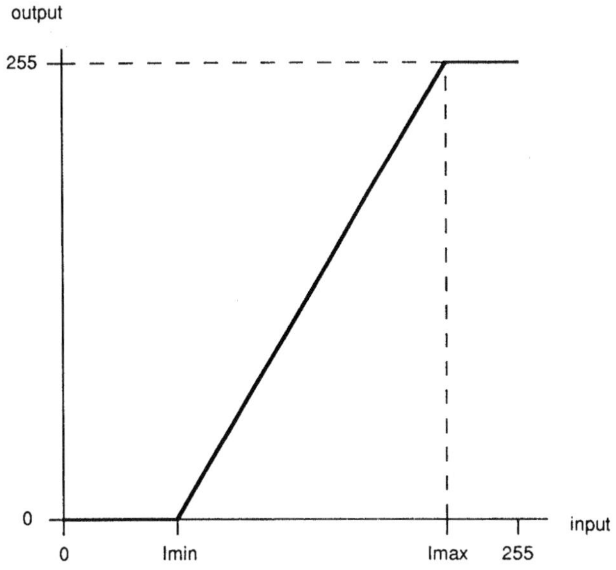

Fig. 1.20 Linear contrast stretch represented as a transfer function.

The simplest form of linear contrast stretch is to map the actual dynamic range onto the maximum available dynamic range. Thus, if the actual dynamic range is $[I_{min}, I_{max}]$ this contrast stretch would be a linear mapping:

$$O(x) = 255 * [I(x) - I_{min}] / [I_{max} - I_{min}] \quad (1.15)$$

where $O(x)$ = output image value
$I(x)$ = input image value
I_{min} = minimum input value
I_{max} = maximum input value

A representation of this contrast stretch in graphical terms as a 'transfer function' can be seen in Fig. 1.20.

A contrast stretch can usually be performed using a GIS by means of a look-up table (LUT). The LUT stores an array of numbers which determine the values displayed for each input value, and by using a LUT the contrast stretch can be performed in real time without having to modify the actual input values. An example of the use of a LUT to store the transfer function can be seen in Table 1.7. This shows a simple linear contrast stretch from an input range of [80, 131] onto the full dynamic range [0, 255], resulting in the transfer function shown in Fig. 1.21.

The simple linear contrast stretch may not prove satisfactory where an image with an overall large dynamic range may have most of its information lying within a relatively small sub-range. This is typified by a fairly dark image with a few bright

Table 1.7 Use of LUT to store a transfer function

Input value	Output value	Input value	Output value
0	0	.	.
1	0	.	.
2	0	.	.
.	.	129	245
.	.	130	250
.	.	131	255
79	0	132	255
80	0	.	.
81	5	.	.
82	10	.	.
83	15	254	255
84	20	255	255
85	25		

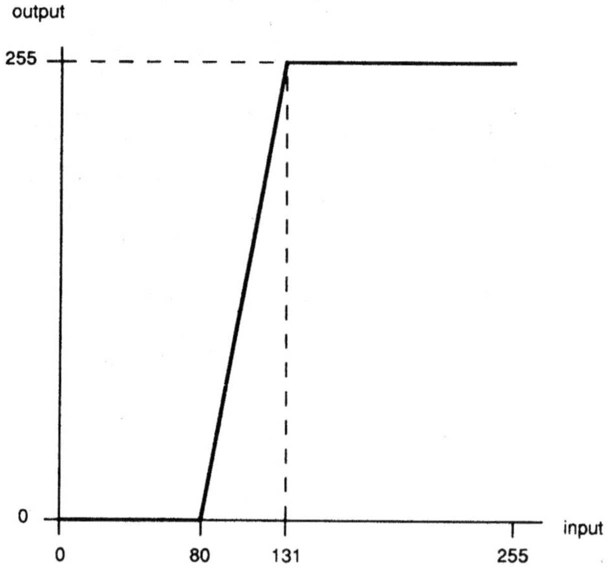

Fig. 1.21 The transfer function for the linear contrast stretch shown in Table 1.7, the range [80, 131] is mapped onto the full dynamic range [0, 255].

'saturated' pixels, caused for example by highly reflective clouds or snow. One way to deal with this problem is to ignore the outlying values near the ends of the histogram and only consider the data lying between the fifth and ninety-fifth percentiles. Another method is to use image-processing functions within a GIS to determine the end points for the stretch interactively.

A more complicated piecewise linear stretch can also be defined by specifying a series of (input, output) pairs. Any input value intermediate to two such points has

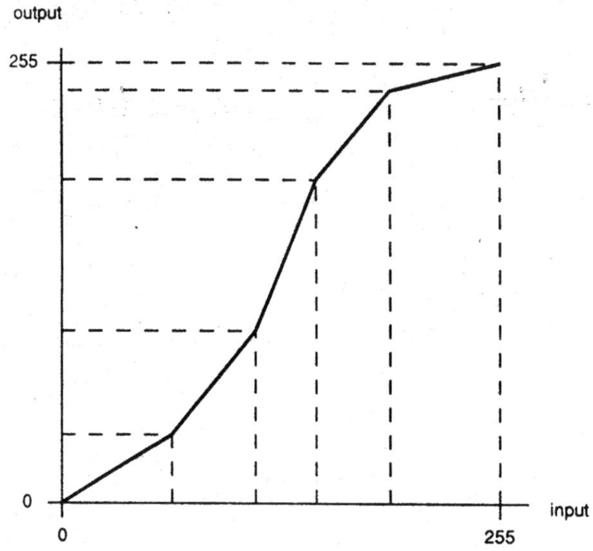

Fig. 1.22 Transfer function for a piecewise linear contrast stretch.

an output value determined by linear interpolation (see Fig. 1.22). This type of contrast stretch may be used to give different emphasis to portions of the histogram corresponding to distinct terrestrial features – for example, sea, low-lying areas and snow-covered upland.

Other automatic contrast stretches are based on transforming the empirical distribution of an image histogram onto a standard histogram whose statistics are better understood. Distributions favoured include the Gaussian (normal) and uniform distribution. In general, better results will be achieved by such a non-linear contrast stretch because the visual response of the human eye is non-linear, i.e. it requires a greater difference in density value near the extremes of black and white for a change to be perceivable, compared to the mid-grey regions.

1.5.3 Filter operations

Various image enhancements can be performed by means of operations with a filter kernel. A kernel is a matrix (often 3 × 3) containing a set of coefficients (or weights). The value of each pixel in the output image is determined by placing the kernel over the corresponding pixel in the input image, multiplying all the pixels beneath the kernel by the appropriate weight and summing the result. The process, known as convolution, can be represented mathematically (for a 3 × 3 kernel) as follows:

$$O(x, y) = \sum_{i=-1}^{i=+1} \sum_{j=-1}^{j=+1} K(i, j) * I(x+i, y+j) + c \tag{1.16}$$

where $O(x, y)$ = output pixel value
$I(x, y)$ = input pixel value

s = scale factor
$K(i, j)$ = filter kernel
c = constant offset

Note that the scale factor and offset are normally selected for convenience of display (for example, to ensure an output in the range 0 to 255) and suitable values for a particular kernel can readily found by experimentation using a GIS. Convolution filters are normally classified as smoothing (or low-pass) filters and edge enhancing (or high-pass) filters.

Smoothing filters
Smoothing filters are used to remove noise in an image, which can produce unwanted high spatial frequency artefacts. The simplest low-pass smoothing filter is the mean filter whereby each pixel is replaced by the average value of its neighbours. In this case

$$K = \begin{bmatrix} 1 & 1 & 1 \\ 1 & 1 & 1 \\ 1 & 1 & 1 \end{bmatrix} \qquad s = 1/9$$

where s = scale factor.

For isolated 'salt-and-pepper' noise a non-linear median filter, whereby each pixel is replaced by the median value of its neighbours, may be preferable to a linear filter as it ignores extreme high or low values, rather than blending them in. (The median filter is, strictly speaking a neighbourhood operation rather than a convolution filter as it is not a linear combination of the input data values.)

Edge-enhancing filters
Most of the information of interest in an image is derived by the human eye from the edges, i.e. the areas where there are sudden changes in the intensity of the image. Several algorithms can be used to enhance these edges to make them more apparent visually. The skill in applying a suitable algorithm is to select one which will emphasis the genuine edges of interest without creating artefacts from spurious edges caused, for example, by noise or data dropout. A general approach to enhancing edges in an image is to emphasise the difference between the central pixel and its neighbours. For example, the Laplacian kernel

$$K = \begin{bmatrix} -1 & -1 & -1 \\ -1 & 8 & -1 \\ -1 & -1 & -1 \end{bmatrix}$$

produces a zero result in homogeneous areas (those with little variation in pixel values) and a large positive or negative result when the pixel is significantly higher (or lower) in value than its surroundings.

By blending a suitable amount of the Laplacian output in with the original image an enhanced version is achieved.

$$O(x, y) = a * I(x, y) + b * L(x, y) \tag{1.17}$$

where $O(x, y)$ = output pixel (x, y)
$I(x, y)$ = input pixel (x, y)
$L(x, y)$ = Laplacian convolution of $I(x, y)$
a, b = constants

Edge-enhancing filters can also be directional – for example, the Sobel filters, K1 and K2

$$\text{K1} = \begin{bmatrix} -1 & -2 & -1 \\ 0 & 0 & 0 \\ 1 & 2 & 1 \end{bmatrix} \qquad \text{K2} = \begin{bmatrix} -1 & 0 & 1 \\ -2 & 0 & 2 \\ -1 & 0 & 1 \end{bmatrix}$$

emphasise horizontal and vertical edges respectively.

It is also possible to combine edge enhancement filters with smoothing filters. For example (Kidwell and McSweeney 1984) the following technique can be used to magnify an image:

- Magnify the image by a factor of 2 using pixel duplication.
- Smooth the image using a low-pass filter.
- Perform edge enhancement using a high-pass filter.

This method avoids the noticeable blocked appearance found when pixel replication is used on its own. The best results are usually found (in the case of Landsat TM data) by using an edge-enhancing kernel twice as large as the multiplication factor (e.g. 4 × 4 filter for a factor of 2 multiplication).

1.5.4 Non-linear filters

For SAR applications the characteristic speckle means that linear convolution filters tend to be ineffective because they destroy edge information. Non-linear filters such as the sigma filter (Lee 1981) are frequently used to overcome these limitations. Over homogeneous regions large window sizes are needed to improve speckle reduction by averaging; however, a large window size reduces the effective resolution of the data – for example, when used on a small bright object artefacts appear at the edge of the filter window (McConnell and Oliver 1996). This means that the background is ill-defined near bright targets and edges where effective discrimination is paramount.

Another approach (Cook *et al.* 1996) uses an objective function derived from a statistical model of SAR imagery. This objective function is then minimised by the method of simulated annealing which is, assuming some weak constraints, guaranteed to give the global minimum. Starting with an initial segmentation, the algorithm proceeds by randomly changing the current state. The annealing then decides whether or not to accept the new configuration by calculating the difference between the likelihoods of the data fitting these segmentations. Simulated annealing can be also be used as the basis of segmentation algorithms to produce 'cartoon images', each region being coloured with the mean of the data values in that region (see Fig. 1.23 colour section). The number of regions may be in the thousands and will need further processing to produce an effective classification into a small number of categories.

1.5.5 Multi-band operations

Band selection

There are two basic colour definitions for image maps: either monochrome or some kind of colour combination. If a multi-spectral sensor such as Landsat TM or SPOT is used as the basis for a monochrome image map it is usual to select an infrared band (such as TM7) as this usually provides better feature discrimination. Alternatively, a panchromatic sensor such as SPOT PA may be used. A colour image map on the other hand will use the sensor's ability to image an area in several portions of the spectrum, not only visible but into the infrared. On a display device, or a photograph, three image bands (one for each of red, green and blue) can be depicted. A SPOT XS image with exactly three bands can be mapped exactly onto the three available display bands. With images possessing more than three bands, several methods can be used to reduce the number to three. With some images it is possible to produce a natural colour image. For example, with Landsat TM three visible wavebands are located approximately at the three primary colours. Thus the combination:

$$\{3, 2, 1\} \rightarrow \{R, G, B\}$$

will produce an image with similar properties to a standard colour photograph of the scene which is simple to interpret as it corresponds to familiar colours – for example, green fields. There are of course many other possible combinations; ignoring the thermal band (TM6) because of its larger pixel size, there are 20 combinations of three bands from six and 120 permutations when they are mapped onto the primary colours. From the multitude of possibilities a few have proved most suitable (Dowman and Peacegood 1988) for the production of image maps, for example:

$$\{5, 3, 2\} \rightarrow \{R, G, B\}$$

This has proved useful where good water body discrimination is required. However, the most suitable combination for an image map will usually depend on the predominant land cover in the image. For example, in areas of dense vegetation band 4 has a characteristically high response, so summer scenes with large areas of vegetation may appear unbalanced, with towns appearing swamped by surrounding countryside. In this case band 5, which has a lower vegetation response, may prove to be a more suitable substitution.

It is also possible to make the best selection of bands mathematically (Chavez *et al.* 1984; Sheffield 1985). This is achieved by examining the variance–covariance matrix for the band combinations of a specific image. The variance of a band gives a measure of the amount of information within it, whereas the covariance between two bands measures how closely the information contained in the first band resembles that in the second. Ideally, therefore, the three bands selected should have large individual variance, but there should be little correlation between any two of them.

The best band combination can be found by taking the variance–covariance matrix $M(i, j)$ and calculating the determinants of each principal sub-matrix – the original values for the thermal band are scaled by a factor of 4 to account for the coarser resolution of 120 metres. There are 35 principal (3×3) sub-matrices corresponding to each possible three-band combination. The triplet $B(i, j, k)$ with the maximum value is the best one to choose but there are still six permutations that map it onto $\{R, G, B\}$. The selection is made by examining the variance in each band (the diagonal entries in M). The eye is most sensitive to green, then to red, then to blue; so the band with most variance is assigned to green, second to red and third to blue. The selected band combination will again depend on the area being imaged but the algorithm has been found to generally pick out combinations with only one visible band, whereas human operators tend to prefer combinations with two.

IHS transform

Another approach to making image maps is to combine imagery from two sources in a way that utilises the best attributes of both of them; for example, the high-resolution of a SPOT Panchromatic Image with the spectral coverage of Landsat TM. One useful tool in the merging of optical imagery is the Intensity Hue Saturation (IHS) algorithm. This is particularly suited for the combination of multi-spectral imagery such as SPOT XS or Landsat TM with higher resolution monochrome images such as SPOT PA, to provide a simulated high-resolution multi-spectral product (Carper *et al.* 1990). A simplified portrayal of IHS space can be seen in Fig. 1.24.

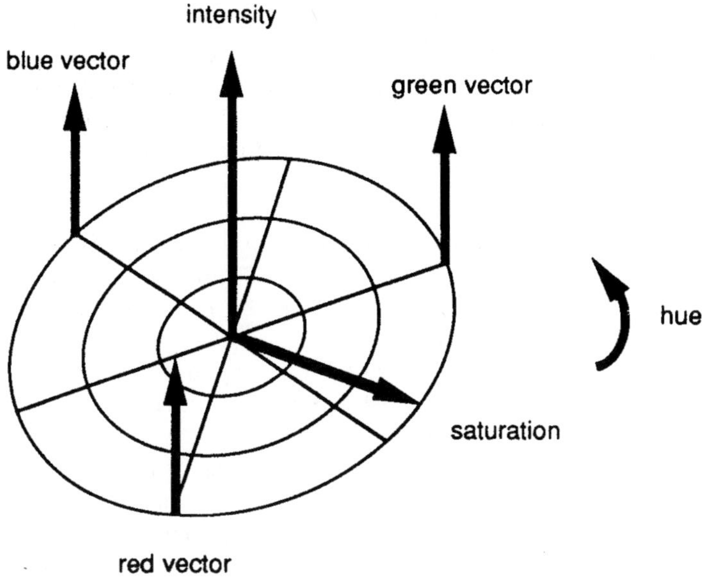

Fig. 1.24 Representation of colours in an IHS coordinate system (adapted from Carper *et al.* 1990).

The IHS transform allows manipulation of spatial information without destroying overall colour balance. The intensity is related to the overall brightness of a pixel, the hue to its colour (that is, the dominant spectral wavelength) and saturation to its purity (relative to grey). A suitable procedure for merging a high-resolution panchromatic image and a lower-resolution multi-spectral image is shown in Fig. 1.25. The first step is to geocode both images and then to transform the low-resolution image from (R, G, B) to (I, H, S) space. It is also necessary to resample the

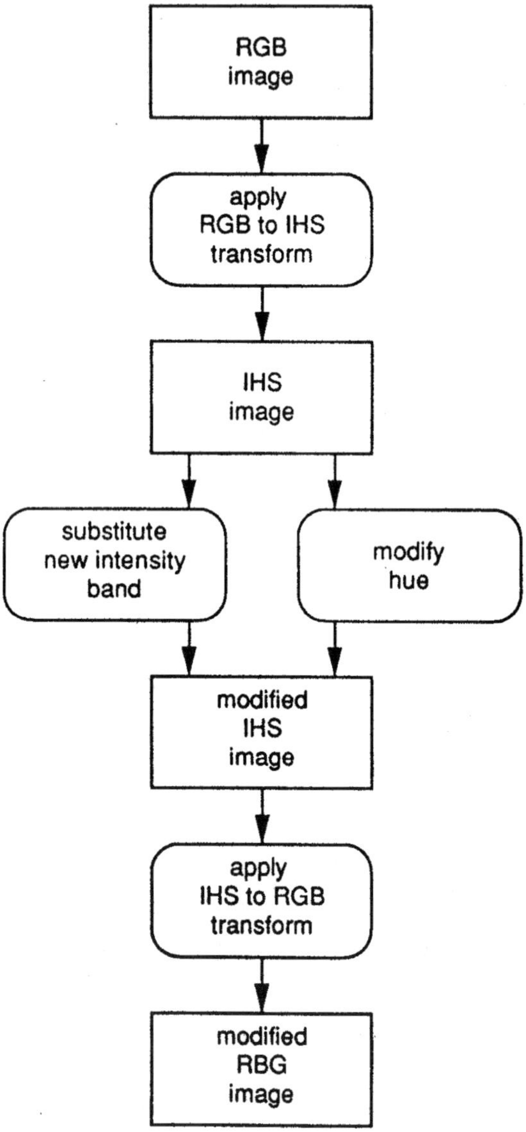

Fig. 1.25 Algorithm for IHS transform.

low-resolution image to the higher-resolution, using, for example, a cubic convolution technique (Welch and Ehlers 1987) or by pixel replication. At this stage the panchromatic image is substituted for the intensity component prior to transformation back from (*I*, *H*, *S*) to (*R*, *G*, *B*). The result is an image that has the high resolution of the panchromatic imagery but retains the spectral characteristics of the multi-spectral imagery (see Fig. 1.26 in the colour section).

It is also possible to perform additional processing in the (*I*, *H*, *S*) space before conversion back to (*R*, *G*, *B*) space, for example:

- histogram equalisation of the hue component to improve the colour balance
- rotation in colour space to move histogram peaks to desired colour (for example, to produce a greener image).

1.6 IMAGE MOSAICS

A mosaic is an image product generated by joining together several geocoded images in a seamless manner to provide coverage of an extended geographic area. The production of image mosaics (Zobrist *et al.* 1983) is a skilled and labour-intensive process, with much intervention from a trained operator required, but the result is often a visually attractive product as well as a valuable source of cartographic information unobtainable by conventional means. Examples of satellite image mosaics are radar maps of Germany and Antarctica being prepared using ERS-1 SAR data (Kosmann *et al.* 1992) and the continental scale of mosaics of Europe (see Fig. 1.27) and Antarctica (see Section 1.6.4).

This section describes:

- Planning the mosaic
- Mosaicking techniques
- Improving mosaic accuracy
- The AVHRR mosaic of Antarctica
- SAR mosaicking.

1.6.1 Planning the mosaic

The first stage in preparing a mosaic product of a particular area is to assess the potential coverage by a particular satellite sensor. For example, using the Landsat World Reference System (WRS) a chart can be drawn up showing all the path/row combinations needed to cover an area of interest. Quick-look prints can then be used to select the best actual images to be used. The area of interest may conform to a country or correspond to a particular map sheet. The images should be as cloud-free as possible, although it is often possible to use one partially cloud-free image to patch in the cloudy areas on another. Ideally the images will be of similar appearance. If for example one image is snow-covered or images correspond to different seasons of vegetation growth it may prove difficult to join images in a seamless manner.

For many image maps (USGS 1986) it is preferable that growing vegetation is

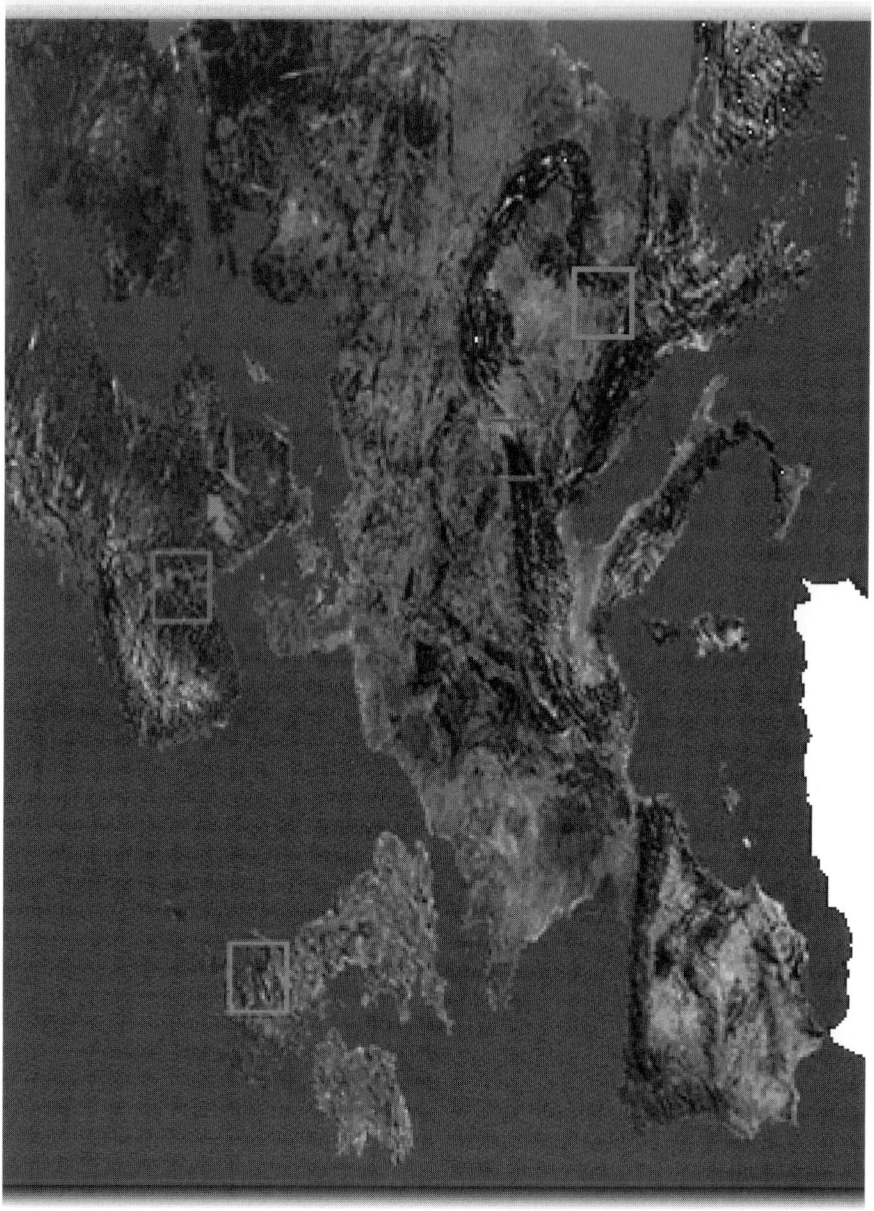

Fig. 1.27 A mosaic of Europe produced from cloud-free Resurs data (courtesy WorldSat, source data copyright Satellus Metria).

Table 1.8 Suitable Sun angles for mosaic production (based on USGS 1986)

Terrain	Sun angle (degrees)
Flat	0–15
Moderate	15–30
Steep	30–50

avoided, as the strong infrared response may overwhelm cultural features. Often autumn imagery is the best – when deciduous leaf-cover has passed its growing season – or, alternatively, spring before the growing season has begun in earnest. The Sun angle should not be too great, as a value of more than 50° will make it difficult to discriminate low-relief features due to lack of terrain shadows; it is also desirable that sun angles are roughly the same for each image to minimise the amount of radiometric correction necessary. A suitable Sun angle will also enhance topographic structure, depending on the local relief. Table 1.8 shows appropriate values suggested by the US Geological Survey.

It is also necessary at this stage to decide on the grid size of the output image. Owing to the large amount of data involved, practical constraints on computer storage may dictate that this be larger than the pixel size of the input images. For example, a 1,000 km square mosaic with a grid size of 25 metres (easily achievable with Landsat TM) would constitute an image of 40,000 lines of 40,000 pixels requiring storage of 1,600 megabytes of data. The storage required for a product will vary inversely with the square of the grid size. In the above example a grid size four times coarser (100 metres) would require one-sixteenth the storage space (100 megabytes), which may be a more manageable proposition.

Before starting the 'production line' it is usually advantageous to construct one or more prototypes – using, for example, sub-sampled images which have been bulk geocoded – to produce a 'mock-up' of the area being covered. This gives an indication of the appearance of the final product which can be used to determine the location of 'cut lines' for the mosaic process. Based on this information a definitive project plan can be drawn up and used as a baseline for the product development.

1.6.2 Mosaicking techniques

The production of an image mosaic can require an immense amount of data: an idea of the effect of data type and hence grid size on the output product size can be seen in Table 1.9, showing the coverage required for an area 1,000 km square.

A common approach to mosaic production is to join one image at a time onto the partially completed mosaic. Thus the first step is to join two overlapping images together (see Fig. 1.28).

In order to balance the radiometric appearance of the component images a linear contrast stretch is applied so that the mean and variance of the overlap region in both images is the same. It can be seen that any point on the mosaic grid falls into one of four categories:

Table 1.9 Output product size as a function of input data type

Sensor	Grid size (m)	Mosaic size			Storage (megabytes)
AVHRR	1,000	1,000	×	1,000	1
Landsat MSS	100	10,000	×	10,000	100
Landsat TM	25	40,000	×	40,000	1,600
SPOT PA	10	100,000	×	100,000	10,000
Ikonos	1	1,000,000	×	1,000,000	1,000,000

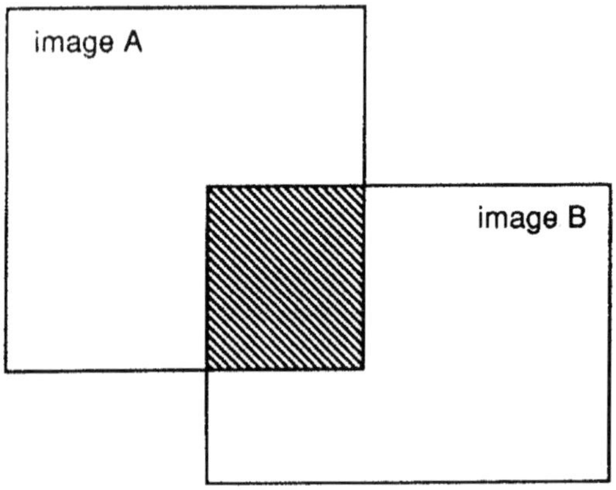

Fig. 1.28 Joining two overlapping images A and B (the overlap area is shaded).

(a) in the first image only
(b) in the second image only
(c) in neither image
(d) in both images.

In cases (a) and (b) the output pixel in the mosaic is set to the value of the corresponding input pixel. In case (c) it is set to a background value (it is customary in mosaicking practice to reserve a value such as 255 as a background value – genuine data with a value of 255 can be reset to 254 with little loss of significant radiometric information). In the overlap area, various simple algorithms can be used to calculate the overlap value, the most often used ones are: 'brute force' (set to the value of the first image, ignoring the second), calculate the average value of the two pixels, or use the maximum (alternately minimum) of the two values.

The disadvantages of the simple approach is that the algorithms used can tend to create an artificial seam at the edge of the overlap region. It is desirable therefore to smooth-out the values in the intersection so that the output value is $A(i, j)$ if (i, j) is on the seam of A, $B(i, j)$ on the seam with B, and a smoothly weighted blending function at intermediate points. A suitable method is based on the mathematical measure of distance, known as the taxi-cab metric. This is a computationally efficient method of defining the distance between two points as the sum of the difference in pixel numbers and the distance in line numbers:

$$d[(x_1, y_1),(x_2, y_2)] = |x_1 - x_2| + |y_1 - y_2| \tag{1.18}$$

where (x_1, y_1) = position of first point
(x_2, y_2) = position of second point

This can be used as the basis of a function z which measures the relative closeness of a point from A and B on a scale of [0, 1] such that $z = 0$ on the join with A and $z = 1$ on the join with B and varies smoothly between them. In terms of the taxi-cab metric:

$$z = d_A / (d_A + d_B) \tag{1.19}$$

where z = relative closeness function
d_A = distance from nearest point on join with A
d_B = distance from nearest point on join with B

The function z can then be used to define the weighting function $w(z)$ used to define the output value:

$$M(i, j) = w * A(i, j) + (1-w) * B(i, j) \tag{1.20}$$

where $M(i, j)$ = mosaic value at point (i, j)
$A(i, j)$ = value at point (i, j) in image A
$B(i, j)$ = value at point (i, j) in image B

Once the first two images have been joined, the same procedure can be repeated for each subsequent image by regarding the partially completed mosaic as 'Image A'. The merging of images is performed by assembling the component geocoded images on a regular grid. Best results are obtained by merging images acquired on the same along-track pass followed by adjacent passes. At this stage any filler data is replaced by genuine image data. A less obvious join between two images can be achieved if, instead of using straight line boundaries, natural features such as field boundaries are used. The first stage is to specify the join line between the two images, which can be done in two main ways:

- using a GIS to draw in the join line
- using the GIS to specify a series of tie points along the desired boundary and use a suitable interpolation method (e.g. linear, cubic spline) to join and complete the line.

It is then a simple matter to process each point in the overlap and to allocate a value depending on which side of the boundary line it falls. This was the method employed by the UK National Remote Sensing Centre to produce a series of regional mosaics from Landsat TM data (Batts, pers. comm.). Where a mosaic encompasses a coastal region it is necessary to mask out the sea to remove the great variation in the sea's appearance from image to image which cannot be removed by blending algorithms. The sea mask may be obtained by using a GIS or by automatically defining the sea boundary by density slicing or histogram thresholding. Care must also be taken with any remaining areas of cloud as it is undesirable to blend in cloud values with genuine surface features. It is difficult, however, to reliably remove clouds automatically, so a skilled operator will 'paint out' clouds interactively.

1.6.3 Improving mosaic accuracy

Improvements in mosaic accuracy fall into two classes: radiometric and geometric. One commonly used method of radiometric correction is to equate the histograms of the two images within the overlap area. This is normally achieved by using the cumulative histograms of the two images and adjusting them so that they agree at selected percentile points. This can be achieved by use of a LUT, interpolating between points as necessary. Another method (Binnie and Colvocoresses 1987) is to perform radiometric matching using a sampling method. Image patches (size 9×9) are analysed along the seam between two mosaic components to compare the average brightness value between the two images; this information is then used as input to a piecewise linear surface fit between the two sets of radiometric values.

To improve geometric accuracy a set of registration tie points in the overlap area can be used to produce a more accurate spatial correspondence where the two images join. This method is used, for example, to automate the production of mosaics from SAR images (Kwok *et al.* 1990). Frame-to-frame registration in the overlap region is performed by selecting image patches and performing one of the following pattern matching operations:

- cross-correlation method
- zero-crossing method.

The zero-crossing method is based on the Marr–Hildreth algorithm (Marr and Hildreth 1980) and may be preferable because changes in SAR image geometry and radiometric characteristics (such as the multiplicative noise evident as speckle) tend to decorrelate the image. A similar method was used for the production of mosaics from Landsat TM images (Binnie and Colvocoresses 1987) with the refinement that tie-points in the seam region are triangulated, and a transformation is defined on each of the triangles in order to form a continuous model for the entire image map.

1.6.4 The AVHRR mosaic of Antarctica

One of the most impressive satellite mosaics is the AVHRR mosaic of Antarctica produced by the UK National Remote Sensing Centre (Merson 1989). The raw material for the mosaic consists of a set of 50 NOAA–AVHRR images of the continent which vary greatly in data integrity and cloud cover. The best 25 scenes were selected to form the relatively cloud-free basis of the mosaic. AVHRR bands 1, 2 and 4 were selected to give a consistent colour scheme throughout. Before the individual images were used they were compressed from their original 10-bit dynamic range [0, 1023] down to 8 bits [0, 255] for subsequent processing, and any data dropouts were replaced by a low-pass line repair algorithm. The images were acquired illuminated by a variety of Sun angles. This led to considerable radiometric inequalities within the image set as the reflectance in bands 1 and 2 tend to be highly correlated with the Sun angle. Because of the large area covered by an AVHRR image it is not reasonable to assume that the Sun angle remains almost constant across the width of an image (as is the case with Landsat), thus a project-specific algorithm was devised to correct for Sun angle effects.

The first stage of processing for each image consisted of producing a Sun angle image derived from Sun–sensor–target geometry, with pixel values proportional to the sine of the Sun elevation. An IIPF was then used to produce scatter diagrams of the Sun image pixel against the intensities from the actual image. The best-fit quadratic curve to the scatter diagram was then used to produce a numeric correction to the raw intensity levels over the whole scene. A study of the map projections commonly used for world mapping reveals that many of them (for example, Mercator) are not suitable for mapping polar regions, typically distorting regions of high latitudes or even disconnecting them completely. For this reason the projection used was Polar Stereographic (centred on the south pole). Each image was geocoded separately. A first approximation was made using satellite orbital ephemeris and scanner parameters to correct for Earth curvature and rotation and variations in satellite orbit and attitude. Another problem with images of the Antarctic is the almost complete lack of suitable accurately located GCPs. This was circumvented by obtaining a digitised Antarctic coastline from the UK Scott Polar Research Institute (SPRI), consisting of 222,000 points compiled from various sources between 1961 and 1981. An extract was taken from the coastline and transformed to raster format for each image to be transformed. This allowed tie-points to be identified on the coastline and on the actual image. The locations of the points in each coordinate system were then stored and the best-fit bilinear matrix calculated to model the transformation. This was then be used to refine the approximate transformation derived from the ephemeris data. The final step was a cubic convolution resampling.

The method used for joining images was slightly different from the overlapping method described previously. In this case, the mosaic was constructed using abutting images, i.e. every pixel in the final output mosaic was derived from a single input image. This method was preferred because the large radiometric variations in the input images made the algorithms that were normally used for blending overlapping images unsuitable. To record the final positions of the constituents of the mosaic a master mask was constructed. This is a grid of pixels corresponding to the size of the final mosaic. Each pixel value is a code indicating which input image will be used to produce the corresponding 1-km square (see Fig. 1.29). The master mask was then used to blank out the pixels in each component that are not going to be used. The scenes used for the final mosaic are shown in Table 1.10.

Various techniques were used to complete the final mosaic:

- adjacent components balanced using a linear contrast stretch to minimise apparent seams
- a smoothing algorithm applied to get rid of any remaining discontinuities
- areas of sea masked out and set to a uniform value.

The sea masking was necessary due to the large variations in sea state and ice margin. The completed mosaic, which can be seen in Fig. 1.30, provides a visually

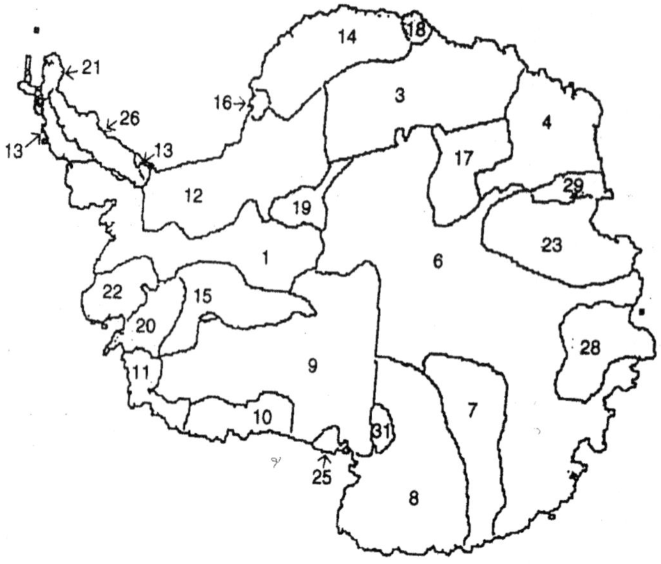

Fig. 1.29 The master mask used for the construction of the AVHRR mosaic of Antarctica (from Merson 1989).

Table 1.10 AVHRR Antarctic scenes used to produce mosaic (based on Merson 1989)

Date	Satellite	Longitude (°E)	Mask key
12 January 1980	NOAA 6	148–184	8
13 January 1980	NOAA 6	230–280	22
14 January 1980	NOAA 7	74–140	6
18 January 1980	NOAA 6	62–121	23, 29
25 January 1980	NOAA 6	246–324	13, 20
11 February 1980	NOAA 6	298–335	12
15 February 1980	NOAA 6	42–67	4
6 November 1980	NOAA 6	210–270	11
8 November 1980	NOAA 6	290–304	15, 26
12 November 1980	NOAA 6	168–230	9
24 November 1980	NOAA 6	280–294	1
27 November 1980	NOAA 6	330–31	16-19
24 January 1981	NOAA 6	300–310	21
29 January 1983	NOAA 7	350–39	3
6 February 1983	NOAA 7	333–10	14
12 February 1983	NOAA 7	140–220	10, 25, 31
16 February 1983	NOAA 7	57–116	28
29 November 1983	NOAA 7	143–172	7

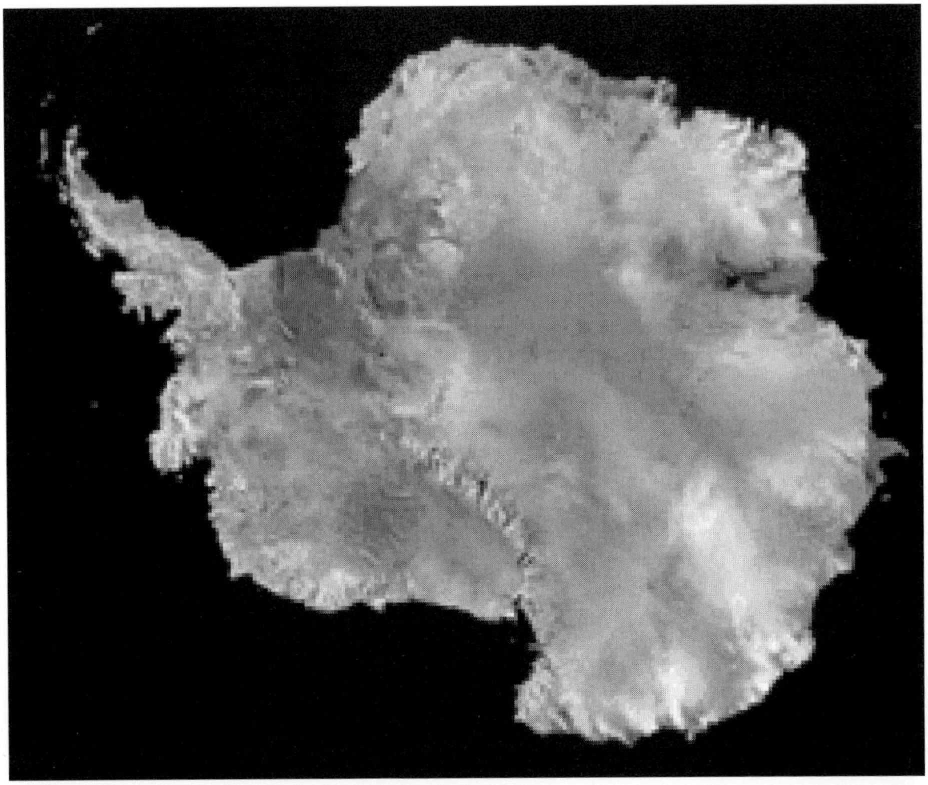

Fig. 1.30 AVHRR mosaic of Antarctica (Courtesy USGS, www.terraweb.wr.usgs.gov)

striking and scientifically useful product with a variety of potential applications in geological and other polar research.

1.6.5 SAR mosaicking

As with optical image mosaicking, the production of SAR mosaics consists of two processing steps:

- geocoding of the input images to a common map projection
- fusion of scenes into a seamless image product.

Unlike optical images the radiometric information content of SAR scenes in rugged areas is primarily due to topographic effects and hence not expected to vary between scenes. However, in the overlap areas radiometric variations due to sensor or processor gain change are noted. To achieve seamless blending, a normalisation procedure based on the grey levels of a set of simulated images is used; the assumption being that these values are directly correlated to the local terrain foreshortening. The matching of real images to simulated images can be fine-tuned using an area-based correlation algorithm.

The use of simulated images (Guindon and Adair 1992) has been applied

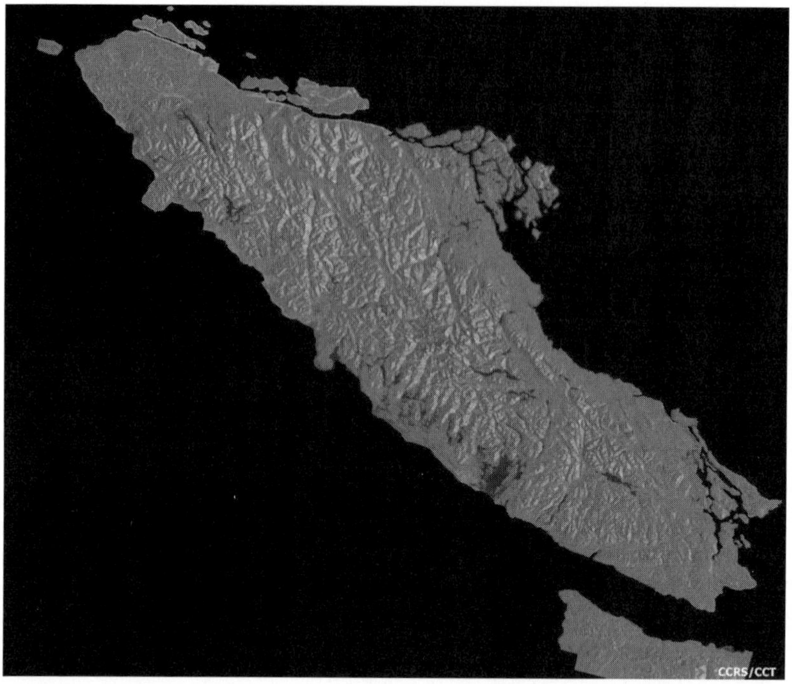

Fig. 1.31 SAR image of Vancouver Island, British Columbia (Canada Centre for Remote Sensing).

successfully to SAR image mosaics by the Canada Centre for Remote Sensing. Using this approach a 53-scene mosaic of the southern British Columbia mainland and a 16-scene mosaic of Vancouver Island, both compiled using ERS data, were produced (see Fig. 1.31).

1.7 IMAGE PRODUCTS AND PROCESSING LEVELS

This section provides comparative details of some of the 'standard products' supplied by the major distributors of satellite data. It describes the major factors that are needed to make informed decisions on the selection of data for a specific project, including spatial resolution, revisit times, spectral bands available and levels of geometric and radiometric processing. The following sources of high-resolution optical and SAR imagery are described:

- Landsat TM
- SPOT
- IRS
- ERS-SAR
- Radarsat

1.7.1 Landsat

A summary of the TM technical specification can be seen in Table 1.11.

Table 1.11 Landsat TM technical specification

Country of origin	United States
Launch	Landsat 4: 16 July 1982 Landsat 5: 1 March 1984 Landsat 6: Failed Landsat 7: 15 April 1999
Orbit	Sun-synchronous (near polar)
Orbit period	99 minutes
Equator crossing	10:00 (Landsat 7)
Inclination	98.2°
Altitude	705 km
Coverage	81°N to 81°S
Repeat cycle	16 days
Swath width	185 × 185 km
Pixel size	30 metres (band 6 is 120 metres and 60 metres for Landsat 7)
Panchromatic (Landsat 7 only)	15 metres
Band 1	0.45–0.52 μm
Band 2	0.52–0.60 μm
Band 3	0.63–0.69 μm
Band 4	0.76–0.90 μm
Band 5	1.55–1.75 μm
Band 6	10.40–12.50 μm
Band 7	2.08–2.35 μm

The base product (Level 0R) for the Landsat ETM sensor is a raw format dataset which has not been radiometrically or geometrically corrected. It does, however, contain the ancilliary data which is needed to subsequently perform these corrections. The next level of product (Level 1R) has been radiometrically corrected either by internal calibration or by applying the gain values stored on the calibration parameter. No additional geometric transformations are applied.

The first level of geometric correction is the Level 1G product. This applies corrections for the spacecraft and sensor geometry that have been generated onboard the satellite. The image data is resampled to a uniform pixel size using a standard resampling kernel (see Section 1.2.2). The product is available in a number of standard map projections including: (1) Universal Transverse Mercator, (2) Lambert Conformal Conic, (3) Transverse Mercator, (4) Polyconic, (5) Oblique Mercator, (6) Polar Stereographic, and (7) Space Oblique Mercator. Although the Level 1G product removes the major geometric distortions such as variations in the satellite orbit and Earth curvature and rotation, it still includes effects due to terrain distortion leaving a residual error of about 250 metres in flat areas at sea level. In areas of moderate relief this can be reduced to circa 30 metres using GCP techniques; in more rugged areas DEMs can be employed to achieved comparable accuracy.

1.7.2 SPOT

A summary of the SPOT technical specification can be seen in Table 1.12.

Table 1.12 SPOT technical specification

Country of origin	France
Launch	SPOT 1: 22 February 1986 SPOT 2: 22 January 1990 SPOT 3: 26 September 1993 SPOT 4: 24 March 1998
Orbit	Sun-synchronous (near polar)
Orbit period	101 minutes
Equator crossing	10:30 (local time)
Inclination	98.7°
Altitude	830 km
Coverage	81°N to 81°S
Repeat cycle	26 days
Swath width	60 km (nadir)
Pixel size	10 metres (panchromatic) 20 metres (multi-spectral)
Panchromatic	0.51–0.73 μm
Band 1	0.50–0.59 μm
Band 2	0.61–0.68 μm
Band 3	0.79–0.89 μm
SPOT 4 only	Short Wave Infrared (SWIR)
Band 4	1.58–1.75 μm

SPOT data is available from SPOT Image with the following standard processing levels:

- *level 1a* – raw image data with detector normalisation in each band
- *level 1b* – basic pre-processing including radiometric and geometric corrections
- *level 2* – precision processing, with or without ground control points, to Lambert, Mercator or other projections
- *level S* – precision level to enable pixel to pixel registration with a reference image for thematic applications.

Of particular interest is the introduction (Pulpin 2001) of the Short Wave Infrared Band (SWIR) introduced as part of SPOT 4. This has the following complementary features:

- It is relatively unaffected by atmospheric corrections, increasing the probability of image acquisition through thin cloud, particularly in tropical regions.
- It is sensitive to surfaced moisture content, providing improved discrimination for land-cover mapping.

- It is less sensitive to scattering by atmospheric water vapour, improving image constrast and shadow definition for example to detect lines of hedgerows.
- It also allows better correlation between images used to generate DEMs.

1.7.3 IRS

A summary of the IRS-1 technical specification can be seen in Table 1.13.

Table 1.13 IRS-1 technical specification

Country of origin	India
Launch	IRS-1a: 17 March 1988
	IRS-1b: 29 August 1991
	IRS-1c: 28 December 1995
	IRS-1d: 29 September 1997
Orbit	Sun-synchronous (near polar)
Orbit period	103 minutes
Inclination	99°
Altitude	904 km
Coverage	81°N to 81°S
Repeat cycle	22 days
Payload IRS-1a, IRS-1b	Linear Imaging Self Scanner (LISS) I
	Linear Imaging Self Scanner (LISS) II
Payload IRS-1c, IRS-1d	PAN
	Linear Imaging Self Scanner (LISS) III
Swath width	148 km (LISS-I)
	145 km (LISS-II)
	142 km (LISS-III)
	70 km (PAN)
Pixel size	72 metres (LISS-I)
	36 metres (LISS-II)
Country of origin	India
	23 metres (LISS-III)
	70 metres (LISS-III; SWIR)
	5.8 metres (PAN)
Band 1	0.45–0.52 μm (not LISS-III)
Band 2	0.52–0.59 μm
Band 3	0.62–0.68 μm
Band 4	0.77–0.86 μm
Band 5 (SWIR)	1.55–1.70 μm (LISS-III only)
PAN	0.50–0.75 μm

The basic IRS product is radiometrically corrected only. In the case of the PAN sensor the three detector arrays give rise to three separate radiometically corrected data stripes. These are stitched together in the geometric correction process. As with Landsat and SPOT there are two distinct levels:

- *path oriented* – geometrically corrected to the orientation as seen by the spacecraft (corrections include Earth rotation, Earth ellipsoid and map projection, satellite attitude and internal sensor distortions)

- *map oriented* – geometrically corrected to orient the images to a 'true north' orientation.

Further details of the IRS programme can be found in Section 7.1.6.

1.7.4 ERS

The main characteristics of the SAR data from ERS-1 and ERS-2 are shown in Table 1.14.

Table 1.14 ERS SAR technical specification

Frequency	5.3 GHz (C band) $\pm$ 0.2 MHz
Polarisation	Linear vertical
Radiometric resolution	2.5 dB at sigma-nought = -18 dB
Swath location	244.5 km to the right of the orbital track
Swath width	80.4 km (nominal within specifications) 102.5 km telemetered
Incidence angle	23° at mid-swath
Data rate	105 Mbit/s

ERS SAR data is available from the European Space Agency in a number of different processing levels.

- SAR *Annotated Raw Data* (RAW) consists of 16 seconds of raw data together with all the auxiliary information necessary for data processing. It is used for testing purposes and for scientific users who require unprocessed data that they can calibrate themselves.
- The SAR *Precision Image* (PRI) is a multi-look image calibrated and corrected for the SAR antenna pattern and range-spreading loss but no geocoding is applied. The product has a location accuracy of about 100 meters in range and 200 metres in azimuth.
- A SAR *Ellipsoid Geocoded* (GEC) product is a multi-look image calibrated for the antenna pattern and range-spreading loss. It is rectified to a map projection but no correction for terrain distortion (either radiometric or geometric) has been applied. The product has location accuracv of better than 100 metres in flat relief areas.
- For higher geometric accuracy it is imperative to use a DEM to remove terrain distortions. The *Terrain Geocoded Image* (GTC) provides a precison product with locational accuracy of better than 50 metres together with a shadow mask and file of layover areas; however, no correction is performed for the effect of terrain on radiometric values.

1.7.5 Radarsat

The main characteristics of the SAR data from Radarsat are shown in Table 1.15.

Table 1.15 Radarsat SAR technical specification

Frequency	5.3 GHz (C band)
Polarization	HH polarisation
Swath width	925 km
Right-looking, steerable antenna	
ScanSAR capability for wide area coverage	
Multi-mode imaging capabilities	7 beam modes
Average radar data rate	73.9 to 100.0 Mbit/s

The Radarsat beam modes can be seen in Fig. 1.32.

Radarsat's SAR does not collect data continuously. The fine, standard, wide, and ScanSAR beam positions are always loaded, while four of the seven extended beam positions are selectively loaded. The ScanSAR beam mode, a feature unique to Radarsat, provides repeat coverage of large for scales in the order of 1:5,000,000 to 1:250,000. Radarsat products are available in three processing levels:

- RAW data has had no standard processing applied.
- Path-oriented is referenced to a standard Earth ellipsoid.
- Map-oriented data has been transformed to a standard map projection.

Table 1.16 provides a comparison between Radarsat (geometric) processing levels and those available from ERS, Landsat and SPOT.

Table 1.16 Comparison of Radarsat products with other EO products (based on Radarsat 2001)

Radarsat product	Code	ERS	Landsat	SPOT
Raw data	RAW	RAW	RAW	1a
Path-oriented data				
Single look complex	SLC	SLC	NA	NA
Path image	SGF	PRI	Precision correction	1b
Path image plus	SGX	NA	NA	NA
ScanSAR narrow	SCN	NA	NA	NA
ScanSAR wide	SCW	NA	NA	NA
Map-oriented data				
Map image	SSG	GEC	Map-oriented systematic correction	2a
Precision map image	SPG	GTC	Map-oriented precisions correction	2b

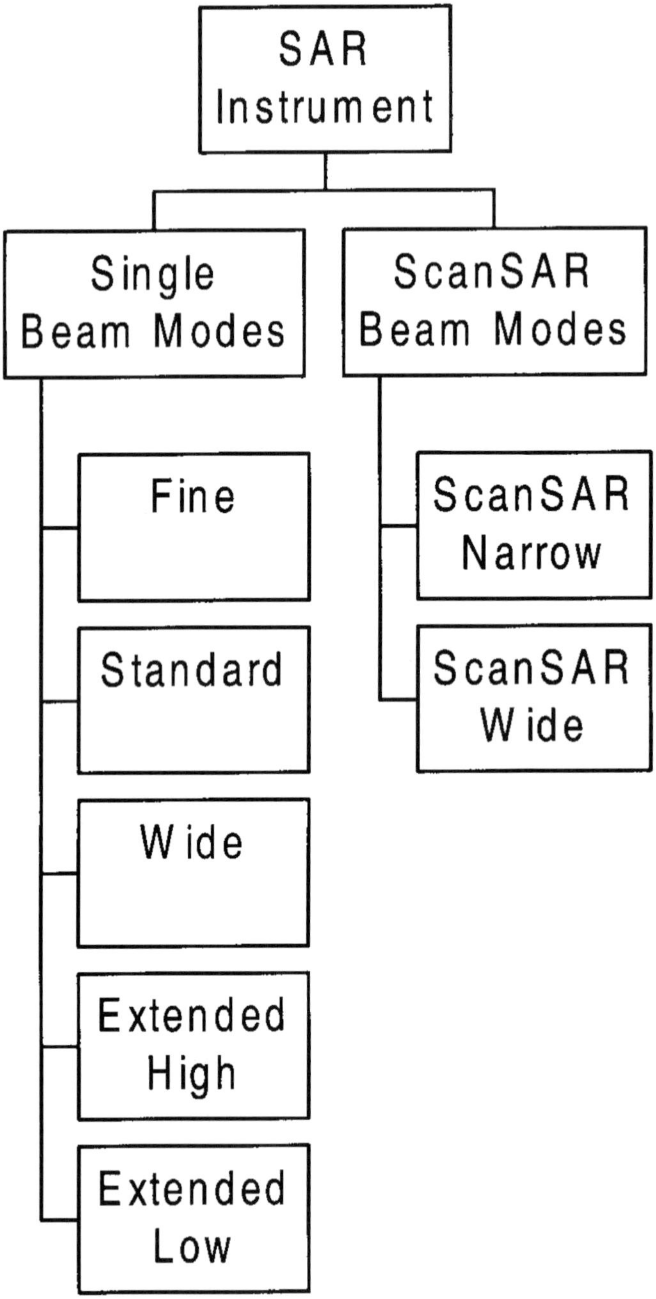

Fig. 1.32 Illustration of Radarsat beam modes (courtesy Canada Centre for Remote Sensing).

REFERENCES

Adamson J, Kerr GW and Jacobs GHP (1988) Rectification quality assessment of Meteosat images. *ESA Journal*, **12**, 467–82.

Albertz J, Lehmann H and Tauch R (1992) The production of satellite image maps – experiments at the Technical University of Berlin. *Proceedings of Satellite Symposia 1 and 2: Navigation and Mobile Communications, and Image Processing, GIS and Space-assisted Mapping, from the 'International Space Year' Conference, Munich, Germany*, 30 March–4 April 1992, ESA ISY-2, pp. 175–80.

Benny AH (1983) Automatic relocation of ground control points in Landsat images. *International Journal of Remote Sensing*, **4** (2), 335–42.

Benny AH (1985) Example of the use of Landsat satellite imagery for the accurate location of offshore islands. *International Journal of Remote Sensing*, **6** (10), 1581–4.

Binnie DR and Colvocoresses AP (1987) The Denali image map. *Photogrammetric Engineering and Remote Sensing*, **53** (3), 307–10.

Brooks J, Kumar R and Levine I (1981) Implementation of the Space Oblique Mercator projection in a production environment. *Proceedings of the Fifteenth International Symposium on Remote Sensing of Environment, Ann Arbor, Michigan*, May 1981, pp. 671–4.

Brush RJH (1988) The navigation of AVHRR imagery. *International Journal of Remote Sensing*, **9** (9), 1491–1502.

Carper WJ, Lillesand TM and Kiefer RW (1990) The use of intensity-hue-saturation transformations for merging SPOT panchromatic and multi-spectral image data. *Photogrammetric Engineering and Remote Sensing*, **56** (4), 459–67.

Chavez P, Guptill SC and Bowell JA (1984) Image processing techniques for Thematic Mapper data. *American Society of Photogrammetry, 50th Annual Meeting, Washington DC*, 11–16 March 1984, pp. 728–43.

Clark BP (1990) Landsat Thematic Mapper data production – a history of bulk image processing. *Photogrammetric Engineering and Remote Sensing*, **56** (4), 447–51.

Coll EC and Pettigrew RG (1988) A high fidelity, high throughput system for geocoding SAR imagery. *Proceedings of IGARSS '88 Symposium, Edinburgh, Scotland*, 13–16 September 1988 (ESA SP-284), pp. 687–90.

Colvocoresses AP (1986) Image mapping with the Thematic Mapper. *Photogrammetric Engineering and Remote Sensing*, **52**, 1499–1505.

Cook R, McConnell I, Stewart D and Oliver C (1996) Segmentation and simulated annealing. In Franceschetti G *et al.* (eds) Microwave Sensing and Synthetic Aperture Radar. Proc. SPIE 2958 (1996), pp. 30–5.

Cracknell AP and Paithoonwattanakij (1989) Pixel and sub-pixel accuracy in geometrical correction of AVHRR imagery. *International Journal of Remote Sensing*, **10** (4 and 5), 661–7.

Curlander JC, Kwok R, Pang SS and Pang AA (1987) *An Algorithm for Generation of Geocoded Data Products from Spaceborne SAR Imagery*. Jet Propulsion Laboratory, California Institute of Technology, Pasadena, California, JPL D-4801, 21 September 1987.

Davison GJ (1986) Ground control pointing and geometric transformation of satellite imagery. *International Journal of Remote Sensing*, **7** (1), 65–74.
Derenyi E and Pollock R (1990) Extending a GIS to support image-based map revision. *Photogrammetric Engineering and Remote Sensing*, **56** (11), 1493–6.
Dowman I and Upton M (1990) A Sun environment for SAR geocoding. *Proceedings of the Third International Workshop on Image Rectification of Spaceborne Synthetic Aperture Radar, Farnham Castle, UK*, 1–3 October 1990, pp. 81–5.
Dowman IJ and Peacegood G (1988) Information content of high resolution satellite imagery. *ISPRS Congress, Kyoto*.
Emery WJ, Brown J and Nowak ZP (1989) AVHRR image navigation: summary and review. *Photogrammetric Engineering and Remote Sensing*, **55** (8), 1175–83.
Ford GE and Zanelli CI (1985) Analysis and quantification of errors in the geometric correction of satellite images. *Photogrammetric Engineering and Remote Sensing*, **51** (11), 1725–34.
Friedmann DE, Friedel JP, Magnussen KL, Kwok R and Richardson S (1983) Multiple scene precision rectification of spaceborne imagery with very few control points. *Photogrammetric Engineering and Remote Sensing*, **49** (12), 1657–67.
Guindon B and Adair M (1992) Analytic formulation of spaceborne SAR image geocoding and 'Value Added' product generation procedures using Digital Elevation Data. *Canadian Journal of Remote Sensing*, **18**, 2–12.
Harrison BA, Jupp DLB, Hutton PG and Mayo KK (1989) Assessing remote sensing technology the microBRIAN example. *International Journal of Remote Sensing*, **10** (2), 301–9.
Heard MI, Mather PM and Higgins C (1992) GERES: a prototype expert system for the geometric rectification of remotely-sensed images. *International Journal of Remote Sensing*, **13** (17), 3381–5.
Ho D and Asem A (1986) NOAA AVHRR image referencing. *International Journal of Remote Sensing*, **7**, 895–904.
Keydel W (1992) Basic principles of SAR. In *Fundamentals and Special Problems of Synthetic Aperture Radar (SAR)*. Advisory Group for Aerospace Research and Development (AGARD), AGARD-LS-182, August 1992, pp. 1–13.
Kidwell KB (1986) *NOAA Polar Orbiter Data User's Guide*. NOAA NESDIS, National Climate Data Centre, Washington DC, USA.
Kidwell RD and McSweeney JA (1984) Art and science of image maps. *Proceedings ASPRS 51st Annual Meeting, Washington DC*, Vol. 2, 10–15 March 1984, pp. 771–82.
Kosmann D, Roth A, Schreier G and Winter R (1992) The ERS-1 radarmaps of Germany and the Antarctic peninsula. *Proceedings of the Central Symposium of the 'International Space Year' Conference, Munich, Germany*, 30 March–4 April 1992, ESA ISY-1, pp. 321–5.
Kwok R and Curlander JC (1987) Automated rectification and geocoding of SAR imagery. *Proceedings of IGARSS '87 Symposium, Ann Arbor*, 18–21 May 1987, pp. 353–6.
Kwok R, Curlander C and Pang SS (1990) An automated system for mosaicking spaceborne SAR imagery. *International Journal of Remote Sensing*, **11** (2), 209–23.

Larijani LC (1998) *GPS for Everyone*. American Interface Corporation, New York.
Laycock J (1990) Resampling algorithms for geocoding. *Proceedings of the Third International Workshop on Image Rectification of Spaceborne Synthetic Aperture Radar, Farnham Castle, UK*, 1–3 October 1990, pp. 15–17.
Lee J-S (1981) Refined filtering of image noise using local statistics. *Computer Graphics and Image Processing*, **15**, 380–9.
Lloyd D and d'Souza G (1987) Mapping NOAA-AVHRR imagery using equal-area radial projections. *International Journal of Remote Sensing*, **8** (12), 1869–78.
McConnell I and Oliver C (1996) Comparison of annealing and iterated filters for speckle reduction in SAR. In Franceschetti G *et al.* (eds) *Microwave Sensing and Synthetic Aperture Radar*. Proc. SPIE 2958 (1996), pp. 74–85.
Mack GAH (1992) Skyfix – a better position through the use of Inmarsat and GPS. *Proceedings of Satellite Symposia 1 and 2: Navigation and Mobile Communications, and Image Processing, GIS and Space-assisted Mapping, from the 'International Space Year' Conference, Munich, Germany*, 30 March–4 April 1992 (ESA ISY-2), pp. 81–6.
Marr D and Hildreth E (1980) Theory of edge detection. *Proceedings of the Royal Society of London*, **B207**, 187–217.
Merson RH (1989) An AVHRR mosaic image of Antarctica. *International Journal of Remote Sensing*, **10** (4 and 5), 669–74.
Moik JG (1980) *Digital Processing of Remotely Sensed Images*. National Aeronautics and Space Administration, Washington DC, NASA SP-431.
O'Brien DM and Prata AJ (1990) Navigation of ERS-1 Along Track Scanning Radiometer (ATSR) images. *ESA Journal*, **14** (4), 447–65.
Pulpin T (2001) Interview with Thierry Phulpin, SPOT Image. http://www.spotimage.fr/home/system/imexplo/goodmir/welcome.htm
Radarsat (2001) *Radarsat Illuminated: Your Guide To Products and Services*, Radarsat International web-site. http://www.rsi.ca/classroom/class.htm
Raggam J (1990) SAR parameter adjustment and related modules for simulated ERS-1 data. *Proceedings of the Third International Workshop on Image Rectification of Spaceborne Synthetic Aperture Radar, Farnham Castle, UK*, 1–3 October 1990, pp. 9–14.
Schreier G (1991) SAR geocoding – operational service for ERS-1 and future sensors. *Proceedings of Spatial Data 2000, Christ Church, Oxford University*, 17–20 September 1991, pp. 29–38.
Schreier G, Kosmann D and Roth A (1990) Design aspects and implementation of a system for geocoding satellite SAR-images. *ISPRS Journal of Photogrammetry and Remote Sensing*, **45**, 1–16.
Sharun C (1993) GPS and GIS: operational use of GPS for cutback layout under forest canopy. *GIS '93 Symposium, Vancouver, British Columbia*, February 1993, pp. 189–93.
Sheffield C (1985) Selecting band combinations from multi-spectral data. *Photogrammetric Engineering and Remote Sensing*, **51** (6), 681–7.
Shlien S (1979) Geometric correction, registration and resampling of Landsat imagery. *Canadian Journal of Remote Sensing*, **5**, 74–87.

Sun W and Takagi M (1987) Geometric distortion correction with high accuracy for NOAA satellite images. *Proceedings of the 1987 International Geoscience and Remote Sensing Symposium, Ann Arbor, Michigan*, pp. 1257–62.

Swann R, Hawkins D, Westwell-Roper A and Johnstone W (1988) The potential for automated mapping from geocoded digital image data. *Photogrammetric Engineering and Remote Sensing*, **54** (2), 187–93.

Tilley DG and Bonwit KS (1989) Reduction of layover distortion in SAR imagery. *Remote Sensing of Environment*, **27**, 211–20.

USGS (1986) *Procedure Manual for Preparation of Satellite Image Maps*. National Mapping Program Technical Instructions, Open-File Report 86–19, Department of the Interior, US Geological Survey, National Mapping Division.

Weiler J, Brule M, Lim K, Ball R and Turner T (1986) An algorithm for automatically acquiring ground control points in SAR imagery. *Proceedings of the SAR Applications Workshop held at Frascati, Italy*, 16–18 September 1986, ESA SP-264, pp. 103–10.

Welch R and Ehlers M (1987) Merging multiresolution SPOT HRV and Landsat TM data. *Photogrammetric Engineering and Remote Sensing*, **53** (3), 301–3.

Westin T (1990) Precision rectification of SPOT imagery. *Photogrammetric Engineering and Remote Sensing*, **56** (2), 247–53.

Williams JM (1995) *Geographic Information from Space: Processing and Applications of Geocoded Satellite Images*. Wiley-Praxis Series in Remote Sensing.

Williamson SJ (1983) *Light and Colour in Nature and Art*. New York, John Wiley & Sons.

Wilson AK (1988) The effective resolution element of Landsat Thematic Mapper. *International Journal of Remote Sensing*, **9** (9), 1303–14.

Zobrist AL, Bryant NA and McLeod RG (1983) Technology for large digital mosaics of Landsat data. *Photogrammetric Engineering and Remote Sensing*, **49** (9), 1325–35.

Further reading

Anon (2000) Landsat 7 Home Page, NASA Goddard Space Flight Center, Greenbelt MD. http://landsat.gsfc.nasa.gov/

Davis WA and Kenue SK (1978) Automatic selection of control points for the registration of digital images. *Proceedings of the Fourth International Conference on Pattern Recognition, Kyoto, Japan*, 7–10 November 1978, pp. 936–8.

Rignot EJM, Kwok R, Curlander JC and Pang SS (1991) Automated multisensor registration: requirements and techniques. *Photogrammetric Engineering and Remote Sensing*, **57** (8), 1029–38.

2

Information extraction

As it becomes increasingly important to provide accurate and detailed inventories of surface cover types, so it becomes correspondingly difficult to provide all the information required by conventional means. Thematic information derived from geocoded satellite images is becoming a major primary source and in some circumstances may provide the only available solution to the problem. This chapter describes the extraction of thematic information from satellite images. It contains the following sections:

- Pre-processing and calibration, using mathematical models of illumination, atmospheric effects, the Earth's terrain and imaging sensors.
- Multi-spectral image classification: basic techniques for extracting information from multi-spectral images including the calibration and pre-processing necessary to interpret 'raw' pixel data, the theoretical basis of classification and the two main approaches to classification (supervised and unsupervised) and models for validating the classification results.
- Other information extraction algorithms: other algorithms for information extraction are described, namely, principal components analysis (for decorrelating image information), vegetation indexes using visible/infrared images and extraction of linear and other regular features.
- Thematic information from SAR: the basic sources of information in a SAR image, calibration and pre-processing necessary, and methods of classification.
- Combining SAR and optical data to make the most of the strengths of both sources of information.
- Multi-temporal analysis, including the use of image data of the same area acquired at different times for enhanced information extraction via the use of image differencing and other change detection algorithms and improvements to classification schemes using multi-temporal data.

This chapter describes information extraction techniques; subsequent chapters describe the determination of height information from space, integration with GIS; land, ocean, atmosphere and cryosphere application areas; applications for global environment and security and the emerging discipline of geomatics.

2.1 PRE-PROCESSING AND CALIBRATION

To appreciate the nature of the information that can be derived from satellite images it is important to understand the physical processes involved and what the digital numbers in an image actually represent. It is assumed that the reader has some knowledge of the physical basis of remote sensing in terms of electromagnetic radiation and its interaction with the Earth's surface, such as can be found in *Computer Processing of Remotely-Sensed Images* (Mather 1987). In the early development of remote-sensing algorithms many researchers manipulated the digital numbers in images directly. This proved satisfactory for qualitative analysis, but for accurate interpretation it is important to be able to calibrate the image values into actual geophysical units. In particular this accurate calibration is an essential prerequisite for the accurate multi-temporal analysis described in Section 2.6. In a perfect system the image values would correspond directly to the radiance of a point on the Earth's surface; however, in reality, factors such as sensor optics, illumination variation, terrain orientation and atmospheric effects all interact to provide a complicated relationship between recorded and actual values.

A model for calibration can be broken down into the following sub-models:

- Illumination model
- Atmospheric model
- Surface interaction model
- Sensor model.

2.1.1 Illumination model

With passive sensors the illumination energy will originate either from the sun (in the case of visible/NIR bands) or from the object itself (in the case of thermal bands). In the case of active instruments such as SAR, or radar altimeter, the instrument itself provides the energy source. The maximum spectral radiant energy (Lintz and Simonett 1976) available for an image during daylight hours can be approximated by the spectral distribution of a black body at 5,800 K (the Sun) combined with that of another black body at 300 K (representing the mean temperature of the Earth). The two main determinants of the illumination available from the Sun will be its distance from the Earth (governed by the time of year) and its elevation angle (governed by the time of day).

2.1.2 Atmospheric model

The atmospheric model must account for a number of effects that the Earth's atmosphere has on the electromagnetic signal – in the case of passive sensing of a terrestrial target a ray of light from the Sun will pass once through the atmosphere on the way to the target, experiencing attenuation effects, then pass through the atmosphere again experiencing further attenuation effects, before finally being detected by the sensor. Depending on the wavelength of the signal, atmospheric gases and aerosol particles will absorb signals, act as scatterers and also as radiant energy sources. One consequence of this is the strong attenuation of electromagnetic

Table 2.1 Main atmospheric windows (based on Lintz and Simonett 1976)

Portion of spectrum	Window (μm)
UV/visible	0.30–0.75
	0.77–0.91
Near-IR	1.00–1.12
	1.19–1.34
	1.55–1.75
	2.05–2.40
Mid-IR	3.50–4.16
	4.50–5.00
Thermal-IR	8.00–9.20
	10.2–12.4
	17.0–22.0

radiation in various spectral wavebands, within which the available energy from a transmitted signal has diminished so considerably after passage through the atmosphere as to be virtually unusable for remote-sensing purposes. This means that there are only certain atmospheric 'windows' (wavebands) in which remote sensing is a practical proposition. The major atmospheric windows are shown in Table 2.1.

The effect of the atmospheric absorption at wavelengths used for Earth observation can be seen in Fig. 2.1.

The two main scattering mechanisms (Chavez 1988) are Rayleigh scattering, which occurs when the wavelength of the radiation is much greater than the scattering particles, and Mie scattering, which is not as wavelength dependent and is largely caused by dust and smoke particles. Detailed atmospheric models which may be used for atmospheric correction of satellite data, such as 'Simulation of the Satellite Signal in the Solar Spectrum' (Tanre *et al.* 1990) have been developed. This model (known as 5S) actually consists of eight separate atmospheric models:

- No absorption
- Tropical
- Mid-latitude summer
- Mid-latitude winter
- Sub-Arctic summer
- Sub-Arctic winter
- US standard 1962 atmosphere
- User defined (using radiosonde or other meteorological data).

The last option reflects the fact that, no matter how sophisticated static models of typical atmospheres may be, it is always preferable to model the atmosphere using accurate and contemporary meteorological data. An updated version known as 6S ('Second Simulation of the Satellite Signal in the Solar Spectrum') is now available

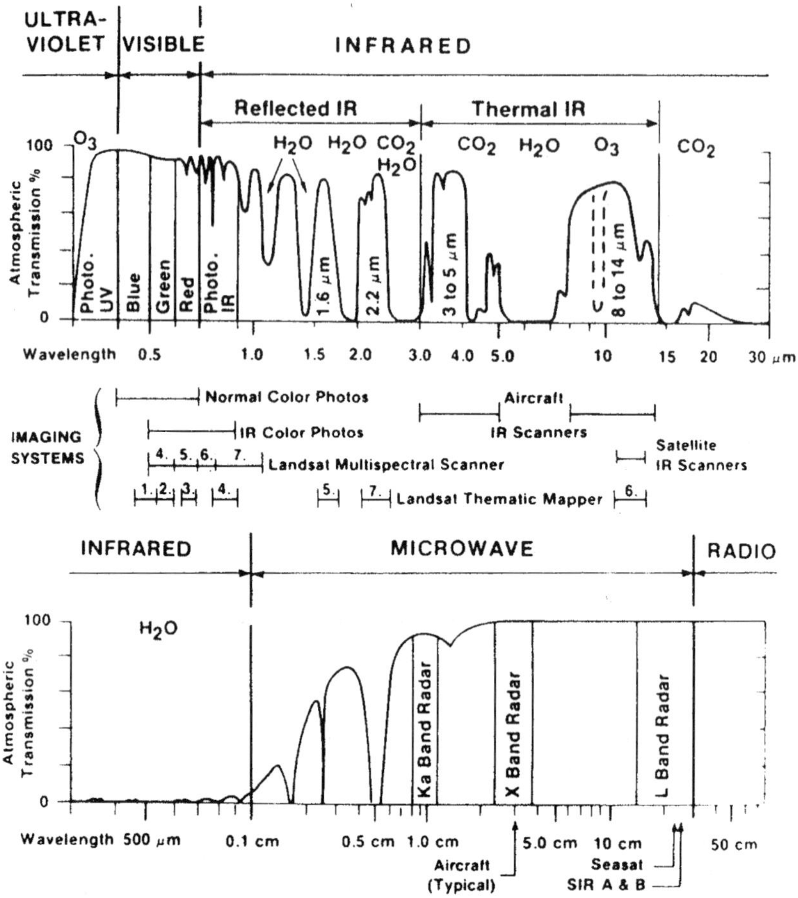

Fig. 2.1 The effect of atmospheric absorption at wavelengths used by various imaging systems (from Sabins 1987).

(Vermote *et al.* 1997). The new version now permits calculations of near-nadir (down-looking) aircraft observations, accounting for target elevation, non-Lambertian surface conditions, and new absorbing species (CH_4, N_2O, CO).

2.1.3 Surface interaction model

The surface interaction model accounts for one of the major contributions to spectral variation in an image – that provided by terrain variation. The basic imaging geometry is shown in Fig. 2.2, the three critical variables being the position of the illumination source (the Sun), the target and the sensor.

For a given image the satellite orbital plane can be regarded as fixed relative to the Sun, thus the major variation is caused by the orientation of the surface facets. Two facets with the same surface attributes but differing slope and aspect will have a

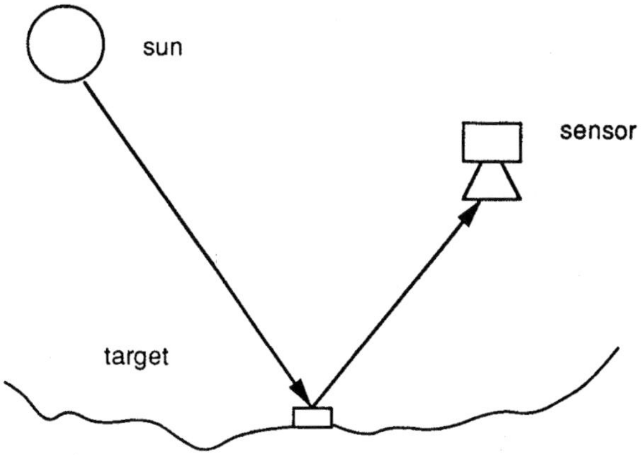

Fig. 2.2 Basic Sun–sensor–target imaging geometry

different imaging geometry, producing distinct spectral signatures. Thus, even in an area with a homogeneous cover type classification based on spectral signatures may not succeed unless the terrain effect is accounted for.

The transmitted solar radiation will eventually be intercepted by the atmosphere, ocean or land surfaces. At this stage one of three types of energy response will occur:

- absorption by electrons and molecular reactions in the medium – some energy may be re-emitted producing a thermal signal, primarily at longer wavelengths
- transmission – radiation penetrating into surface media, particularly water
- reflection and scattering at different angles.

The simplest form of radiometric correction is to assume that the Earth's surface reflects electromagnetic radiation as a Lambertian model. In such a model the only geometrical parameters required are the Sun angle (azimuth and elevation), the sensor look angle and the terrain slope and aspect. In general the last two will be unknown for a particular scene; they can, however, be derived from a numerical analysis of a Digital Elevation Model (DEM) since the slope represents the rate of change of height at a given point and the aspect represents the direction of greatest slope. Mathematically, slope and aspect can be derived from the partial derivatives of a continuous (differentiable) surface. As the DEM is a discontinuous surface (sampled at grid points) it is necessary to estimate the derivatives by numerical interpolation.

2.1.4 Sensor model

The sensor model will be specific to the particular instrument in question (unlike the more generic illumination and atmosphere models). On the other hand, since the sensor is a precisely engineered artefact it is very amenable to mathematical modelling. Typically, pre-launch calibration in a laboratory will give a nominal function to convert from digital numbers to radiance, and this calibration will be modified during a sensor's lifetime and coefficients will be transmitted as part of the spacecraft housekeeping data. Possible methods for in-orbit calibration include:

- ground and atmospheric measurements
- reference to another sensor
- a combination of the above methods.

An example of sensor calibration models are those used for the Landsat ETM+ (US Geological Survey 2001) for post-launch radiometric calibration. These make use of three onboard calibration devices:

- *Internal calibrator* using a blackbody source
- *Full aperture solar calibrator*
- *Partial aperture solar calibrator*, which consists of a small passive device that allows the ETM+ sensor to image the Sun while viewing the darker Earth.

By combining the models in Sections 2.1.1–2.1.4 (illumination, atmosphere, surface interaction and sensor) for a particular image it is possible to calibrate an image more accurately into geophysical units, and this in turn provides a potential for a more accurate classification of the image. This is described in more detail in the next section.

2.2 MULTI-SPECTRAL CLASSIFICATION

The derivation of feature types based on the multi-spectral information contained within an image is known as *image classification*. This section describes the theory of multi-spectral classification based on the spectral signature of an object; simple classifier techniques; supervised and unsupervised classifications; collection of ground data; mixture modelling; and validation.

2.2.1 Theory of multi-spectral classification

The colour of an object discriminated by the human eye is determined by the way that it reflects light in the visible spectrum, and it is the eye's sensitivity to light in this fairly narrow waveband that allows it to perceive objects as having different colours. Objects reflect or radiate energy at all wavelengths in the electromagnetic spectrum, so sensors capable of detecting other spectral bands such as infrared and ultraviolet will have a greater potential for discrimination of objects than by using visible light alone. Figure 2.3 shows the spectral signatures of five typical surface classes – clear water, turbid water, bare soil and two types of vegetation.

The spectral signature is the typical intensity response of an object as a function of wavelength, and the principle of image classification is to distinguish between different classes on the basis of this signature. Note that the single curves drawn on the diagram are highly schematic, in practice there can be great spectral variation between objects in the same class; thus the simplified curves are better considered as a range of values – an envelope within which spectral values can be expected to lie. To simplify the mathematics of classification, various assumptions are generally made, such as:

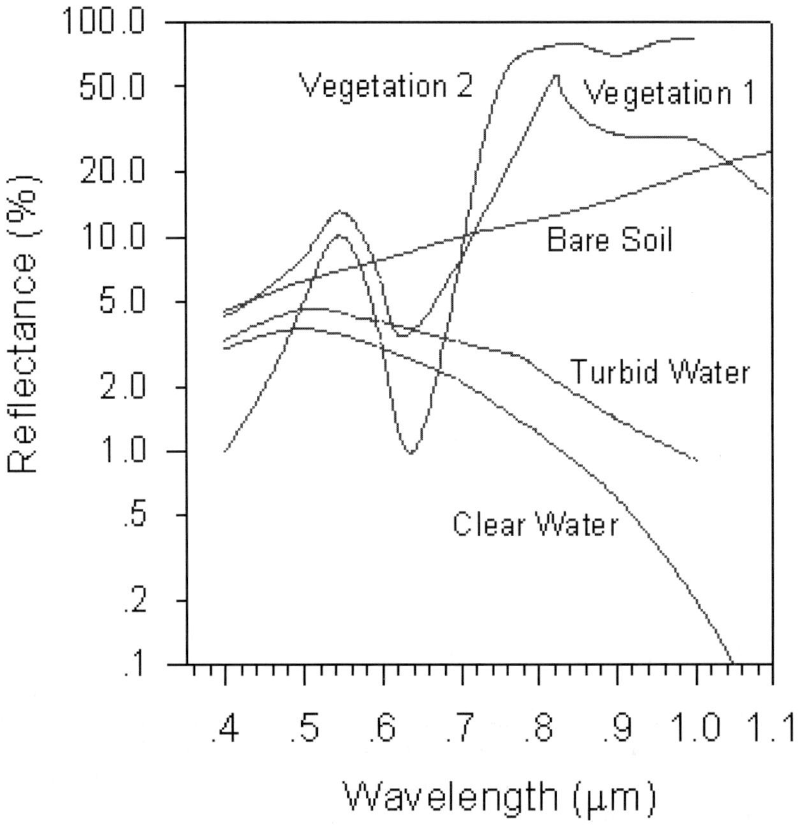

Fig. 2.3 Spectral curves for clear water, turbid water, bare soil and two types of vegetation (courtesy Singapore Science Centre).

- the sensor response is uniform across a spectral band
- each pixel corresponds exactly to a well-defined square on the ground
- the pixel area is covered with a homogeneous feature type
- the histogram of each band has a normal (gaussian) distribution.

These assumptions will not be totally applicable in practice – for example, a particular sensor will not respond uniformly to all wavelengths within its spectral band as reflectance from neighbouring areas will contribute to each pixel, and one pixel could contain the information from several different classes within its spatial extent, i.e. it could be part corn-field, part road and part river (the interpretation of such 'mixels' is considered in Section 2.2.6).

The assumption about the image histograms having a normal distribution is crucial for the operation of various parametric classification techniques described in Section 2.2.4. Since a typical image histogram will rarely have a normal distribution it may be necessary to pre-process the image by applying a Gaussian contrast stretch.

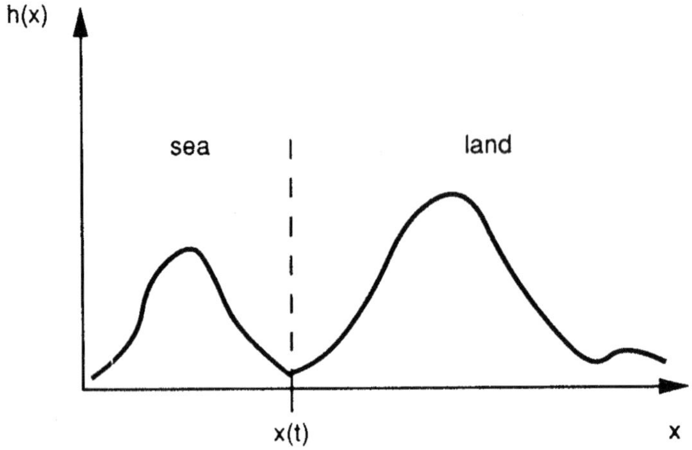

Fig. 2.4 A histogram divided into two image classes by threshold $x(t)$.

2.2.2 Simple classification algorithms

Histogram classification

A simple form of classification of land and sea can be performed by density slicing, that is, setting a number of thresholds in an image histogram to divide classes on the basis of their response in a single spectral band. For example, a single threshold can be used to divide the image into two classes based on simple decision criteria (see Fig. 2.4):

- If the pixel value is less than or equal to threshold, assign it to class 1 (for example, sea).
- If the pixel value is greater than threshold, assign it to class 2 (for example, land).

This is an example of a binary decision classification. The statistical model used is that the histogram represents the combined effects of two pure populations of pixels each with a characteristic statistical distribution. The classification problem can then be stated in terms of probability: 'Find the threshold T such that the probability of classifying a pixel as the wrong class is minimised.' If the two histograms do not overlap then any value between the two histogram ranges is suitable and the probability of misclassification is zero. In practice, there will usually be some overlap and thus a possibility that values in the overlap area could lead to misclassification (for instance, land as sea, or sea as land). In some circumstances improved results may be obtained by using a contour-threading technique (Benny 1980). This makes use of the fact that some interfaces between classes (such as land–sea or river boundaries) can be regarded as a continuous boundary and techniques analogous to those used to thread contours through spot-height data can therefore be used.

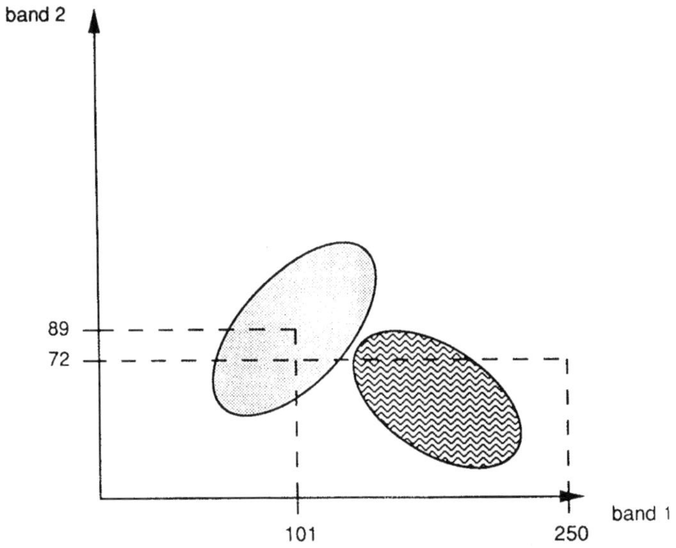

Fig. 2.5 Example of a two dimensional scatter diagram: image band 2 is plotted against image band 1.

Classification using scatter diagrams

The probability of misclassification can be reduced by using more than one spectral band and by examining the histograms of the pixel classes in each spectral band. It is most convenient, however, not to display the image statistic as histograms, but as scatter diagrams. Figure 2.5 shows a typical scatter diagram of two image bands. The x axis represents all the possible values in band 1 [0, 255] and the y axis represent the values possible in band 2. The dots on the scatter diagrams are interpreted as indicating combinations of pixel values in the two spectral bands coincidentally. Thus the dot indicated at (101, 89) indicates that there is at least one pixel in the image with a value of 101 in band 1 and 89 in band 2; whereas the blank at (250, 72) indicates that there are no pixels in the image with this particular combination.

A more sophisticated scatter diagram can be produced in the form of a two-dimensional array, I, which is used to count exactly how many co-occurrences of the various pixel values actually occur in the image. Thus, for example, $I(101, 89) = 57$ indicates that there are exactly 57 pixels in the image with values of 101 in band 1 and simultaneously 89 in band 2. The potential of multi-spectral methods can be judged by comparing a two-dimensional scatter diagram with either of the one-dimensional histograms (Fig. 2.6). In neither case is it possible to separate completely the two classes as there is considerable overlap between them – however, it is a simple matter to separate the classes on the scatter diagram as they form two separate regions (or clusters).

It is thus the task of the image classification algorithm to construct decision boundaries between the clusters to mathematically divide the classes. The methods

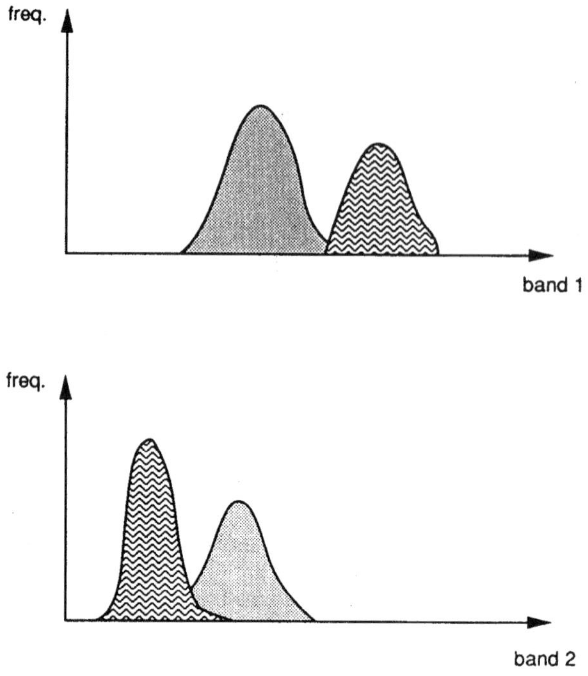

Fig. 2.6 Comparison of the scatter diagram shown in Fig. 2.5 with the two underlying one-dimensional histograms.

used can quite easily be generalised to more than two classes, but in this case there may again be overlap between clusters. In this circumstance the algorithm must select decision boundaries which minimise the probability of a pixel being misclassified. The methods are easily extended algorithmically to three or more spectral bands but the resulting n-dimensional clusters separated by $(n-1)$-dimensional decision boundaries cannot easily be visualised.

2.2.3 Unsupervised classification

There are many different approaches to the cluster analysis needed to partition the feature space into distinct classes. Algorithms which use external information (such as incorporation of known classes based on ground data) are known as *supervised* classifiers, whereas those algorithms which do not, and rely solely on the use of the statistical information inherent in the image are known as *unsupervised* classifiers. The aim of an unsupervised classifier is to divide the feature space into clusters which are individually dense yet collectively well separated. In mathematical terms the objective is to minimise the average distance from points within a cluster to the local cluster centroid while at the same time maximising the distance between cluster centroids. A typical algorithm to achieve this is as follows:

1. Choose the number of clusters to divide points into.

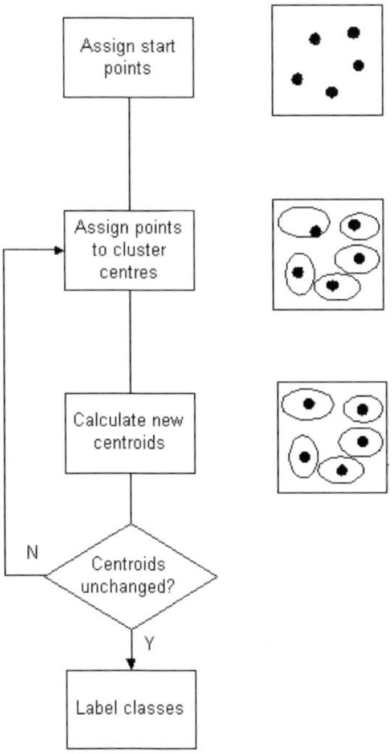

Fig. 2.7 Unsupervised classification flow diagram.

2. For each cluster select an arbitrary starting point to act as centroid.
3. For each starting point assign every other point to the nearest cluster centre.
4. For each cluster calculate the new centroid based on points in the cluster.
5. Repeat steps 3 and 4 until the position of the cluster centroids is unchanged.
6. Assign a label to each class.

Figure 2.7 presents a flow diagram of this process.

The labelling of classes is performed by a skilled operator relating the resultant classifications to actual ground-cover-types based on 'ground truth' data. In practice (Wilkinson 1991) an implementation of the unsupervised classification algorithm will include extra criteria for refining the classification, for example:

- merge rules, which join two clusters together if the distance between their means is less than a specified threshold
- split rules, which break a cluster into two smaller ones if its variance exceeds a specified threshold
- abandoning a cluster and reallocating its members if there are less than a threshold number of them.

Unsupervised classifications are rarely used as the sole means for information extraction from an image, because they use an entirely algorithmic approach, without the introduction of real-world knowledge. They can, however, provide great insight into the feature space representation of an image and provide a good basis for the subsequent use of supervised classification techniques.

2.2.4 Supervised classification

An alternative method (supervised classification) is to provide some information about the pixel classes before the classification takes place. A typical processing flowline for supervised classification can be seen in Fig. 2.8.

The information about the pixel classes is provided by the operator in a process known as 'training' the classifier. Based on ground information the operator will know the classes to which some of areas of the image correspond. The collection and interpretation of ground data is a specialist area in itself and is described in more detail in Section 2.2.5. The training algorithm uses the designated areas to extract the sample pixel values in each class. By working out the statistical distribution of the sample values (for example, mean, variance) the classification algorithm can then assign class probabilities to all the remaining pixels in the image.

The flowline shown in Fig. 2.8 also indicates a number of 'feedback loops' used to refine or validate the classification process, for example:

- edit/revise training sites
- cross-validate and edit statistics
- test accuracy
- post-classification filtering and presentation.

The classification algorithm itself is then applied to the whole image. The result of this will be another image, the same size as the original but with each pixel value being one of the feature codes from the classification hierarchy (for example, 321 = natural grasslands). It is often beneficial to edit the final output – for example, to remove isolated pixels or to correct inaccuracies discovered in the classification. This can either be performed interactively using graphics tools or, in some cases, automatically. For example, an algorithm can be used to remove all groups of pixels less than a certain area (since small groups of pixels are more likely to suffer from spectral contamination, making accurate classification more unlikely). If the output classification is intended for a vector-based GIS or digital mapping system the final stage will be the conversion of the classified image from raster format to vector format; this involves taking each connected block of pixels within a class and calculating the polygon that represents the boundary.

The box classifier

The simplest supervised classification algorithm is the parallelepiped or box classifier. The idea behind this is shown in Fig. 2.9.

In two dimensions it may be possible to separate clusters by drawing a rectangle around each one. The training algorithm calculates the size of the rectangle for each sample class, which can be achieved simply by calculating the maximum and minimum values within the sample distribution. A statistically more accurate result is obtained, however, by removing outliers – for example, using the 5–95 percentile range instead of the entire distribution. The box classification algorithm is 'non-

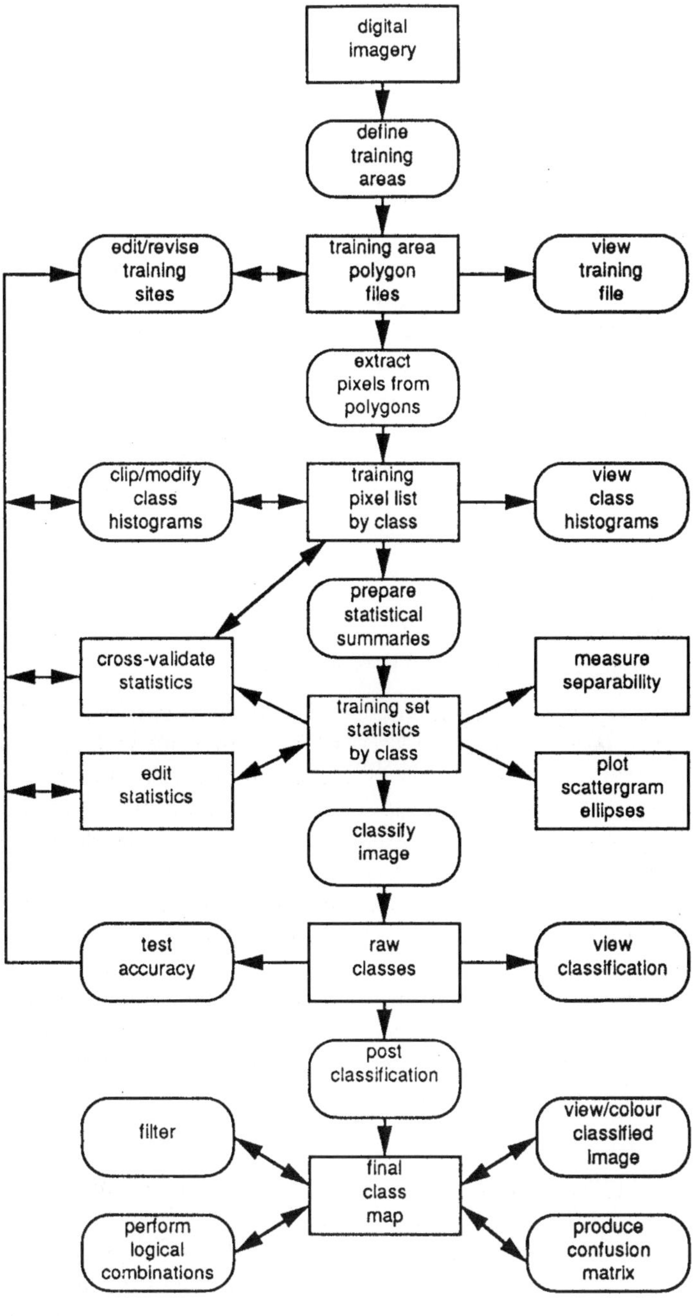

Fig. 2.8 A typical processing flowline for supervised classification (adapted from van Genderen and Uiterwijk 1987).

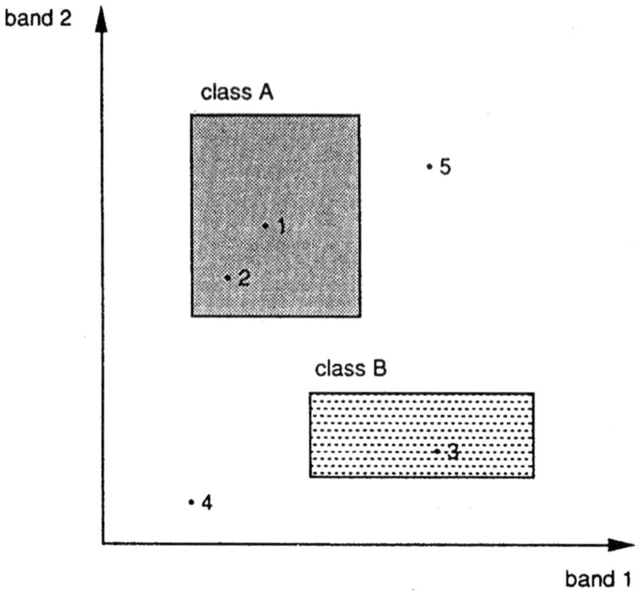

Fig. 2.9 Multi-spectral classification using a parallelepiped (box) classifier: points 1 and 2 lie in class A, point 3 lies in class B, points 4 and 5 are marked as unclassified.

parametric' in the sense that it does not employ statistical parameters to calculate class separation. Its operation is very simple: to classify the entire image it goes through on a pixel-by-pixel basis examining the value in each spectral band used for training. If the pixel value lies within the multi-dimensional box, it is classified as being that feature. In cases where a pixel lies in the overlap between two boxes, a decision rule has to be used; a typical rule is to allow the operator to specify which class should have priority.

Parametric classifiers

The box classifier uses a simple decision rule to decide whether a pixel should be included in a particular class or not. There is, however, far more information available in the training samples which can be used to provide better classification estimates. For example, a pixel with a spectral value near a class centroid is more likely to be a member of that particular class than one far away from the centroid, or in the overlap of more than one cluster. Classifiers that make use of the statistical distribution of the training samples are known as parametric classifiers. A simple example of a parametric classifier is the 'minimum distance' classifier, which calculates the centroids of each class in feature space (using the training data); the classification then proceeds by assigning each pixel to the class whose centroid is nearest (see Fig. 2.10).

This method has the disadvantage (similar to the box classifier) of assigning a pixel to a class even when it is 'obvious' that there is a greater probability that it

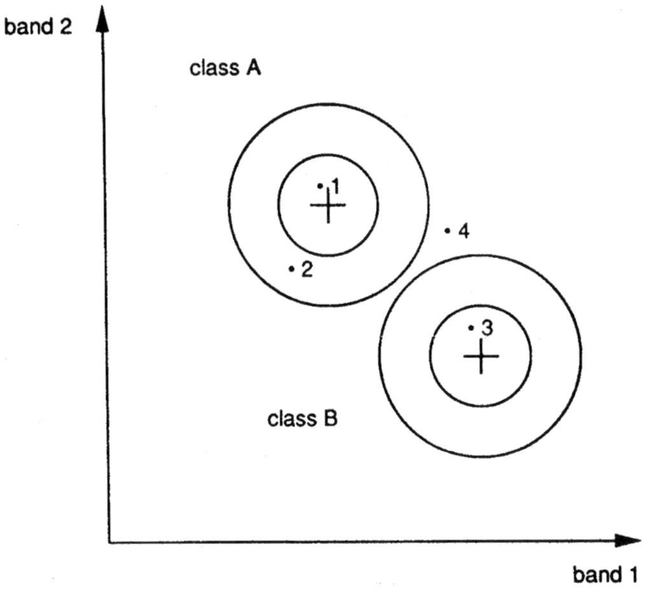

Fig. 2.10 Multi-spectral classification using a minimum distance classifier: points 1 and 2 lie in class A, point 3 lies in class B, point 4 lies marginally in class A, point 5 lies marginally in class B but with a large probability of misclassification.

belongs to another class. Neither algorithm takes into account the probability density functions that describe the shape of a cluster.

One of the most-used types of parametric classifier is that known as 'maximum likelihood'. This assumes that each class has a multi-variate normal distribution, which means that the distance from the cluster centroid, calculated in terms of the variance in each band, can be used as a measure of the probability that the pixel belongs in that class (see Fig. 2.11). To classify the image each pixel is located in feature space and the maximum likelihood distance to each of the class centroids is calculated. The one given the lowest value is the class to which the pixel is assigned. The distance can also be stored as it gives some indication of the confidence that can be placed in the classification.

2.2.5 Collection of ground data

For any successful image classification it is important to be able to collect sufficient ground data to accurately calibrate and provide confidence in the subsequent classification. Sources of data can vary from existing maps, aerial photographs and airborne scanner data to field surveys carried out using portable radiometers. The plan for the collection of ground data will be based upon the classes that will be needed for the final image classification. These could vary from a simple single class (for example, inland water) through variants in types of woodland and forestry to a complicated hierarchical classification scheme. In all cases it is important to select

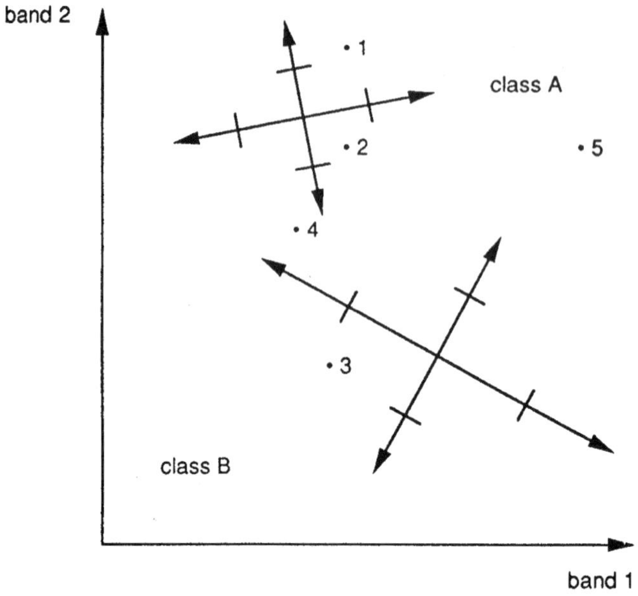

Fig. 2.11 Illustration of a maximum likelihood classification: classes A and B are portrayed by the principal axes of the corresponding feature-space clusters, annotated by 'one unit of distance' tick marks. According to the classification, points 1 and 2 lie in class A, point 3 lies in class B, point 4 lies marginally in class B, point 5 lies marginally in class B but with a large probability of misclassification.

the ground survey sites using a well-defined and statistically valid scheme, a typical approach (Taylor *et al.* 1997) being:

- Divide the region of interest into a grid of blocks, say 10 km by 10 km.
- Choose a random sample segment, say 1 km by 1 km, within one block.
- The systematic random sample is created by selecting segments at the equivalent locations in each of the blocks.
- Repeat the process to produce several replicated samples in each block.
- Additional restrictions can be included to ensure that samples are not too close together and to avoid 'forbidden' areas.

Figure 2.12 shows a typical sample grid for a systematic sample.

The choice of an appropriate classification scheme is essential to the successful classification of the image and maximising the information derived. A typical hierarchy for classifying land cover types is shown in Table 2.2. This shows the 44 classes (not all of them derivable solely from satellite imagery) used for the Commission of the European Communities CORINE (Coordination of Information on the Environment) project used to define an information system on the state of the environment throughout the European Union.

Figure 2.13 (colour section) shows how SPOT satellite image data was combined

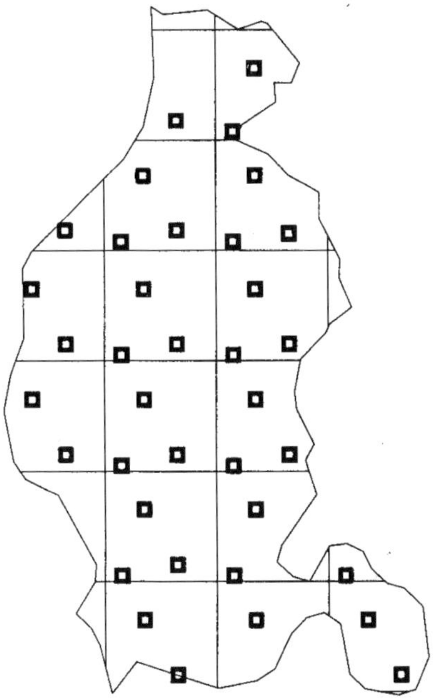

Fig. 2.12 Systematic sample within a distance threshold (based on Taylor *et al.* 1997).

Table 2.2 Land cover hierarchy (from Cornaert and Maes 1992)

1 ARTIFICIAL SURFACES
- 11 Urban Fabric
  - 111 continuous urban fabric
  - 112 discontinuous urban fabric
- 12 Industrial, commercial
  - 121 industrial or commercial units and transport
  - 122 road and rail networks
  - 123 port areas
  - 124 airports
- 13 Mine, dump and construction sites
  - 131 mineral extraction sites
  - 132 dump sites
  - 133 construction sites
- 14 Artificial vegetated areas
  - 141 green urban areas
  - 142 sport and leisure facilities

2 AGRICULTURAL AREAS
- 21 Arable Land
  - 211 non-irrigated arable land
  - 212 permanently irrigated land
  - 213 rice fields
- 22 Permanent crops
  - 221 vineyards
  - 222 fruit trees and berry plantations
  - 223 olive groves
- 23 Pastures
  - 231 pastures
- 24 Heterogeneous areas
  - 241 annual crops associated with permanent crops
  - 242 complex cultivation patterns
  - 243 principally agriculture, some natural vegetation
  - 244 agro-forestry areas

3 FOREST AND SEMI-NATURAL AREAS
- 31 Forests
  - 311 broad-leaved forest
  - 312 coniferous forest
  - 313 mixed forest

32 Scrub and/or herbaceous vegetation
- 321 natural grasslands
- 322 moors and heathland associations
- 323 sclerophyllous vegetation
- 324 transitional woodland-scrub

33 Open Spaces
- 331 beaches, dunes and sands
- 332 bare rocks
- 333 sparsely vegetated areas
- 334 burnt areas
- 335 glaciers and perpetual snow

4 WETLANDS

41 Inland wetlands
- 411 inland marshes
- 412 peat bogs

42 Maritime wetlands
- 421 salt marshes
- 422 salines
- 423 intertidal flats

5 WATER BODIES

51 Inland waters
- 511 water courses
- 512 water bodies

52 Marine waters
- 521 coastal lagoons
- 522 estuaries
- 523 sea and ocean

---

with geographic databases and local data to provide a classification of the region around Toulouse, France, following the above hierarchy. This was used to monitor the implementation of management measures designed to protect an area for ecological reasons or for its flora or fauna, and to assess the vulnerability of such areas to urban growth.

The result of the data collection exercise will be a database of ground information consisting of a number of training areas together with a feature class assigned from the appropriate hierarchy. These training areas will either be represented in vector format (as a collection of polygons) or in raster format (as a collection of pixels). There are four principal methods of inputting the data required:

- *direct coordinate input* – the ground data is recorded as a list of coordinate pairs (map references) which can be keyed in by hand
- *interactive graphic input* – the image is displayed on the screen and an operator visually identifies each training area, using a mouse, or similar device to draw an appropriate polygon for each training area, or to paint-in its raster extent
- *digitiser input* – the ground data is presented as overlays on a conventional map-sheet; these can then be digitised with a tablet in the conventional manner
- *direct digital input* – ground data is increasingly available in digital format; as such data will often be geocoded it can be put into the system directly.

Following the input and display of the training data the operator may carry out further addition, modification and deletion of the displayed areas. The specification of training areas is greatly simplified if they are already available in a digital format within a Geographic Information System (GIS). This is described in more detail in Chapter 4.

The first step in the 'training session' is for the operator to specify which bands will be used as the basis for spectral discrimination. A typical GIS set-up will allow the image bands to be displayed along with the training area boundaries defined as polygons. Based on this collection of information the training session statistics are calculated. The statistics derived from each class will depend on the classification algorithm to be employed. For example, in the case of the box classifier the range and selected percentiles will be calculated; for maximum likelihood the feature centroids and variance vectors will be calculated.

2.2.6 Mixture modelling

One possible source of error in the classifier can occur if more than one land cover type contributes to a particular pixel – either the pixel is marked as unclassified or is spuriously attributed to the wrong class. The technique known as 'mixture modelling' (Foody and Cox 1991) attempts to 'unmix' a pixel into its components to allow more accurate classifications and areal estimates of cover types. Mixture modelling is based on three assumptions:

- surface components are spectrally distinct
- the spectral mixing process can be described by a linear equation
- the only contribution to changes in spectral reflection is that due to the surface components alone.

If the contribution of each component to the mixed pixel is denoted by f_i then two numerical constraints apply (Drake and White 1991):

$$f_i > 0 \text{ for all values of } i \quad (f_1 + f_2 + \cdots + f_n = 1) \tag{2.1}$$

The linear mixture model can be solved either by standard least-squares estimation or by using a regularised estimator (Settle 1990) which introduces a prior estimate of pixel mixture throughout the image. For example, a sub-pixel spectral analytical process was used to classify Bald Cypress and Tupelo Gum wetland in Landsat Thematic Mapper imagery in Georgia and South Carolina (Huguenin *et al.* 1997a). The sub-pixel process enabled the detection of Cypress and Tupelo trees in mixed pixels. Field investigations revealed that both cypress and tupelo trees were successfully classified when they occurred both as pure stands and when mixed with other tree species and water. Large areas of wetland where cypress was heavily mixed with other tree species were correctly classified by the sub-pixel process and not classified by the traditional classifiers. Sub-pixel classification (Huguenin *et al.* 1997b) has also been used for automated bathymetric analysis. The approach adopted was to identify and remove unwanted spectral contributions from background materials in the pixels. This provides a means for automatically identifying and removing terrain and surface reflection (sky and Sun reflection) contributions from water pixels. It also allows composite depths and bottom materials within pixels to be resolved into individual components, such as shallow coral and deep sand. This enables more accurate determinations of the water column and bottom reflectance characteristics to be made. In addition the algorithmic approach provides the ability to calculate atmospheric correction factors for the scene being processed. This allows the attenuated bottom radiance to be converted from units of digital number into units of calibrated reflectance, providing a means for automatically calculating depth using a standard regression analysis of logarithmic reflectance.

2.2.7 Validation models

The accuracy obtainable in a classification process is dependent on a number of factors. These include the state of the original image (geographic location, illumination, time of year), quality and timeliness of the ground data and whether the image has been geometrically transformed. The measure of classification accuracy will also depend on the classification scheme employed; it is a lot simpler to divide features into two broad classes (such as land/sea) than into 30 or 40 different classes separated only by relatively subtle distinguishing attributes.

In terms of the validation models introduced in Section 1.1.10, there are two main approaches to validation of the classification process which are generally used in a complementary fashion:

- *process validation* – using the statistics generated by the classification process to produce quality control parameters
- *product validation* – quality control of the final classification product; for example, comparing classification output with known land cover categories at sample sites.

The third category of validation, *blunder detection*, can be difficult to apply in practice as it is not simple to define what constitutes a 'gross error or physically impossible result' without detailed prior knowledge of the area being classified.

Process validation

Process validation consists of the analysis of the statistics generated from the training classes and is thus best suited for use with parametric classifiers. The primary consideration is the 'separability' of the clusters that represent each class in feature space. Separability is based on two parameters:

- clusters should be widely separated in feature space (as defined by the distance between cluster centroids)
- each individual cluster should be 'compact' (as defined by the variance of the distance of members of a cluster set from its centroid).

The separability of the training clusters can be ascertained interactively by displaying the training data as a scatter diagram. This will indicate 'problem' categories with large variances or which overlap the distributions of other categories. A more rigorous approach is to calculate statistical measures of between-cluster and within-cluster distributions which will indicate, for every possible combination of pairs of categories, the separability between them.

Product validation

The most common method used for product validation is the confusion matrix (Foody 1992), which is a comparison of the image classes (as determined by multispectral classification) of a sample of points against their actual classes as determined

		Ground Data			
		conifer	decid.	agric.	water
Image Classification	conifer	317	23	0	0
	deciduous	61	120	0	0
	agriculture	2	4	60	0
	water	35	29	0	8

Fig. 2.14 Confusion matrix for four classes: conifer, deciduous, agriculture and water (adapted from Foody 1992).

from ground data. A typical confusion matrix can be seen in Fig. 2.14, which shows, for example, that 317 sample points were correctly classified as conifer, but that 61 genuine conifer samples were misclassified as deciduous and 23 deciduous samples were classified as conifer. The confusion matrix can be used to measure overall classification accuracy by summing the correctly classified samples (located on the leading diagonal of the matrix) and dividing by the total number of samples. The confusion matrix can also be used to calculate the marginal accuracy attained by a particular class.

Another useful product validation technique is 'residual image analysis' (Jupp and Mayo 1982). This involves the construction of a 'mean image' where each pixel is replaced by the mean value of the class to which it has been assigned. The residual image is constructed by subtracting the mean image value from the raw image values on a pixel-wise basis, for each band:

$$R(i, j) = I(i, j) - I^*(i, j) \tag{2.2}$$

where $R(i, j)$ = value of pixel (i, j) in the residual image
$I(i, j)$ = value of pixel (i, j) in the original image
$I^*(i, j)$ = mean value of spectral class containing (i, j)

The result is an image which displays the variability within each spectral class. Areas of great variability in the residual image will suggest classes for which further selection of training data and possibly refinement of classes may lead to a more accurate overall classification. More advanced 'knowledge-based' techniques for post-classification sorting and validation are described in Section 4.2.

2.3 OTHER INFORMATION EXTRACTION ALGORITHMS

This section describes some of other information extraction algorithms that can be applied to optical imagery, including:

- Principal Components Transform
- vegetation indexes
- extraction of linear features

2.3.1 Principal Components Transform

Statistical analysis of satellite images reveals that there is often a great deal of correlation between image bands – thus, for example, in a scatter diagram plot of Landsat TM band 4 value against those of band 5, the characteristic appearance of the cluster will tend to be an elongated ellipse at an angle to the two main axes (see Fig. 2.15).

The objective of a Principal Components Transform (PCT) is to produce a decorrelated linear combination of the original image bands which maximise the information content available in a small number of 'principal' bands. In the simplified two-dimensional case this is equivalent to defining two new axes – one along the direction of maximum pixel variability and the second representing variability from this axis. This results in a new 'principal component' containing most of the information (variance) from the original image data and a second component which is not correlated with the first and contains less information. With more than two image bands the PCT is not simple to visualise but is derived by considering the variance–covariance matrix, $C(i, j)$, where the entries represent the covariance between band i and band j of the image. For example:

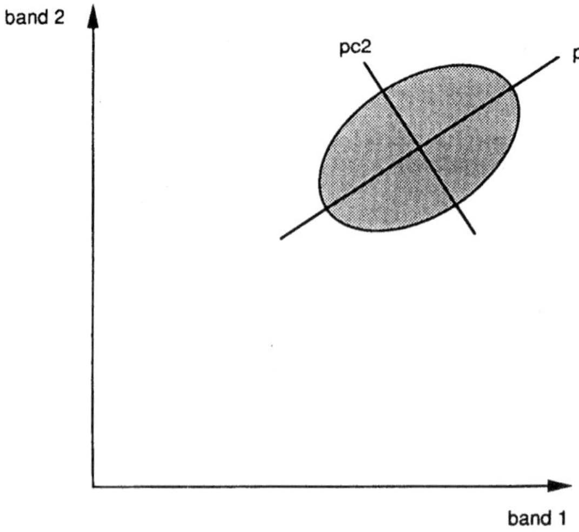

Fig. 2.15 Principal Components Transform: two-band data is mapped onto principal axes pc1 and pc2.

$C(2, 3)$ = covariance between band 2 and band 3
$C(4, 4)$ = variance of band 4 (equivalent to covariance between band 4 and itself)

An alternative approach is to utilize the correlation matrix where all entries are normalized by the band variances so that they lie in the range $[-1, 1]$. The PCT is a linear transformation which, when applied to the image band values, results in a variance–covariance matrix whose non-diagonal entries are zero, implying that there is no correlation between the bands of the transformed image. In mathematical terms this requires finding the eigenvectors of the matrix $C(i, j)$, and further details of this technique can be found in specialist works such as *Digital Image Processing* (Pratt 1985).

A similar approach to the PCT is the 'tasseled-cap' transform (Kauth and Thomas 1976; Crist and Cicone 1984) so-called because of its cap-shaped appearance for vegetated areas. Instead of deriving the coefficients of the transformation statistically, typical values which correspond to attributes of land cover ('brightness', 'greenness' and 'wetness') are defined. The point of the cap (which lies at low red reflectance and high NIR reflectance) represents regions of high vegetation, and the flat side of the cap directly opposite the point represents bare soil.

Values of tasseled-cap parameters for Landsat TM are shown in Table 2.3. The following should be noted:

- the coefficients define a linear combination of the input bands, for example: brightness $= 0.30 \times$ (band 1) $+ 0.28 \times$ (band 2) $+ \cdots + 0.19 \times$ (band 7)
- the thermal band (band 6) is omitted because of its different spatial resolution
- the (unnamed) fourth, fifth and sixth tasseled cap bands are usually not used because of their lack of information content.

Table 2.3 Tasseled cap coefficients (after Wilkinson 1991)

Name	TM band coefficient					
	1	2	3	4	5	7
Brightness	0.30	0.28	0.47	0.56	0.51	0.19
Greenness	−0.28	−0.24	−0.54	0.72	0.08	−0.18
Wetness	0.15	0.20	0.33	0.34	−0.71	−0.46
(Four)	−0.82	0.08	0.44	−0.06	0.20	−0.28
(Five)	−0.33	0.05	0.11	0.19	−0.44	0.81
(Six)	0.11	−0.90	0.41	0.06	−0.03	0.02

2.3.2 Vegetation indexes

Multi-spectral classification is a general-purpose tool which can be used to classify an entire image into land cover types. More specialised information extraction algorithms can also be used. One example of these are vegetation indexes which make use of the fact that information in a multi-spectral image is typically concentrated within a small number of spectral bands – for example (Sheffield 1985), after a PCT 99% of the information in a 7-band Landsat TM image is usually contained within the first three principal components. Vegetation indexes generally make use of two spectral bands – one in the red part of the spectrum and one in the near-infrared as these portions of the spectrum are optimal – to distinguish vegetation from background soil.

A basic assumption behind vegetation indices (Ray 2000) is that an algebraic combination of spectral bands can provide a useful quantative measure of vegetation. Collected empirical evidence over the years tends to support this view. A further assumption is that all bare soil in an image will form a line in spectral space. For example (Richardson and Wiegand 1977) the Perpendicular Vegetation Index (PVI) defines a characteristic soil line in the two-dimensional feature space defined by a red band (Landsat MSS band 5) and a near-infrared band (Landsat MSS band 7). The line can be found by locating two or more patches of bare soil in the image having different reflectivities and finding the best fit line in spectral space. The vegetation content of a pixel is then defined in terms of its distance from this line.

Since their original development for Landsat data a number of vegetation indexes are have also been used successfully to estimate vegetation from AVHRR data. Although originally intended for meteorological purposes such as cloud detection and measuring of sea surface temperatures, the AVHRR instrument has several interesting properties that lends itself to land cover monitoring, including a much wider area coverage than Landsat, a rapid revisit capacity (often daily) and a large pixel size leading to cheaper processing costs per unit area. AVHRR bands provide a suitable basis for vegetation mapping because the red band (band 1) is sensitive to absorption by the chlorophyll in vegetation up to about 0.7 μm. At longer wavelengths lack of absorbing pigments in the green vegetation together with the leaf structure leads to a much higher reflectivity. Soils, on the other hand, are characteristically more reddish or neutral in colour, so spectral analysis based on the ratio of near-infrared (band 2) to red (band 1) can be used to determine the amount and categories of vegetation cover and subsequently to derive indicators such as Leaf-Area Index (LAI), green-leaf biomass and total biomass. Some typical AVHRR vegetation indexes making use of AVHRR bands 1 and 2 are shown in Table 2.4.

Of the bands shown in the table, the most widely used (Ray 2000) is the NDVI. It has the best dynamic range of any of the indexes and the best sensitivity to changes in vegetation cover. It is moderately sensitive to the soil background and to the atmosphere except at low plant cover. The principal application of AVHRR vegetation indexes is described in more detail in Section 7.2.3.

2.3.3 Extraction of linear features

Another important source of information in a satellite image is the location of linear features which can typify roads and communication features, geological lineaments or boundaries between land cover types. A simple approach to detection of linear features is to use linear convolution filters (of the type described in Section 1.5.3) in

Table 2.4 AVHRR vegetation indexes (based on Hayes 1985)

Name	Acronym	Formula
Environmental Vegetation Index	EVI	NIR − red
Infrared over Red	IR/R	NIR/red
Normalized Difference Vegetation Index	NDVI	(NIR − red)/(NIR + red)
Transformed Vegetation Index	TVI	$(\text{NDVI} + 0.5)^{1/2}$
Vegetative Sponge Index	VSI	sponge * NDVI

Notes: red = band 1 of AVHRR (visible); NIR = band 2 of AVHRR (near infrared) sponge is an environmental moistness variable.

order to detect the edges in an image (that is, pixels that exhibit a marked change in grey level from their neighbours). Typical filters used are the Laplacian kernel (which responds strongly to isolated points) and the series of Sobel filters (which respond strongly to edges in a given direction). The application of a filter to an entire image results in a map of the edge strength at each pixel. By applying a threshold value only the strongest edges are retained, resulting in an edge-map of the image. This method does, however, have a number of drawbacks:

- Edges may not be detected in areas of low-pixel values because the correspondingly subtle changes in relative value at an edge-pixel are not detected by a linear filter which measures absolute changes in value.
- If the threshold is set too high, many edge-pixels may be excluded unnecessarily.
- If the threshold is set too low, spurious edges (due, for example, to noise in the image) may be inadvertently included.

Solutions to these problems include the use of non-linear filtering, the use of adaptive local thresholds and the use of constraint rules to link neighbouring edge candidates together. Further details on the use of knowledge-based approaches can be found in Section 4.3.5.

Another approach to the detection of spatial features (Cross 1988) is the use of the Hough transform. The Hough transform is concerned with the shape of features rather than their spectral properties and has been used widely in image-processing applications (such as robot vision) for the location of features whose general shape (for example, line, circle or 'V') is known but not the exact size, position and orientation. The application of the Hough transform consists of three phases:

- construction of an edge map
- transformation into parameter space
- extraction of features from parameter space.

The edge map is produced by using linear filtering followed by thresholding to produce candidate edges with the strongest response. The parameter space used depends on the type of features it is desired to detect – for example, a straight line can be represented by the equation:

$$y = mx + c \tag{2.3}$$

where m = gradient of the line
c = intersect on the y axis

In this case the two parameters defined are m and c and any line in (x, y) space can be represented as a single point in (m, c) space. The Hough transform works by using an accumulator array in parameter space: for each edge candidate there is a whole family of lines which could pass through it which correspond to a number of bins in the parameter accumulator that are incremented. The bins containing the most entries at the end of the process represent the strongest linear features.

2.4 THEMATIC INFORMATION FROM SAR

The processing of SAR for thematic mapping applications differs fundamentally from that of optical imagery and is still in a far less understood state of research. This section describes:

- sources of information in a SAR image
- calibration and pre-processing
- classification methods
- SAR classification models

2.4.1 Sources of information in a SAR image

Spaceborne SAR images are (generally) single frequency, thus they cannot be classified by multi-spectral means and have a characteristically different statistical distribution. The mechanism whereby incident microwave radiation interacts with a surface area is complex and imperfectly understood, and the aim of much research is to provide better models of the fundamental imaging process which can subsequently be used as the basis for more advanced classification methods. There are some characteristic relationships between various types of target and their appearance in a SAR image – for example, specular reflectors such as lakes and rivers will appear very dark and good corner reflectors such as bridges will appear as very bright tones. The basic interaction between radar waves and a surface target depends on how rough the target is (Churchill and Sieber 1991). In mathematical terms roughness can be defined by threshold equations such as the Rayleigh criterion which defines a rough surface as one that obeys the equation:

$$\Delta h \cos \theta > \lambda/8 \tag{2.4}$$

where Δh = average vertical displacement of the surface
θ = incidence angle of the radar beam
λ = wavelength of radar

If, according to the above equation, a surface is smooth, then mirror-like specular reflection will occur, whereas for rougher surfaces scattering will occur in many different directions. Examples of surface types of varying roughness and the effect on image appearance can be seen in Table 2.5.

Table 2.5 Example of surface roughness and consequent image appearance (after Stone 1986)

Roughness	Image appearance	Target description
Very smooth	Black	Absolutely smooth and flat, open water, fine-gravel mud or silt
Smooth	Dark grey	Smooth and flat with minor perturbations, sand, no vegetation
Intermediate	Mid grey	Gravelly/stony areas, ploughed fields with shallow furrows, low vegetation
Rough	Light grey	Deeply ploughed fields, small cobbles and boulders, grass and low scrub vegetation
Very rough	White	Gullying, boulders, tall widely spaced vegetation, blocky rock outcrops

There are however, apart from surface roughness, a number of other interdependent influences on the radar signal (Radarsat 2001):

- the amount of moisture in the soil or on the vegetation
- man-made structures such as buildings and ships that strongly reflect radar energy back to the sensor, hence appearing as bright 'point targets'
- topography has a major influence on the radar signal with slopes inclined away from the sensor creating significantly less backscatter than those facing it.

The dependence of backscatter on radar geometry shows that two of the most important radar parameters for target discrimination are incidence angle and resolution, the value of which for the major SAR missions are shown in Table 2.6. The number of looks shown in the table is that required to achieve the azimuth and range resolution indicated.

Table 2.6 Incidence angle and resolution for SAR sensors (based on Churchill and Sieber 1991)

Resolution sensor (m)	Incidence angle (°)	Azimuth (m)	Range (m)	Looks
Seasat	20	25	25	4
SIR-A	47	40	40	4–7
SIR-B	15-60	25	17–58	4
ERS-1	23	30	30	4
JERS-1	35	18	18	3
ERS-2	23	30	30	4
SIR-C	15–55	30	10-60	4
Radarsat	20–50	4–100	4-100	1–8

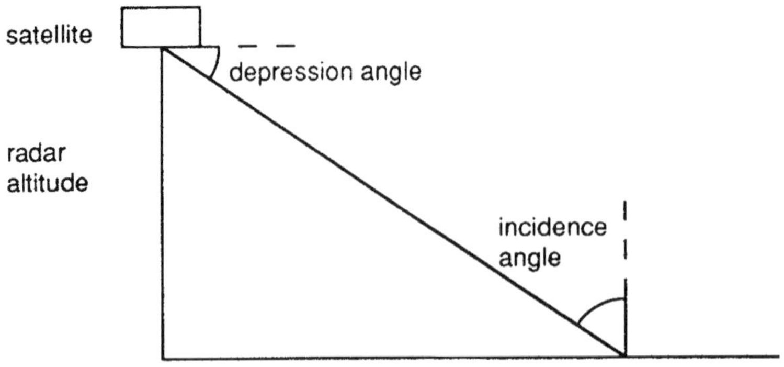

Fig. 2.16 Relationship between depression and incidence angle.

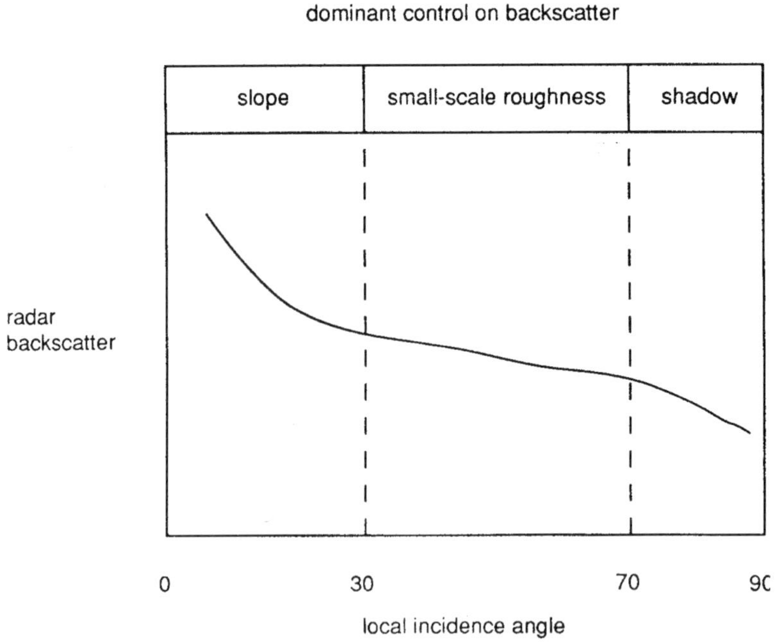

Fig. 2.17 Dominant control on backscatter at various incidence angles (based on Stone 1986).

Figure 2.16 shows the definition of incidence angle which is the complement of the depression angle. The dominant control on backscatter at various incidence angles is shown in Fig. 2.17.

In general (Stone 1986) shallow incidence angles such as that obtainable with SIR-A (47°) are better for surface roughness discrimination; steeper incidence angles such as Seasat (20°) and ERS-1 (23°) are more appropriate for areas of high relief. Another important parameter is wavelength (Churchill and Sieber 1991): at radar

Table 2.7 The main radar bands for spaceborne SAR (adapted from Churchill and Sieber 1991)

Band name (GHz)	Frequency (mm)	Wavelength	Sensors
L	1–2	150–300	Seasat SIR-A SIR-B SIR-C JERS-1
C	4–8	37.5–75	SIR-C ERS-1 ERS-2 Radarsat SRTM
X	8–12	25–37.5	SIR-C SRTM

wavelengths the signal recorded is greatly determined by scatterers of characteristic size 1–10 cm such as leaves, stems and branches of the target. The main bands used for spaceborne SAR, using the standard IEEE nomenclature (IEEE 1976), are shown in Table 2.7.

The last important parameter for SAR imagery is polarisation which will either be horizontal or vertical. Since either type of polarisation may be specified for transmit or receive, there are a total of four polarisation options (see Table 2.8).

Table 2.8 Polarisation options

Designator	Transmit	Receive	Type	Sensors
HH	Horizontal	Horizontal	like	Seasat SIR-A SIR-B JERS-1 Radarsat
VV	Vertical	Vertical	like	ERS-1 ERS-2 SIR-C (X-band)
VH	Vertical (polarimetric C and L bands)	Horizontal	cross	SIR-C
HV	Horizontal (polarimetric C and L bands)	Vertical	cross	SIR-C

2.4.2 Calibration and pre-processing

The two main complications which preclude the use of simple image analysis techniques for thematic analysis of SAR data are:

- speckle, which arises from coherent, monochrome microwave radiation and results in high variance of image tone resulting in the likely failure of pixel-based methods
- the viewing geometry.

The removal of speckle can be attained by the overlaying of several looks of the same area and hence averaging out the speckle values or the use of non-linear filters, such as median filters or the Lee filter (Lee 1981), to remove speckle without smoothing the genuine image data unduly. See Section 1.5.4 for further details on non-linear filters.

Topography can have a significant effect on image tone. Experiments with airborne SAR (Foody 1986) indicate that variations in topography can account for 5–25 per cent of the variance in image tone. This results in an increased range of possible values for each class of land-cover type leading to overlaps between classes and possible misclassification. The brightness of the target in a SAR image is dependent on the angle between the surface facet and the incoming radiation, hence it is a function both of the incidence angle and the local terrain slope. It is highly desirable that the effect of slope is removed as part of the calibration process before any quantitative classification is made, and this is most effectively performed with the use of a Digital Elevation Model. Another useful pre-processing technique (Foody 1987) makes use of fitting a smooth curve to a radiometric profile of the data in order to eliminate local effects.

2.4.3 Classification methods

The classification of SAR images can be divided broadly into four categories (in increasing order of sophistication):

- *cerebral* – visual photo-interpretation of SAR images as photographic products
- *image processing* – use of simple enhancement techniques to improve the interpretation using a GIS
- *information extraction algorithms* – use of statistical and other mathematically based tools to help automate the interpretation process
- *model based* – use of specific models of SAR geometry, target type and surface interaction.

Cerebral processing

The cerebral approach to classification of SAR images is to examine them visually, using a hard copy image, or an GIS in the manner of a photo-interpreter. The two main objectives for manual interpretation of a feature (Quegan *et al.* 1985) are tone (which can be identified on a subjective scale – very light, light, medium, dark and very dark) and texture (on a similar scale – very smooth, smooth, mid-texture, rough and very rough). The basic approach is shown in Fig. 2.18.

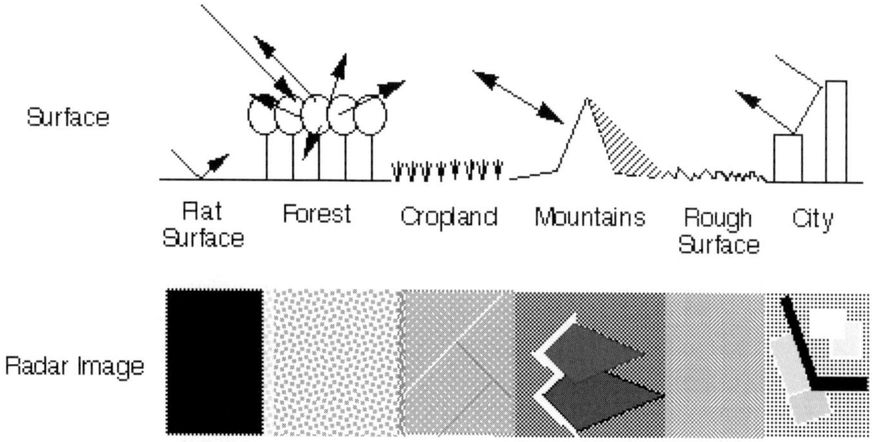

Fig. 2.18 Basis of Cerebral Processing (Courtesy NASA, Jet Propulsion Laboratory).

Image processing

Image-processing tools can be of use to remove the effects of speckle and image geometry. Standard image-processing techniques such as contrast stretching will also improve interpretability, but the use of linear edge-enhancing filters will often create spurious detail as the result of speckle, and therefore are not normally applicable.

Information extraction

The main information extraction algorithms are concerned with the statistics of a SAR image (such as mean, variance and higher order textural statistics of backscatter) and may be obtained on a per-pixel or per-region basis. Again, the speckle inherent in the images makes image segmentation techniques difficult to apply, and a better result may be obtained using digitised map information such as field boundaries (Wooding 1985). An alternative approach, and one suited to minimising the effect of image geometry on classification accuracy (Foody 1987), is to divide the image into sectors, either crudely based on equal divisions of ground-range distance or more accurately based on breaks in slope on the radiometric profile.

A recent innovation is to make use of 'tandem pairs' acquired during a mission where ERS-1 and ERS-2 were configured to follow the same orbit, and to acquire images with the same geometry and good temporal separation (tandem pairs are also important for derivation of height information from SAR interferometry – see Section 3.3.2). Different information layers can be extracted from tandem pairs including coherence, the backscattering intensity, the ratio between the two backscattering intensity images and the backscattering intensity texture. For multiple tandem pairs, the time average of coherence, backscattering intensity and texture and the temporal variability of the backscattering intensity are also calculated. A recent project (Strozzi *et al.* 2000) showed that ERS SAR interferometry can map typical landuse classes with the exception of urban areas. The use of image pairs with one to three days acquisition time intervals is required;

however, the presence of topography reduces the performance, as do variations in meteorological conditions.

2.4.4 SAR classification models

The model-based approach to SAR interpretation is based on identifying the main parameters of a surface type that will effect the radar backscatter. Models used for canopy backscatter range from regression estimators to complex ray-tracing algorithms (Seker and Schneider 1998). As described previously, roughness is an important parameter in such a model and so is the target's dielectric constant; for example, the moisture content of crops will have a significant effect on the backscattered signal. A detailed model for a particular target type will also need to include the background material and the geometrical relationships between individual scatterers that comprise the target. For example, a model used for radar classification of forestry (Simonett *et al.* 1987) has the following major components:

- direct backscattering from the soil surface
- backscattering from large branches and trunks
- volume scattering from foliage, canopy and leaf litter
- interaction caused by corner reflection from tree trunks
- other interaction such as multiple scatterings between foliage, ground surface, shrub canopy and branches.

Another important aspect of the modelling approach is temporal evolution of the backscatter. For example (Graham 2000), a semi-empirical 'water cloud' model was used to estimate crop moisture in the potato crop based on the backscatter of ERS radar data at the key points of the growing season: cultivation, planting, growth, saturation and harvest. The conceptual model can be seen in Fig. 2.19.

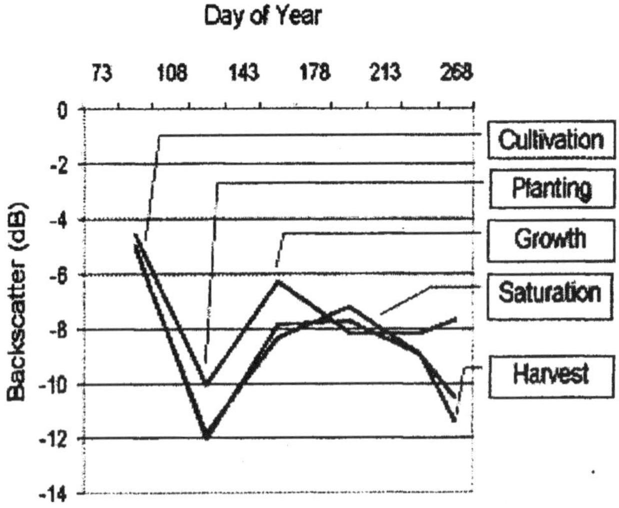

Fig. 2.19 Semi-empirical radar model for crop development (based on Graham 2000).

2.5 COMBINING SAR AND OPTICAL IMAGERY

One of the most interesting areas of development recently has been the combination of SAR and optical imagery.

> '*It is becoming clearer and clearer that the retrieval of bio- or geo-physical parameters using only SAR data from existing or planned spaceborne systems will be somewhat limited...*' (Borgeaud 1999)

There are many possible approaches and we present a few of them here. The first example (Williams *et al.* 2000) uses combined SAR imagery from ERS-1 with optical imagery from SPOT to help to calculate the area of the UK potato crop. Although the use of optical satellite data for crop classification is well established in Europe, the prevalence of cloudy conditions makes it difficult to guarantee image acquisition during the growing season even with specially programmed SPOT image acquisition. Another limitation of optical imagery is the difficultly of discriminating crop types based on multi-spectral classification alone – for example, potatoes often have a very similar spectral response to sugar beet and other crops growing in the same region. Radar images, on the other hand, can always be acquired for a particular growing season, but it can be difficult to discriminate crop types based on SAR data alone. Thus, the following approach was adopted to attempt to maximise the information derivable from the various data sources:

- A multi-temporal dataset of five ERS-2 SAR images and 3 SPOT XS images was acquired for each test area. Where possible the SWIR band from SPOT 4 was requested as it provides improved discrimination of crop types.
- In areas of low relief, such as East Anglia, it proved possible to directly overlay SAR and optical images using control point techniques. In more rugged areas the images were ortho-corrected using a DEM.
- The SAR data is processed using an iterative 'simulated annealing' technique to reduce speckle and to smooth the image data. This is then combined with the SPOT data as multi-dimensional 'layer stack' and a multi-spectral classification applied.
- A commercial GIS was used to present the imagery, derived 'potato maps' and associated statistics. The GIS was also used to combine imagery and maps of field boundaries to assist local field staff with the collection of ground truth information according to a predefined nested random sampling scheme.
- By using maps of field boundaries in the GIS it proved possible to produce information on a per-field rather than a per-pixel basis, which provides more accurate classification results and area statistics.

The project has shown classification accuracies of up to 95 per cent of potato field correctly identified using this approach, compared with typical results of 85 per cent using SPOT imagery alone and 75 per cent using SAR imagery alone.

The second example (Merlin *et al.* 1999) is based on the clear relationship between SAR imagery and the underlying topography. In this case the study area (the

Mustang kingdom in Nepal) is so mountainous it is difficult to overlay SAR images and optical data due to the extreme topographic distortion. The alternative approach is to use a DEM to generate a synthetic SAR image based on the expected response from topography alone. In most cases the synthetic images are sufficiently similar to the actual SAR images of the region to allow them to be geometrically registered using control points. Because the SPOT image can also be registered with the DEM it is then possible to create a dataset where all images are co-registered. Another piece of research (Manninen *et al.* 1998) showed that the seasonal variation of intensities in SAR images is typically higher on agricultural fields than in forests. This means that multi-temporal use of SAR data enables the distinction of fields and forests. The actual classification of crop types can then be carried out using multi-temporal optical imagery.

Finally (Mangolini and Arino 1996) developed a methodology to derive crop statistics from ERS SAR data and Landsat TM. This uses two separate processing chains for each type of data rather than the combined approach described previously for the UK potato crop. Firstly, the fields are segmented by a combination of an automated algorithm and visual segmentation using Landsat TM images. As the objective is to derive field boundaries it is possible to do this using imagery from the previous year if insufficient cloud-free data is available from the current season. Once the fields have been defined, the backscattering coefficients for each field can be derived from a time series of ERS SAR images. This is then used as the basis of a multi-spectral classification of crop types.

2.6 USING MULTI-TEMPORAL INFORMATION

This section describes the use of multi-temporal information – that is, a sequence of images acquired at different times. Multi-temporal datasets can be used for change dectection or for multi-temporal classifications.

2.6.1 Change detection

Change detection is concerned with identifying and quantifying areas of change, for example deforestation or urban growth. Other potential applications include study of agricultural changes, monitoring of coastal processes and hazard and disaster assessment. To compare two or more images of the same area it is essential that they can be related to a single coordinate system. For interactive or cerebral processing it is often sufficient that the images are registered together using one image (typically the earliest one) as the base image (see Section 1.1.7 for details). For more accurate multi-temporal analysis, it is desirable that the images are all geocoded to the same map projection. Ideally the geocoding process should result in an accuracy of better than one pixel in registration. If this is not achieved then small areas or edges of larger areas may be detected as spurious change due to the inaccuracy of the registration. Even when using accurate models of satellite and terrain geometry it is not always possible to achieve the desired accuracy in a first-pass geocoding process and it may be necessary to fine tune the registration using further tie points or correlation-based methods. It is also important to produce a multi-temporal series whose radiometric values are directly comparable – change must be attributable to genuine surface changes and not to systematic effects such as varying illumination conditions, atmospheric attenuation or changes to sensor gain or bias. A simple method of doing this is by histogram matching, and this will work fairly well providing that the areas of change are not large enough to bias the histograms significantly. However, as with all modelled interpretation of satellite data, the most accurate results will be obtainable if the data is accurately calibrated and converted to geophysical units.

After the images have been precision geocoded and calibrated it can be hypothesised that the changes between the two images can be attributed to genuine terrestrial changes. An initial estimate of the degree of change between the two images can be performed using a GIS. One technique (for example) is to display one image in red and the second in green. Areas of no change will appear as grey-tones with areas of great change being visible in one of the primary colours. Other techniques include 'swiping' from one image to the other, whereby areas of great change appear as obvious discontinuities in tone.

2.6.2 Image differencing

More quantitative results can be achieved by using algorithmic methods, the simplest being to subtract two images which have been geometrically registered and radiometrically corrected. This results in a 'change image' whose values are defined as:

$$C(i, j) = I_2(i, j) - I_1(i, j) \tag{2.5}$$

where $C(i, j)$ = value of pixel (i, j) in change image
$I_1(i, j)$ = value of pixel (i, j) in oldest image
$I_2(i, j)$ = value of pixel (i, j) in youngest image

A typical histogram of such a change image can be seen in Fig. 2.20. This can be interpreted as follows – the mode present around zero is contributed by the areas with no significant change; the tails of the histogram correspond to the areas of greatest change.

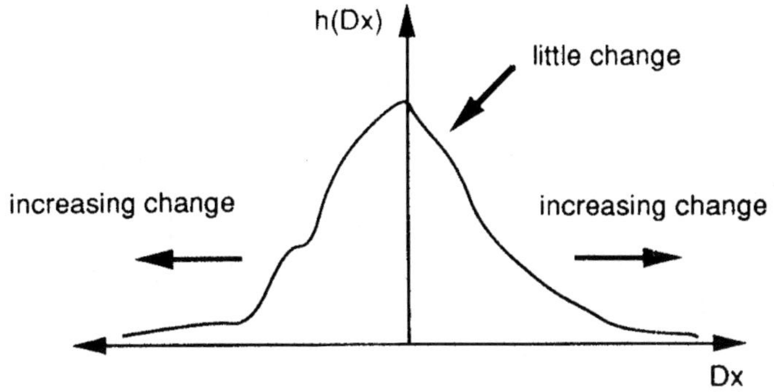

Fig. 2.20 Schematic histogram of a change image derived by subtracting two registered multi-temporal images.

The change image itself can be interpreted using two filtering techniques:

- *thresholding of the image histogram* – this removes the pixels near the centre of the histogram which are not associated with change
- *cleaning of small areas* – following thresholding to remove (a) isolated pixels and edge-pixels where the detected change may be caused spuriously, for example, by inaccurate registration, or (b) pixels that suffer from mixture effects.

The result is a number of areas whose size and spatial relationships can be measured, and although this may not in itself be enough to identify change, it may at least provide suitable 'triggers' for future examination by further EO sources or ground survey.

2.6.3 Other change detection algorithms

A number of other change detection techniques have been proposed and a survey can be found in the review article by Ashbindu Singh (1989). The first class of algorithms are essentially variants on the image differencing technique and include image ratioing and regression analysis. The advantage of image ratioing is that it detects large relative changes in low pixel values that would be thresholded out by image differencing; regression analysis uses statistical analysis to find the best-fit line between pixel values in the two images, effectively eliminating differences in the mean value and variance from the two sets of data.

A second class of algorithms is concerned with comparing transforms of the image data, and suitable choices include vegetation index transforms and the Principal Components Transform (PCT). Vegetation index methods can make use of any of the algorithms described in 'Vegetation indexes' (Section 2.3.2) such as the Perpendicular Vegetation Index (PVI) or the tasseled cap transform. The comparison between the two transformed images from the different dates can then be performed by the use of simple image differencing. A similar approach is used for the PCT images; or, alternatively, PCT (see Section 2.3.1) can be performed on the entire dataset amalgamated from the two images. The expected result (Byrne *et al.* 1980) is that major changes associated with scene illumination and atmospheric effects should appear in the major components with the 'interesting' changes being present in the minor components; it is, however, relatively difficult to analyse such a dataset.

The third major approach is to perform a multi-spectral classification on each image individually and to compare the results, producing a class-transition for each pixel in the image. This approach is best used within the context of a Geographic Information System (see Chapter 4 for further details).

2.6.4 Multi-temporal classification

Multi-temporal classification is based on the observation that the spectral signature of different land cover types, particularly those of crops, will vary over time. This leads to the possibility of more accurate classification using a series of 'snap shots' taken throughout the season than is possible using images taken at a single date.

For example, Price *et al.* (1997) used a multi-date Landsat Thematic Mapper image set to map land cover in a high plains agro-ecosystem in Kansas. Use was made of four spectral bands for each of three dates. An unsupervised ISODATA (Iterative Self-Organizing Data Analysis) technique was used as a preliminary processing step before submitting to maximum likelihood classification. This provided 100 output classes which were then reclassified as either cropland or grassland. Prior to this study, overall classification accuracy of single-date imagery for grassland and cropland was 70 per cent. Whereas the multi-temporal approach provided classification ranging from 92.0 to 99.5 per cent.

As we have already seen in Section 2.5, multi-temporal techniques are very important for the classification of SAR imagery where the variation of a single parameter (backscatter) over time can be used as a surrogate 'spectral signature' for multi-spectral classification. For example, the seasonal variation of intensities in SAR images is typically higher on agricultural fields than in forests (Manninen *et al.* 1998). Therefore, multi-temporal use of SAR data enables the distinction of fields and forests. Single SAR images are not as easy to use for forest mapping, although at certain times of the season the intensities of fields and forests are clearly different. Owing to the speckle, direct classification is not usually successful. In an earlier example (Cihlar *et al.* 1992) a variety of land cover types (including agricultural, urban, forestry and golf courses) displayed distinctive annual sinusoidal variations in backscatter and consistent year-to-year stability in backscatter profiles. Figure 2.21

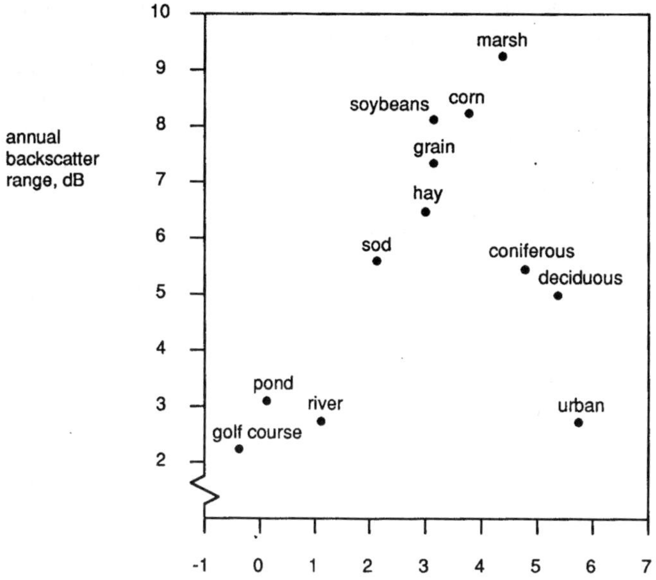

Fig. 2.21 Distinguishing land cover classes on the basis of annual backscatter range combined with annual mean backscatter – experimental results derived from aircraft C-band SAR (based on Cihlar *et al.* 1992).

shows experimental results using C-band aerial SAR which demonstrate how it is possible to distinguish land cover classes on the basis of annual backscatter range combined with annual mean backscatter.

REFERENCES

Belward AS and Valenzuela CR (eds) (1991) *Remote Sensing and Geographical Information Systems for Resource Management in Developing Countries*. Kluwer Academic Publishers, Dordrecht, The Netherlands.

Benny AH (1980) Coastal definition using Landsat data. *International Journal of Remote Sensing*, **1** (3), 255–60.

Borgeaud M (ed.) (1999) *Second International Workshop on Retrieval of Bio- and Geophysical Parameters*. Special Issue of ESA's *Earth Observation Quarterly*, EOQ No. 62, June 1999.

Byrne GF, Crapper PF and Mayo KK (1980) Monitoring land cover change by principal component analysis of multi-temporal Landsat data. *Remote Sensing of Environment*, **10**, 175–84.

Cihlar J, Pultz TJ and Gray AL (1992) Change detection with Synthetic Aperture Radar. *International Journal of Remote Sensing*, **13** (3), 401–14.

Chavez PS (1988) An improved dark-object subtraction technique for atmospheric scattering correction of multi-spectral data. *Remote Sensing of Environment*, **24**, 459–79.

Churchill PN and Sieber AJ (1991) The current status of ERS-1 and the role of radar remote sensing for the management of natural resources in developing countries. In Belward AS and Valenzuela CR (eds), pp. 111–43.

Cornaert M-H and Maes J (1992) Land cover: an essential component of the CORINE information system on the environment. GIS Implications. *Proceedings of the Central Symposium of the 'International Space Year' Conference, Munich, Germany*, 30 March–4 April 1992, ESA ISY-1, pp. 473–81.

Crist EP and Cicone RC (1984) A physically-based transformation of Thematic Mapper data – the TM tasseled cap. *IEEE Transactions on Geoscience and Remote Sensing*, **GE-22** (3), 256–63.

Cross AM (1988) Detection of circular geological features using the Hough transform. *International Journal of Remote Sensing*, **9** (9), 1519–28.

Drake N and White K (1991) Linear mixture modelling of Landsat Thematic Mapper data for mapping the distribution and abundance of gypsum in the Tunisian Southern Atlas. *Proceedings of Spatial Data 2000, Christ Church, Oxford*, 17–20 September 1991, pp. 168–77.

Foody GM (1986) An assessment of the topographic effects on SAR image tone. *Canadian Journal of Remote Sensing*, **12** (2), 124–31.

Foody GM (1987) A method for thematic classification with Synthetic Aperture Radar data. *Geocarto International*, **2**, 31–6.

Foody GM and Cox DP (1991) Estimation of sub-pixel land cover composition from spectral mixture models. *Proceedings of Spatial Data 2000, Christ Church, Oxford*, 17–20 September 1991, pp. 186–95.

Foody GM (1992) Classification accuracy assessment: some alternatives to the Kappa coefficient for nominal and ordinal level classifications. *Proceedings of the 18th Annual Conference of the Remote Sensing Society, University of Dundee, Scotland*, 15–17 September 1992, pp. 529–38.

van Genderen J L and Uiterwijk U (1987) A practical procedure for classifying digital imagery. *Proceedings of the Annual Conference of the Remote Sensing Society, Nottingham*, September 1987, pp. 287–96.

Hayes L (1985) The current use of TIROS-N series of meteorological satellites for land cover studies. *International Journal of Remote Sensing*, **6** (1), 35–45.

Huguenin R, Karaska M, van Blaricom D and Jensen J (1997a) Subpixel classification of bald cypress and tupelo gum trees in Thematic Mapper imagery. *Photogrammetric Engineering and Remote Sensing*, **63**, 717–25.

Huguenin R, Boudreau E and Karaska M (1997b) Adaptation of the AASAP (n.k.a. subpixel classifier) analysis software for automated bathymetry mapping. Paper presented at the *ERIM Fourth International Conference on Remote Sensing for Marine and Coastal Environments, Orlando, Florida*, 17–19 March 1997.

IEEE (1976) *Standard Letter Designation for Radar Frequency Bands*, IEEE Standard 521-1976.

Graham AJ (2000) On the use of synthetic aperture radar to explore moisture issues related to *solanum tuberosum*. *Aspects of Applied Biology: Remote Sensing in Agriculture*, **60**, 173–7.

Jupp DLB and Mayo KK (1982) Use of residual images in Landsat image analysis. *Photogrammetric Engineering and Remote Sensing*, **48**, 595–604.

Kauth RJ and Thomas GS (1976) The tasseled cap – a graphic description of the spectral temporal development of agricultural crops as seen by Landsat. *Proceedings of the Second Annual Symposium on Machine Processing of Remotely Sensed Data, Purdue University, Indiana*, 1976, pp. 4B41–4B49.

Lee J-S (1981) Refined filtering of image noise using local statistics. *Computer Graphics and Image Processing*, **15**, 380–9.

Lintz J and Simonett DS (eds) (1976) *Remote Sensing of Environment*. Addison-Wesley, Reading, Mass.

Manninen AT, Häme TP and Lohi A (1998) Comparison of ERS SAR and Landsat TM forest classification. *Retrieval of Bio- and Geophysical Parameters from SAR Data for Land Applications Workshop, European Space Agency ESTEC, Netherlands*, 21–23 October 1998.

Mangolini M and Arino O (1996) ERS SAR and Landsat TM multi-temporal fusion for crop statistics. *Earth Observation Quarterly*, **51**, March.

Mather PM (1987) *Computer Processing of Remotely-sensed Images: An Introduction*. John Wiley & Sons, Chichester.

Merlin L, Nicolas JM and Blamont D (1999) Using layovers in fusion of optical and SAR data: application to Mustang Landscape analysis. *CEOS SAR Workshop, Toulouse*, 26–29 October, 1999.

Pratt WK (1985) *Digital Image Processing*. John Wiley, New York.

Price KP, Egbert SL, Nellis MD, Lee R-Y and Boyce R (1997) Mapping land cover in a High Plains agro-ecosystem using a multi-date Landsat Thematic Mapper modeling approach. *Transactions of the Kansas Academy of Science*, **100** (1/2), 21-33.

Quegan S, Churchill PN, Wright A, Lamont J, Rye AJ and Trevett JW (1985) SAR for agriculture and forestry. *Proceedings of a Workshop on Thematic Applications of SAR Data, ESRIN, Frascati, 9-11 September 1985*, ESA SP-257, pp. 7–14.

Radarsat (2001) Radarsat illuminated: your guide to products and services. http://www.rsi.ca/classroom/class.htm

Ray TW (2000), A FAQ on vegetation in remote sensing. http://www.yale.edu/ceo/Documentation/rsvegfaq.html

Richardson AJ and Wiegand CL (1977) Distinguishing vegetation from soil-background information. *Photogrammetric Engineering and Remote Sensing*, **43**, 1541–2.

Sabins FF (1987) *Remote Sensing Principles and Interpretation* (2nd edition). WH Freeman & Company, New York.

Seker SS and Schneider A (1988) Electomagnetic scattering from a dielectric cylinder of finite length. *IEEE Transactions of Antennas and Propagation*, **36** (2), 303–7.

Settle JJ (1990) Contextual models and the use of Dirichlet priors for mixing problems. *Proceedings of the Fifth Australian Remote Sensing Conference, Perth, Western Australia*, 8–12 October 1990, pp. 314–23.

Sheffield C (1985) Selecting band combinations from multi-spectral data. *Photogrammetric Engineering and Remote Sensing*, **51** (6), 681–7.

Simonett DS, Strahler AH, Sun G and Wang Y (1987) Radar forest modelling: potentials, problems, approaches, models. *Proceedings of the Annual Conference of the Remote Sensing Society, Nottingham*, September 1987, pp. 256–70.

Singh A (1989) Digital change detection techniques using remotely-sensed data. *International Journal of Remote Sensing*, **10** (6), 989–1003.

Stone RJ (1986) Radar system and ground parameters – an interactive case study of semiarid terrain in Tunisia. Mapping from modern imagery. *Proceedings of a Symposium held by Commission IV of the International Society for Photogrammetry and Remote Sensing, Edinburgh, Scotland*, 8–12 September 1986, pp. 261–70.

Strozzi T, Dammert P, Wegmüller U, Martinez J-M, Askne J, Beaudoin A and Hallikainen M (2000) Landuse mapping with ERS SAR interferometry. *IEEE Transactions on Geoscience and Remote Sensing*, March 2000.

Tanré D, Deroo C, Duhaut P, Herman M, Morcrette JJ, Perbos J and Deschamps PY (1990) Description of a computer code to simulate the satellite signal in the solar spectrum: the 5S code. *International Journal of Remote Sensing*, **11**, 659–68.

Taylor JC, Sannier C, Delincé J and Gallego FJ (1997) *Regional Crop Inventories in Europe Assisted by Remote Sensing: 1988–1993*, Joint Research Centre, European Commission, EUR 17319 EN.

US Geological Survey (2001) Landsat-7 Level-0 and Level-1 Data Sets Document. http://eosims.cr.usgs.gov:5725/DATASET_DOCS/landsat7_dataset.html

Vermote EF, Tanré D, Deuzé JL, Herman M and Morcette JJ (1997) Second simulation of the satellite signal in the solar spectrum, 6S: an overview. *IEEE Transactions on Geoscience and Remote Sensing*, **35**, 675–86.

Wilkinson GG (1991) The processing and interpretation of remotely-sensed satellite imagery – a current view. In Belward AS and Valenzuela CR (eds), pp. 71–96.

Williams J, Parker D, Harris R, Turner R and Baker D (2000) Area estimation of the British potato crop using earth observation and GIS. *Aspects of Applied Biology: Remote Sensing in Agriculture*, **60**, 195–8.

Wooding MG (1985) SAR image segmentation using digitized field boundaries for crop mapping and monitoring applications. *Microwave Remote Sensing Applied to Vegetation*. European Space Agency, SP-227, pp. 93–8.

Further reading

Bird AC (1991) Principles of remote sensing: interaction of electromagnetic radiation with the atmosphere and the earth. In Belward AS and Valenzuela CR (eds) pp. 17–30.

Crosetto M and Mroz M (1998) Optical-Radar Data Fusion for Land Use Classification. *Proceedings of the ISPRS–Commission VII Symposium, Budapest (Hungary), Int. Arch.* Vol. XXXII, Part 7, pp. 698–705.

Malingreau J-P (1986) Global vegetation dynamics: satellite observations over Asia. *International Journal of Remote Sensing*, **7** (9), 1121–46.

Saich P and Borgeaud M (1998) Interpreting ERS SAR signatures of agricultural crops in Flevoland 1993–1996. *Retrieval of Bio- and Geophysical Parameters from SAR Data for Land Applications Workshop, European Space Agency ESTEC, Netherlands*, 21–23 October 1998.

Townshend JRG and Justice CO (1986) Analysis of the dynamics of African vegetation using the Normalized Difference Vegetation Index. *International Journal of Remote Sensing*, **7** (11), 1435–45.

Vieira CAO and Mather PM (2000) *An Examination of the Effectiveness of Multitemporal Crop Classification, RSS 2000*. University of Leicester, UK.

Wiegand CL, Gausman HW, Cuellar JA, Gerbermann AH and Richardson AJ (1973) Vegetation density as deduced from ERTS-1 MSS response. *Third Earth Resources Technical Satellite Symposium*, NASA SP-351, 1, pp. 93–116.

3

Height information from space

One of the most important data structures used for spatial analysis is a Digital Elevation Model (DEM), which is a digital representation of the Earth's topography – for example, a matrix containing the height of points on the Earth's surface above some reference datum. DEMs have traditionally been produced by digitising map contours derived from stereo-pairs of aerial photographs.

Unfortunately, DEMs of usable detail are still not available for much of the Earth's surface (such as deserts, mountain ranges and other inaccessible regions), and even when available may not be of sufficient accuracy, or may suffer from problems of inconsistency of data acquisition and processing methodology. DEMs have great importance in the geocoding of both optical and SAR images, in the calibration of images (for example, to provide an input to a reflectance or backscatter model) and in the visualisation of images as well as a wide variety of other applications. Faced with the increased global demand for DEMs, satellite data provides a realistic source for all but the finest grid DEMs – and with the advent of Very High Resolution (VHR) images 1 metre grid spacing is available (see Section 8.1).

Chapter 3 describes topographic mapping from space and contains the following sections:

- photogrammetry using satellite images: the SPOT geometry that makes it suitable for derivation of height information, alternative along-track geometries and photogrammetric methods of height extraction
- digital stereo matching: correlation, feature-based and hybrid stereo matching methods and the validation of the DEM produced
- production of DEMs from SAR using interferometric techniques
- DEM validation
- the Shuttle Radar Topography Mission (SRTM)
- applications of DEMs: the importance of DEMs as an adjunct to information extraction from satellite images and for a variety of other applications

3.1 PHOTOGRAMMETRY USING SATELLITE IMAGES

This section describes photogrammetry using satellite images, starting with the SPOT stereo geometry and then looking at photogrammetric methods of height extraction and along-track stereo.

3.1.1 SPOT stereo geometry

The most common method used to derive DEMs from satellite images is to use the parallax information present in a pair of images viewing the same scene at different angles. This is similar to the way that the human visual system obtains three dimensional information (see Fig. 3.1).

A single human eye is essentially a two-dimensional sensor – it can move elevationally (up and down) and azimuthally (side to side) relative to its position in

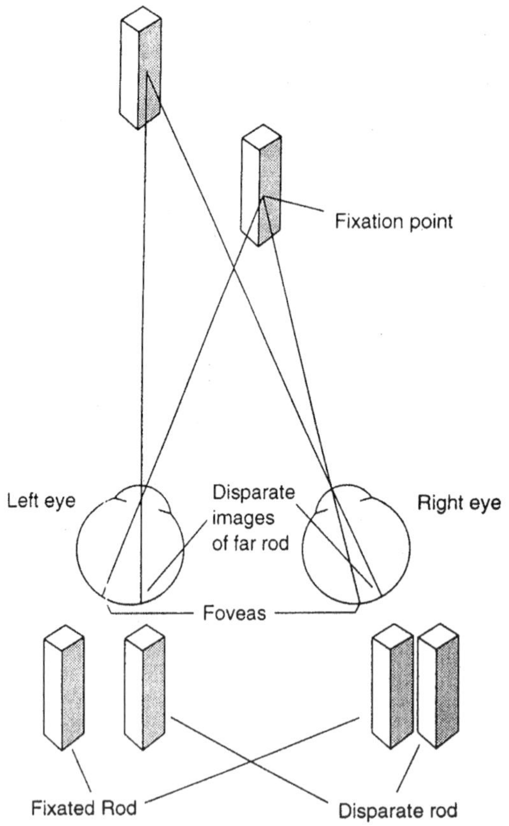

Fig. 3.1 The human visual system: the two eyes are fixated on the nearer rod, the more distant rod produces a binocular disparity (a larger distance between the image of the rod and the fovea of the left eye compared to that of the right eye). This effect is illustrated by the corresponding pair of rods below each eye (from Frisby 1979).

the head. The angular location of any object viewed by a single eye can thus be specified by the two angles describing the ray of light from the object to the eye. This, in itself, is not sufficient to retrieve three-dimensional information, however the information from the right eye also determines another ray that the point must lie on. The combined attitude information from both sensors together with their known configuration in the baseline of the human skull enables the brain to calculate the true three-dimensional location of the point.

The basic requirement for satellite stereo viewing is that the sensor is capable of viewing the same point on the Earth from two different angles. In the case of Landsat (either MSS or TM) the orbit ensures that the repeat path is consistent to within a few kilometres. Thus two images from the same Landsat (path, row) taken at different dates will have virtually the same viewing angle and will not be suitable for use as a stereo pair. Certain areas of the world, however, will lie in the overlap region between two adjacent Landsat paths – one of the satellite tracks passing to the east and one to the west. The overlap between adjacent Landsat paths varies with latitude and can be seen in Table 3.1.

Table 3.1 Overlap between adjacent Landsat paths (from Hill 1991)

Latitude (°)	Overlap (%)	Latitude (°)	Overlap (%)	Latitude (°)	Overlap (%)
0	7.3	30	19.7	60	53.6
10	8.7	40	29.0	70	68.3
20	12.9	50	40.4	80	83.9

Images from adjacent paths acquired in this region will view a point from two different angles and can thus be used as a stereo pair. However, in comparison with the somewhat limited acquisition of suitable stereo pairs from Landsat, the SPOT sensor is designed specifically for stereo viewing, and it is possible to acquire stereo pairs in a variety of angular configurations simply by altering the angle of the sensor mirror prior to image acquisition (see Fig. 3.2).

The ability to steer each HRV instrument up to 27° off-nadir allows images of a given area to be acquired more frequently than the nominal 26-day repeat cycle; for example seven passes per 26 days at the equator, and 11 passes at latitude 45°. This provides the ability to acquire stereo pairs of SPOT images the same area from different viewing angles (see Fig. 3.3).

Using SPOT stereo pairs it is possible to produce a DEM with height accuracy of better than the equivalent of one pixel; for example, one of the first tests of stereoscopic accuracy of SPOT data using analytic plotters (Konecny *et al.* 1987) achieved RMS errors of 6.5 metres in elevation and an experiment in the Aix-en-Provence area of France, also using an analytical stereoplotter (Rodriguez *et al.* 1988), achieved accuracies under 10 metres in plan and 7 metres in elevation, with a best result of 3 metres in elevation (using a base/height ratio of 1).

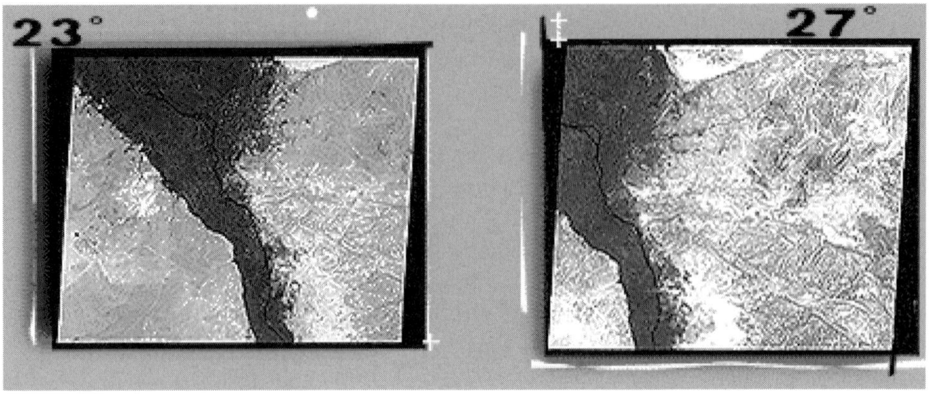

Fig. 3.2 SPOT acquisition angles (courtesy SPOT Image).

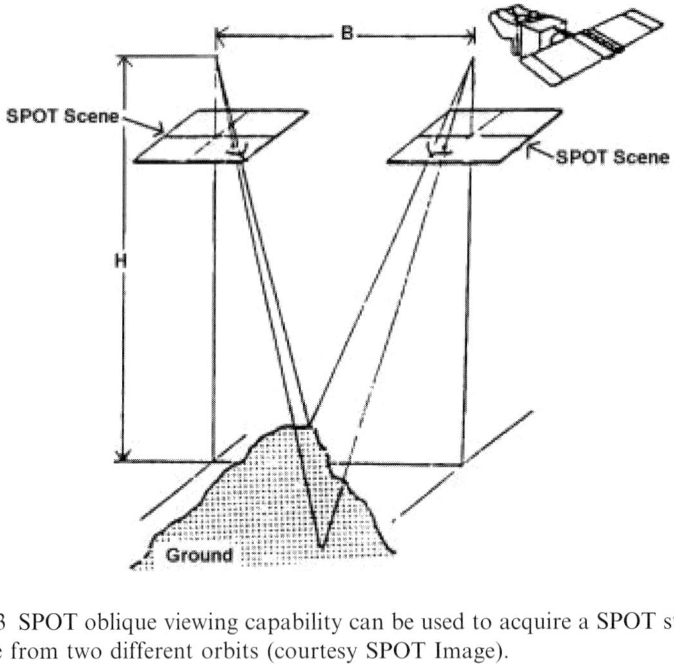

Fig. 3.3 SPOT oblique viewing capability can be used to acquire a SPOT stereo pair of a scene from two different orbits (courtesy SPOT Image).

3.1.2 Photogrammetric methods of height extraction

One simple method of deriving height and contour information from a SPOT stereo pair is to reproduce the left- and right-hand images photographically and use standard photogrammetric techniques, such as the use of an analytic stereoplotter to produce the desired information. Unfortunately the geometry of a SPOT image is somewhat different from that of an aerial photograph and such techniques are not directly applicable (see Section 1.2.3). One solution (Trinder and Donnelly 1988) is to divide the image into a number of segments and to assume that a central point perspective is a fair approximation within that segment (that is, the standard photogrammetric collinearity equations are applicable); alternatively a programmable stereoplotter can be used to relate the image coordinate system to the geographic coordinate system in use to drive the stage plates to their correct position for stereo viewing.

The dynamic orbital exterior orientation for a SPOT image is first calculated by modelling the sensor geometry then calibrated by the use of a few ground control points. These are then used as the input to the process controlling the analytical stereoplotter and form the basis of the inverse collinearity equations used to calculate the instantaneous transformation required to position the right- and left-hand photographs.

Another possible hybrid approach is the digital stereoplotter – in a typical system the right- and left-hand images are presented on a split screen image display, but control over the photogrammetric process is still effected manually. The advantages of such a system are that it avoids the image degradation when digital data is written to film and allows contour and digital elevation data to be generated, edited and output in an entirely digital format. A digital stereoplotter uses similar software to calculate the image acquisition geometry and the resultant exterior orientation, but instead of mechanically positioning the two SPOT photographs the digital images can be repositioned graphically. The operator views both screens through a mirror stereoscope device and controls the motion via an interactive input device.

3.1.3 Along-track stereo

The main advantage of along-track stereo is that images are acquired with a very small temporal separation, typically of an order of minutes. This means that the images will have very similar atmospheric and surface terrain characteristics, as well as an easily modelled geometry, leading to more effective stereo-matching and potentially more accurate generation of DEMs. Along-track stereo (Durand 2000) is derived using views acquired at different angles along the line of the orbit. These can be forward, vertical or backward. These views may be acquired in two different ways:

- from a single viewing instrument that is tilted
- by using several instruments pointed at different angles in the plane of the orbit.

The first method has the advantage that the base/height ratio can be varied (in the same way, for example, that is achievable for SPOT across-track stereoscopy) and also provides the operational benefit of a lighter payload. This was the method originally proposed for SPOT 5, however it was considered that the additional time needed for instrument stabilisation following tilting would compromise the number of acquisitions that could be made, so the second approach was eventually adopted.

Using fixed pointing sensors means that only a fixed base/height ratio is obtainable from any pair of instruments. However by using three or more sensors a selection of different ratios is possible. The OPS optical sensor used on the Japanese

JERS-1 satellite uses two fixed sensors: one nadir-looking and the other 15.3° forwards, which gives a base/height ratio of 0.3.

It is planned that SPOT 5 will carry three identical HRG instruments pointed along-track at −19.2°, 0° and +19.2° respectively. This allows the following configurations:

- before–after stereoscopy with a base/height ratio of 0.8 – suited to areas of moderate relief
- before-nadir or after-nadir stereoscopy with a base/height ratio of 0.4 – useful to minimise concealed area such as cliffs
- the use of all three instruments (tri-stereoscopy): this combines the advantages of the first two methods for a variety of terrains but requires more complicated processing.

The tri-stereo combination has previously been used experimentally on the German MOMS-2P mission flown on the PRIRODA module of the Russian Space Station MIR. This featured a nadir-looking sensor and fore and aft instruments at ±21.4°.

Even with the small gap between image acquisitions some compensation for Earth rotation is necessary. This can either be done by controlling the attitude of the satellite depending on the latitude (at the equator the effect is more pronounced than near the poles) or by using lateral calibrations of the view-changing mirrors for each instrument.

3.2 DIGITAL STEREO MATCHING

We have seen how, with a certain amount of ingenuity, terrain height information can be derived by standard photogrammetric techniques. It is desirable, however, both in terms of operator involvement and the ultimate accuracy of the product, to generate the DEMs entirely by digital means. There are two main aspects to this: the first is the imaging geometry, the model-based solution of which has been described in Section 1.2.3; the second is image correspondence, which is concerned with the location of the same feature in both images and the calculation of the stereo disparity (and hence the terrain elevation) between them. There are two classes of algorithms for solving the correspondence problem: the first class is based on mathematical correlation of image areas; the second is based on featured techniques such as those used for computer vision. In addition, various hybrids of the two approaches are possible.

3.2.1 Correlation methods

The correlation approach to stereo matching (Greenfeld 1991) relies on certain assumptions about the problem domain, primarily that:

- it is possible to pre-define a coarse mapping between the two images (for example, by means of seed points)
- the radiometric properties of the two images are similar

- there are no unmodelled geometric distortions between the image geometries
- two images patches to be matched have distinct textures (smoothly responding surfaces such as ice-sheets can prove very difficult to correlate).

The steps necessary for production of DEMs using the correlation approach can be described in broad terms as follows:

1. Acquire a suitable stereo pair (cloud-free, good base/height ratio).
2. Calculate the imaging geometry for the left-hand image.
3. Calculate the imaging geometry for the right-hand image.
4. For each point, determine the stereo disparity.
5. Calculate the height from the stereo disparity.
6. Perform any interpolation necessary.
7. Reformat onto the final DEM grid.

The use of correlation techniques is broadly similar to that used for the relocation of GCPs (see Section 1.1.9). However, the relocation of GCPs has a number of simplifying aspects; for example, the image will be acquired on the satellite's repeat cycle so that a GCP will be viewed with essentially the same geometry. It is then a simple matter to use one chip as a template and move it in column and row pixel increments until the best fit position is found. The correlation of SPOT stereo pairs is more problematic as they are non-epipolar (an epipolar image geometry means that, given a point in the first image, the match point must lie on a known line in the second image). With an epipolar geometry the search space can be restricted to one dimension and is the basis of many stereo-matching algorithms, such as those used for robot vision. Unfortunately, due to their dynamic nature described previously, SPOT stereo pairs are not acquired in epipolar format and cannot be converted to epipolar format without further geometric rectification; thus any correlation algorithm must use a two-dimensional search space with a resulting increase in computational complexity. The algorithm is further complicated by two additional factors:

- *radiometric* – the radiometric properties of patches to be matched can vary greatly between the pair, in particular off-nadir images can have noticeable non-linear intensity differences
- *geometric* – off-nadir viewing combined with terrain variations mean that patches will not always have the same size, shape and orientation.

It is therefore apparent that any correlation scheme used to solve the image correspondence problem must use a two-dimensional search space and be capable of modelling a variety of geometric and radiometric distortions. One of the commonly used algorithms (Gruen 1982, 1986) is an adaptive least-squares correlation algorithm. Gruen's algorithm is highly accurate (up to 0.05 pixel has been obtained on aerial photography), however it has a limited pull-in range of about two pixels, thus must be provided with known seed points from which the matched sheet can grow out. The least-squares criterion is a measure of how well the two images match and is based on minimising the sums of squares of the differences between the corresponding grey levels. The adaptive attribute of the algorithm is that a set of parameters can be specified to allow for geometric and radiometric distortions between the two patches. The radiometric constraint models the transformation between the patches as a linear contrast stretch:

$$I_2(x, y) = a * I_1(x, y) + b \tag{3.1}$$

where $I_1(x, y)$ = value at (x, y) in first image
$I_2(x, y)$ = value at (x, y) in second image
a, b = constants

The geometric constraints are modelled as affine (linear) transformations. This allows a wide variety of scale, translation and perspective changes to be modelled; dealing with all the cases where the patches in both images are small enough to be considered as facets of plane surfaces. Note that, to be effective, the two constraints must be chained together – that is, the radiometric constraint can only be used after the point is mapped into its 'true' position by the geometric constraint. The parameters of the model are altered throughout the search space until the best fit position is found. The large number of parameters in the model make it impracticable to use a large search space, therefore it is essential that each attempt at correlation starts from a good initial estimate of the corresponding patch. This estimate is known as the seed point of the algorithm and can be derived by various methods such as manual definition, coarse matching on a reduced resolution version of the stereo pair or by use of a feature-based matching algorithm (see Section 3.2.2). As with any searching algorithm the result of an attempt to establish corresponding points can have one of three outcomes:

- *success* – algorithm converges to correct match point
- *failure* – algorithm fails to converge
- *failure* – algorithm converges but to wrong point.

The first case is the desired outcome; the algorithm in the second case 'knows' that it has failed and can attempt to find the correct point using a new seed point. The third case is the most problematic because the algorithm 'thinks' that it has converged correctly (Otto and Chau 1989). It is very difficult to counter false convergence, thus it is essential to have good quality control over the production of a DEM (described in Section 3.4). The stereo-matching algorithm can be speeded up by various means. For example (Norvelle 1992) different shape windows can be used to constrain the search space. This uses the results of matches on neighbouring pixel pairs to determine the size and shape of the area that it is worth searching within. This can also produce more accurate correlations and reduce the averaging effect that may occur when a large window is used over areas of varying terrain. The result of the stereo-matching phase will be a measure of the stereo disparity throughout a control network. The stereo disparities will represent the height relative to a floating datum; that of zero parallax. They can now be converted to a fixed datum by incorporating the information contained in the sensor geometry model and the Earth model. The result of this is a coarse-gridded DEM.

It is also possible to use an iterative method to refine the DEM produced; for example, the 'Iterative Orthophoto Refinements' method (Norvelle 1992) uses the following algorithm:

1. Reduce full-scale images by a factor of 4.
2. Perform stereo matching on reduced-scale images.
3. Calculate approximate DEM from stereo disparities.
4. Produce orthoimages of full-scale input images (using the approximate DEM).
5. Perform stereo matching on orthoimage pairs.
6. Update DEM based on elevation errors by stereo matching.
7. Iterate steps 4 to 6 until DEM is sufficiently accurate.

This method uses the fact that if the DEM is accurate it can be used to produce a pair of orthoimages which are virtually identical in their geometry. Thus, if used as a stereo pair there will be no noticeable stereo disparities. If however, there are inaccuracies in the DEM these will be reflected in mismatches in the stereo-matching process which can be used to iteratively refine the DEM. It is important that the orthoimages generated at each iteration are based on the original image data – this removes the problem of image degradation due to repeated resampling.

3.2.2 Feature-based methods

The other possible approach to the stereo-matching problem is to use feature-based methods, based on techniques of computer vision. The use of computer vision techniques to model 'real-world' features has certain potential advantages. Correlation algorithms can be restricted by their underlying radiometric and geometric constraints and can suffer from 'loss of lock' when they fail to converge. In human vision the most important low-level component is generally considered to be the detection of edges, since the boundary between two features will often be accompanied by a significant change in image intensity over a very small distance; and the problem of edge detection is also fundamental to computer-based image interpretation. The 'edge detector' thought to be employed in the human visual system can be approximated by several Difference of Gaussian (DoG) filters applied to an image (Kauffman and Wood 1987). Conceptually each DoG filter is used to detect edges that occur at a different spatial frequency and thus can be used to detect edges at a wide variety of separations; unlike the Laplacian or other kernel-based filters which are often restricted to edges at a 'one pixel' frequency.

Various approaches to feature matching are possible (see Fig. 3.4 for one example), the key components in such a matching system being (Greenfeld 1991):

- selection of matching primitives (such as points or edges)
- method for extracting primitives (for example, the particular edge detection algorithm used)
- the selection of a suitable search/prediction strategy
- methods for assessment of results
- quality-control and consistency measures.

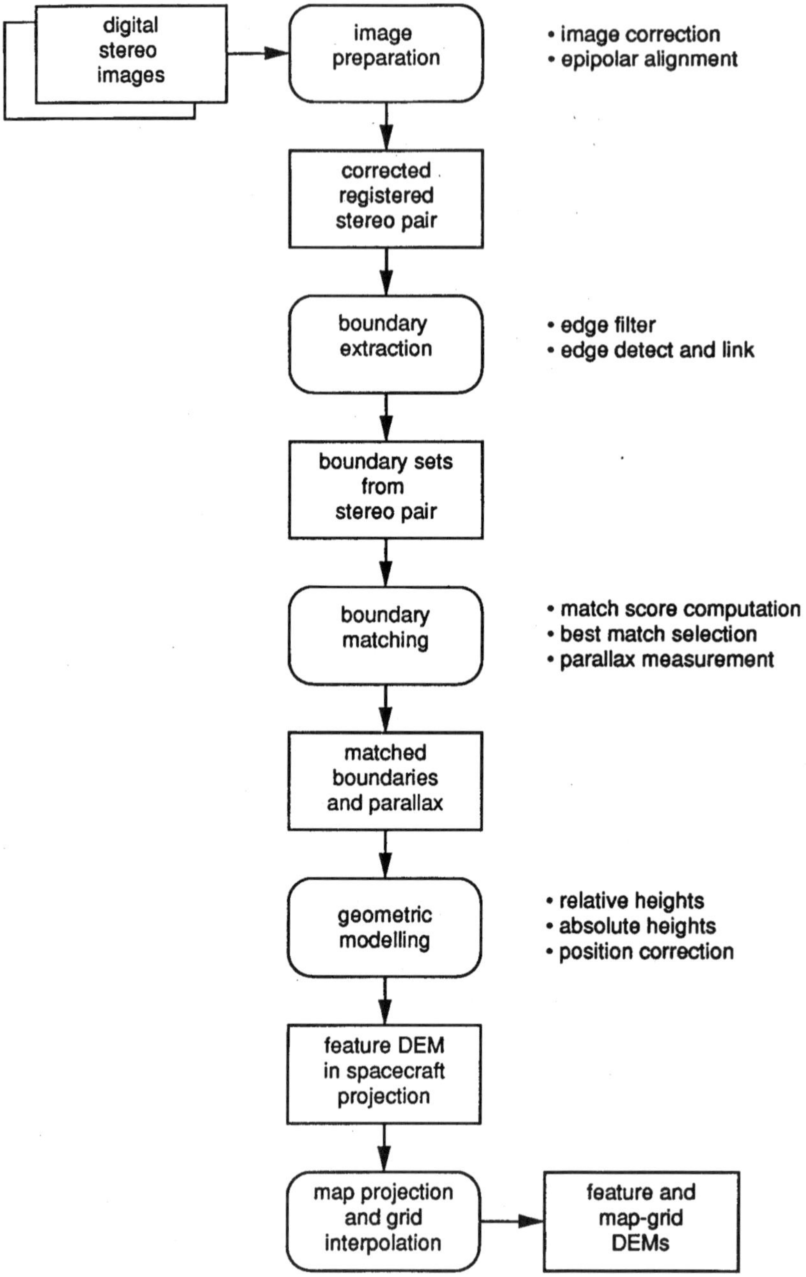

Fig. 3.4 A possible approach to feature-based DEM generation (adapted from Cooper *et al.* 1987).

A typical first step is an image preparation stage which takes the raw digital stereo images and performs the necessary image rectification to produce a corrected vertically registered stereo pair. This is then followed by a boundary extraction phase (Cooper *et al.* 1987) to extract features from the image pair which are invariant to radiometric effects, such as differences in illumination conditions. One approach is to use DoG filters, applied to the entire image, to detect a candidate set of edge locations to sub-pixel accuracy by the positions of zero crossings (it is generally much easier to determine the position where a function crosses an axis than to determine the maximal location of a characteristically flat correlation peak). The edges are subsequently validated to remove effects of sensor noise, scene noise and characteristics of the DoG filters.

Another approach (Brockelbank and Tam 1991) is to use a context-sensitive filter. This compares the result of an edge extraction filter with an adaptive threshold (defined to be the local intensity gradient). In areas of high edge-gradients the correspondingly high threshold means that only the strongest edges are found, whereas in low-gradient areas weak edges can be found (this avoids the problem of having no edges in some regions of the image which may occur with global thresholding). This should result in a more even distribution of edges over the image. Neighbouring edges are then linked to produce a set of boundary objects. This is performed by a traversal of the candidate boundaries which locates and links the position of zero-crossings, measures their shape, and discards those below significance thresholds for shape and value. The result of this is a series of image features represented in a symbolic vector format which makes the subsequent feature-matching process far more rapid than retaining all the original image information. Symbolic representation (Greenfeld 1991) may also be employed at this stage. This involves establishing a list of attributes to characterise a specific edge, such as orientation, strength and relative location. This can be a useful precursor to a boundary-matching phase ensuring that only similar edges get considered.

The objective of the boundary-matching phase is to determine, for each boundary in the left-hand image, the corresponding boundary in the right-hand image. The algorithm starts with the boundaries produced by the largest DoG filters which will provide the coarsest measure of boundary location. For each feature detected in the first image the best match feature in the second image is determined. Because both features are stored in a symbolic vector format rather than as raster patches, the matching process is far less computationally intensive. Once the matching boundaries have been detected the stereo disparity between the two boundaries is calculated, typically at separations of one pixel.

Once the initial coarse stereo disparity has been calculated, the same process can be repeated at each resolution level as the algorithm focuses into the disparity at the finest resolution. This hierarchical 'coarse-to-fine' matching system exploits the relationship between adjacent levels to provide a natural description of image features proceeding from the largest attributes down to the finest discernible image detail; this again is a reasonable approximation of how the human visual system operates on a number of hierarchical scales depending on the symbolic nature of the feature being processed. The actual matching criteria are based on a statistical similarity score between the two boundaries being compared; this is followed by the selection of the best such match.

When this has been achieved at the finest level of detail, the stereo parallax for each boundary-pair is calculated. Thus the result of the boundary-matching phase is the set of matched boundaries and the stereo parallax associated with them. Depending on the algorithms used these will be a measure of the stereo disparity at the edges of significant features or alternatively throughout a control network. As with correlation-based methods geometrical modelling can now be used to generate the final DEM – it may also be necessary to use an interpolation algorithm to convert from a feature-based to a grid-based product.

3.2.3 The hybrid approach

Either a correlation or feature-based approach may successfully produce a DEM, but possible disadvantages are inherent in each method. For example, the correlation method may fail to find sufficient successful matches to produce a DEM, or the feature-based approach may produce good results for identified features – but produce no elevation values at intermediate points. Other problems encountered with correlation-based methods (Kauffman and Wood 1987) include:

- problems with non-linear grey-level effects encountered with off-nadir viewing
- the effect of different lighting conditions
- use of algorithms with limited pull-in range, which rely on accurate seeding.

To date there has been a certain reluctance to use hybrid approaches with most practitioners firmly in the correlation or feature-based camps; for example, David Marr (1982) in his classic book *Vision* asserts that

> '. . . *items will be matchable if they correspond to things in space that have a well-defined physical location. Gray-level pixel values do not. Hence, gray-level correlation fails.*'

Despite this point-of-view and the corresponding opinions of the 'correlationists', it would seem beneficial to investigate approaches which make the most of the strengths of each class of algorithm. There are two main types of hybrid-processing schemes:

- *one-step methods* where (for instance) a preliminary DEM is produced by feature matching and the fine details are determined by correlation
- *adaptive methods* where an 'intelligent' controller determines the most appropriate method for a particular situation.

One example of the simple hybrid approach is to use a feature-based method (Barnard and Thompson 1980) to provide the seed points for a Gruen algorithm (Day and Muller 1988). The Barnard and Thompson feature matcher uses an 'interest operator' which can produce results to the order of a pixel accuracy. This is sufficient to be within the pull-in range of the Gruen algorithm which completes the job to sub-pixel accuracy. A similar approach was used for an experimental system to compare the three approaches to stereo matching (Brockelbank and Tam 1991); this system consisted of an area-based module, a feature-based module (using a second directional derivative edge detector and a process of matching stronger features first) and a hybrid module which used the feature-based method to seed the area-based module. Experiments using images of the Dinosaur National Monument (Colorado, USA) and Red Deer (Alberta, Canada) suggested that the hybrid module performed as well as the area-based one, with both outstripping the feature-based approach.

An example of the second method is the approach proposed in a stereo-matching system which uses a variety of techniques (Greenfeld 1991). The basic structure of the system can be seen in Fig. 3.5.

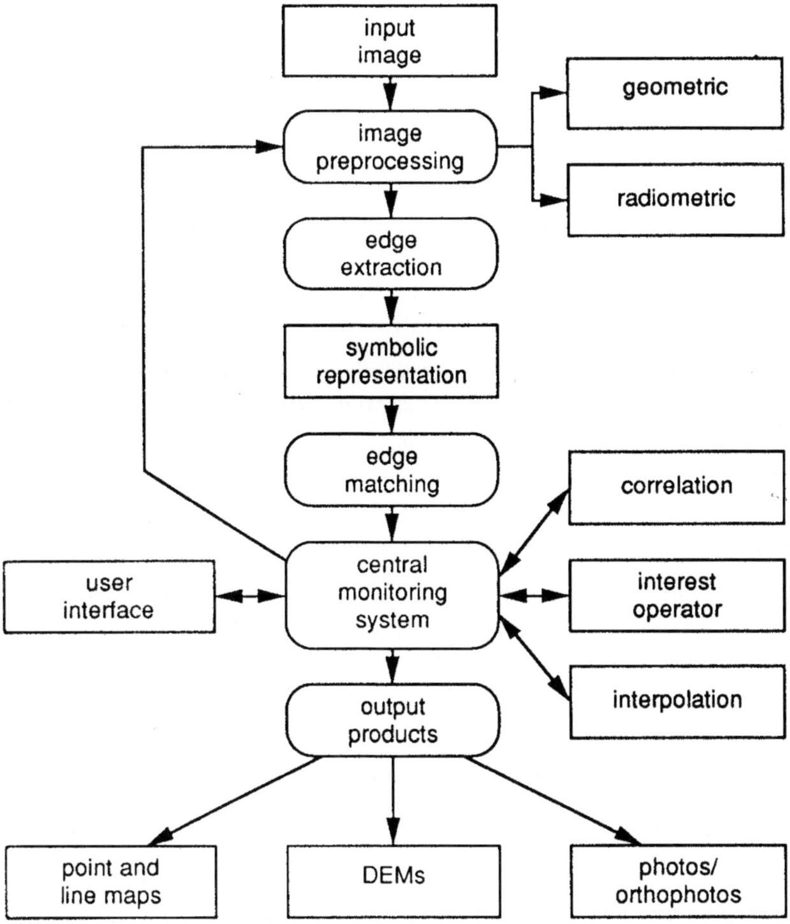

Fig. 3.5 Use of hybrid techniques in a photogrammetric system (from Greenfeld 1991).

The first phase of the matching is performed by an edge-matching process. Subsequently a central monitoring system can invoke one of three additional matching processes:

- a correlation-based algorithm
- an 'interest operator'
- an interpolator.

This approach '... is an enormous challenge in digital photogrammetry' (Greenfeld 1991) but promises to yield an effective, robust stereo-matching system which requires a minimal amount of operator intervention.

3.3 DERIVING HEIGHT INFORMATION FROM SAR

Although the state-of-the-art may not be as advanced as that for optical images, there is great potential in the production of DEMs from spaceborne radar. One of the benefits in using radar is that it measures distances directly, whereas optical sensors measure angles subtended by targets which have to be subsequently converted into distances. The two radar instruments that have potential for DEM production are Synthetic Aperture Radars and Radar Altimeters. Altimeters have been successfully flown on missions such as Seasat and ERS-1 and are capable of measuring height to within a few centimetres; however, due to their large footprint they are not suited for production of local-scale DEMs, and most research and development to date has concentrated on height-extraction from SAR. There are three categories of information inherent in a SAR image (Polidori 1992):

- geometric information in terms of range and azimuth
- information from the phase of the radar signal (this is possible because the signal is a complex wave having both real and 'imaginary' parts)
- radiometric information from the backscattered signal.

Each source of information gives rise to a different method of height determination:

- stereo matching (radargrammetry)
- SAR interferometry
- shape from shading (radar clinometry).

Operational flowlines for the derivation of height using each of the above methods are shown in Fig. 3.6.

3.3.1 Radargrammetry

In common with optical imagery, radar images acquired from two different view angles can be used to form a stereo pair. In order to retain the parallax effect the images are geometrically transformed (Simard *et al.* 1986) to an intermediate projection based on an ideal geoid – that is, no correction is made for terrain effects. An additional processing step which is necessary with SAR images is to remove the effect of speckle. This is a random statistical process, due to the coherent nature of

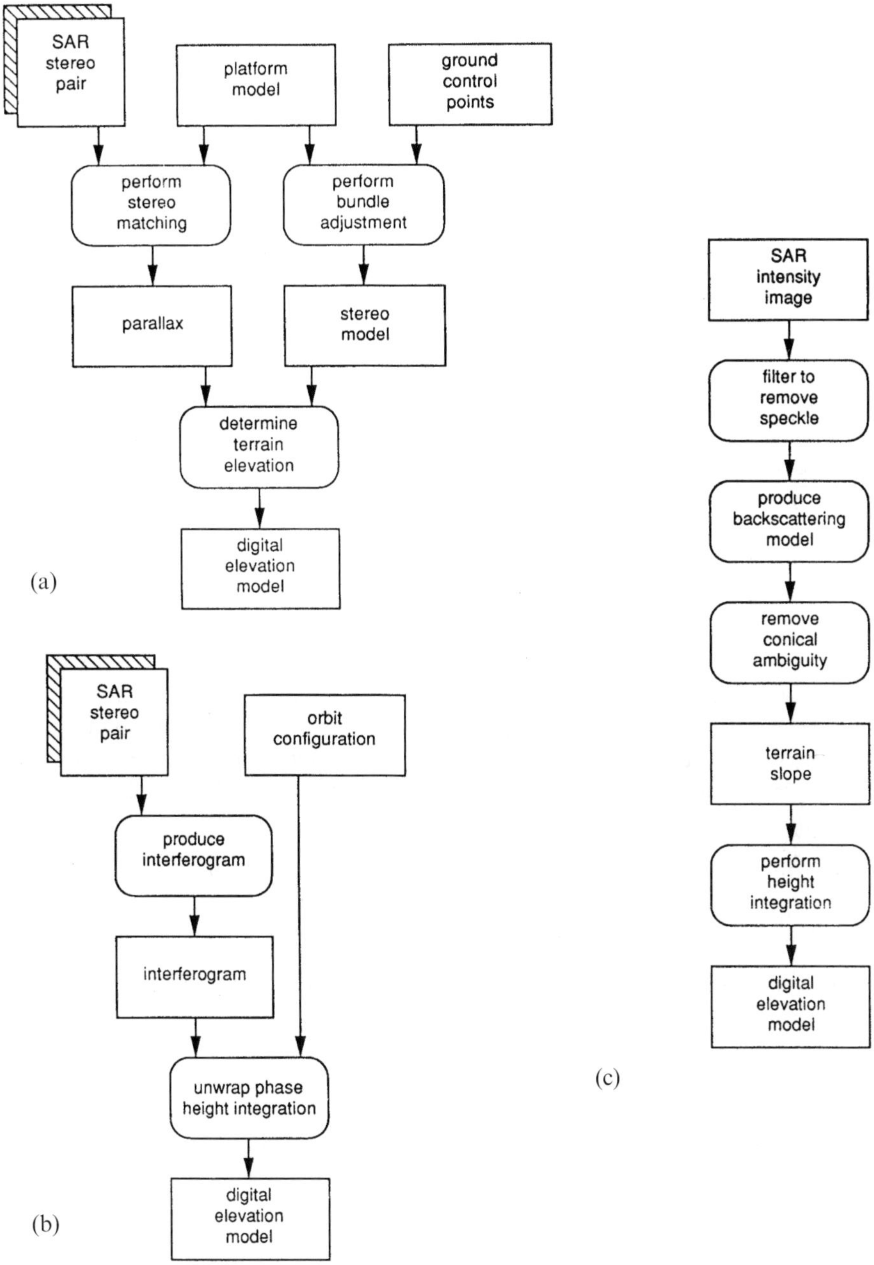

Fig. 3.6 Operational flowlines for DEM generation from SAR (based on Polidori 1992): (a) radargrammetry; (b) interferometry; (c) clinometry.

the SAR illumination, which causes noise to appear in the SAR image. Most research into stereo matching using SAR concentrates on using correlation techniques rather than feature extraction as the whole topic of edge/feature extraction from SAR images is not as well advanced as that for optical images. Stereo matching using radar imagery is more difficult than in the optical case (Simard *et al.* 1986) due to the inherent geometric and radiometric characteristics of radar data.

In general the production of DEMs from stereo SAR imagery using radargrammetry can be accomplished in four phases:

(1) geometric and radiometric pre-processing
(2) automatic stereo matching
(3) three-dimensional modelling
(4) production of the final DEM.

The first stage is to apply a low-pass filter to the input images to remove speckle; however, convolution with a low-pass linear filter may have an undesired side-effect by smoothing in the speckle, thus reducing high-frequency information (local height detail) in the resultant DEM. To avoid this, non-linear adaptive filters such as the Lee filter (Lee 1981) and the Frost filter (Frost *et al.* 1982) which remove speckle – while preserving high-frequency information – are preferred. There are various possible approaches to the correlation process. For example, a method investigated by the Canada Center for Remote Sensing (Simard *et al.* 1986) uses an iterative matching algorithm at different spatial resolutions. The matching of homologous points is performed on a series of stereo pairs with each successive pair being derived by reducing the resolution of the preceding dataset. This acts to constrain the search space algorithm by using the coarse-level locations as starting points for each successive level of the search. This also allows a parallel approach to be adopted for validation as anomalous points can be detected and filtered out at each level of resolution. Interpolation of missing values and low-pass filtering can also be performed at each level of the algorithm.

Stereo reduction is used to optimise the correlation process by using the digital parallax model (DPM) to transform one of the images onto the geometry of the other. This will reduce the overall parallax and, consequently, the search space needed for fine tuning. Parallax values can be converted to elevation values using a line of sight intersection with the instantaneous platform positions derived from the orbital model. The final stage of processing, as with optical imagery, is to correct the floating height values obtained from the parallax effects into a DEM by comparing them with the reference datum used for the geoid.

3.3.2 SAR interferometry

SAR interferometry represents yet another approach to the derivation of height information from spaceborne sensors. A radar antenna (Massonet 1997) illuminates its subject with 'coherent' radiation: the electromagnetic wave emitted following a regular sinusoidal pattern. Thus radar instruments measure both the amplitude and the exact point in wave cycle (the phase) of the returned signal. The basic principle of

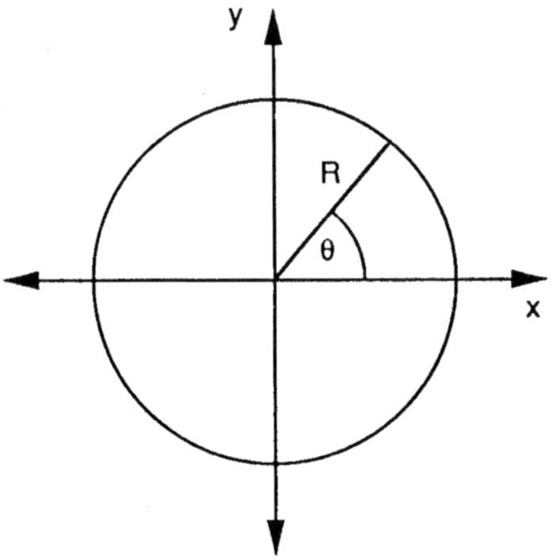

Fig. 3.7 Representation of a complex signal ($x + iy$) in polar coordinates.

SAR interferometry is to use two focused complex images gathered by the same SAR sensor in two passes along parallel or crossing orbits such as those produced by the ERS-1/2 Tandem Mission. The mathematical definition of a complex signal can be seen in Fig. 3.7.

Because the signal is complex it can be represented as a 'real' component and an 'imaginary' component (or quadrature). It is usually more convenient to convert these values to a magnitude-phase representation. The phase component of a SAR image (Rocca and Prati 1992) is the result of two mechanisms:

- basic phase, proportional to the length of the two-directional path of the radar signal trip
- a phase shift introduced by scattering at the target.

Thus if there is no significant change in terrain scattering effect (for example, by a change of surface cover) or elevation (for example, due to snow or vegetation changes) between two SAR surveys of the same target from two different locations, then there will be a direct relationship between the terrain height and the phase difference between the two signals. The main problem with a phase difference approach (Lin *et al.* 1992) is that the phase difference calculated is the principal value of the phase, that is, a value in the range $[-\pi, \pi]$. An infinite number of phase angles can have the same principal value, the relationship being given by the formula below:

$$\theta = \theta_p + 2n\pi \tag{3.1}$$

where θ_p = principal angle in range $[-\pi, \pi]$
θ = actual angle
n = any integer

Thus one of the fundamental tasks of SAR interferometry (also known as InSAR) is to be able to 'unwrap' this phase ambiguity onto its actual value and hence onto the height value. If the phase unwrapping is not performed successfully, local errors in calculating the phase difference will tend to propagate, causing global errors in the derived DEM. The basic conditions to allow successful phase unwrapping are:

- the magnitude must be non-zero everywhere (a zero magnitude means that the phase is undefined)
- the difference in phase between two sample points must be in the range $[-\pi, \pi]$ otherwise there will be an ambiguity in the unwrapping algorithm.

To achieve this it is necessary to account for both phase noise (speckle) and the effects of the phase aliasing. Phase noise is caused by the unpredictable combination of the reflections from the many hundreds of small reflectors – rocks, leaves, trees, grass, buildings – that contribute to a single pixel. The phase aliasing effect is equivalent to several height differences being mapped onto the same phase difference – for instance, in areas of rapidly varying terrain it becomes problematic to determine the true height difference. In particular, pixels with low backscatter are ignored. The low values correspond to a return with a low signal-to-noise ratio, with the noise destroying phase information. One possible approach to removing phase ambiguity is to use a low-resolution DEM to produce seed information about local topography and likely height differences.

Typical steps in producing a DEM from interferometric SAR (Small and Nüesch 1996) are shown in Fig. 3.8. The preparatory steps are to determine the orbit parameters of candidate images, and to acquire a pair of undistorted, coherent images free of phase discontinuities (to maintain coherence it is necessary to generate both images using the same Doppler centre frequency). Following this the images are registered using standard correlation techniques to find the best-fit location – in this case a coarse–fine registration scheme is used to 'pull-in' the images. For the fine registration two types of tie-points may be used: height tie-points (measured in areas of uniform unwrapped phase) and standard Easting and Northing tie-points.

The next stage is to generate the phase interferogram – the phase differences are calculated using the complex form of each SAR image and multiplying the two images using the complex conjugate (the complex conjugate of $a+bi$ being $a-bi$). Figure 3.9 (colour section) shows an interferogram of Mt Etna in Italy acquired by the Space Shuttle X-SAR sensor in 1994. Each of the coloured interferometric fringes represents an altitude change of about 45 metres.

It is then necessary to remove the phase ambiguities. This is the most important stage as it converts the phase differences (which are only known modulo 2π) to actual phase differences and hence to height. Using an accurate imaging model the height differences are converted to absolute heights and mapped onto a grid in the desired map projection. Either forward or backward geocoding (Small and Nüesch 1996) may be used. The forward method is used when there is no reference DEM; the backward method is used once an initial DEM is available, either to refine the DEM or for the generation of additional geocoded products. As well as its use for DEM generation SAR interferometry has proved very successful for detection of very small surface deformations. Applications of this to subsidence measurement and earthquake monitoring are described in Sections 5.3.1 and 7.3.3 respectively.

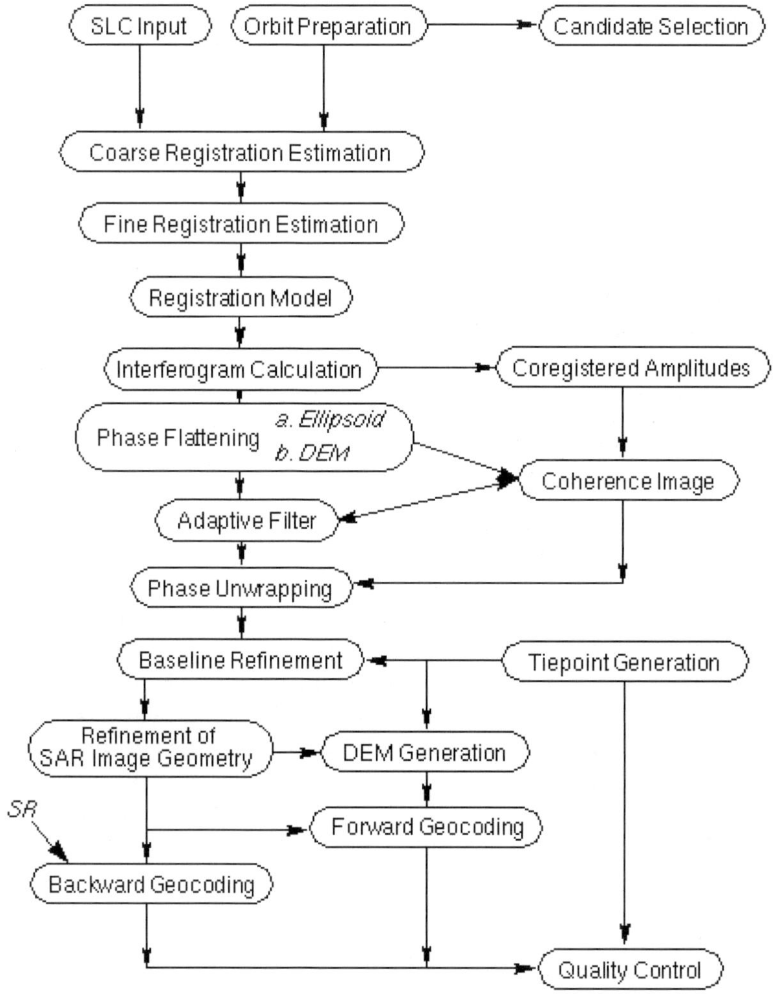

Fig. 3.8 InSAR processing steps (Department of Geography, University of Zurich).

3.3.3 Radar clinometry

The human visual system uses a variety of other 'algorithms' apart from stereo information processing, including:

- *motion* – how the object being viewed changes its angular location with time
- *occlusion* – whether the object is behind or in front of other objects in the visual field
- *shading* – variations in illumination response caused by orientation of the object with respect to the light source
- *context* – acquired knowledge about the scene being viewed.

In a similar way secondary cues could be used to derive three-dimensional information from a satellite image. However, a typical image tends not to be as rich in cues as a normal scene viewed by the brain (for example, short timescales will generally preclude analysis of motion), and the algorithms needed to determine the information are not fully understood. Nevertheless it is possible to obtain some information about an image from the supplementary information present. For example, in optical images an indication of terrain relief comes from the shadow cast from solar illumination. Depending on the Sun-elevation angle, areas of pronounced relief will have variations in spectral response depending on whether a point is on the bright side or the shadow side of a raised feature. Other information will come from the texture of a region with low-lying areas being characteristically homogeneous, and more mountainous areas showing a more ragged 'crinkly' appearance.

With radar images it may be possible to obtain information about the topography of a region from the effect it has on an image's appearance; however, in the case of SAR images there will be no solar illumination or shadowing effect. Radar clinometry, also known as 'shape from shading' (Guindon 1989), is an attempt to derive three-dimensional information from a single radar image by relating the local terrain slope to the radar backscatter cross-section (Polidori 1992). This is based on a mathematical model of the backscattering process together with certain constraints on the geometry of the surface being imaged. The result of radar clinometry (Thomas *et al.* 1991) is a measurement of slope (local differences in height) which must then be referred to some local datum in order to produce the absolute height values required for a DEM. Research so far has not succeeded in obtaining particular accuracy from this method because of potential sources of error, including:

- calibration errors
- effect of speckle
- inaccurate model for backscatter
- propagation of errors/problems converting to height values.

However if DEMs can be obtained in this way they do not require the acquisition of a stereo pair and, because they are likely to be accurate in their high-frequencies, there may be a role for radar clinometry, either in combination with other methods or for validation purposes (such as checking consistency with DEMs obtained otherwise).

3.3.4 Complementary approaches

In many circumstances the best results may be obtained by combining different SAR DEMs with high spatial resolution and good height accuracy. However, in areas of complex topography problems arise in the phase unwrapping and DEM quality becomes compromised. Another problem occurs if there are changes in atmospheric humidity between the two acquisitions, leading to changes in refractive index. This in turn leads to changes in phase, and artefacts which are misinterpreted as terrain relief. DEMs coming from radargrammetry tend to be intrinsically less precise than InSAR DEMs, their quality is best in regions of low relief but are less susceptible to the atmospheric conditions during the image acquisitions. Thus the fusion between these kinds of data can improve both the precision and the completeness of the generated DEMs. For example, according to Crosetto and Aragues (1999):

- radargrammetric height data can be used to compensate for the atmospheric distortions in the InSAR DEMs
- radargrammetric height data can support the phase unwrapping
- InSAR and radargrammetric height data can be fused in order to perform the joint estimation of the terrain surface.

Another complementary approach (Crippa and Crosetto 1998) is to use the combined strengths of optical and SAR. For example, in rugged areas stereo SPOT can be used to 'fill in the holes' caused by radar foreshortening, and a similar approach can be used where there is low coherence in the interferogram. Conversely, radar can be used to fill in gaps in moderate terrain where optical imagery is cloud-covered. Finally, even a crude DEM obtained by optical means can be used as the seed for phase unwrapping or a guide to areas where layover and shadowing are likely to effect the utility of radar.

3.4 DEM VALIDATION

For such a complicated undertaking as the generation of a fine-gridded DEM a substantial degree of quality control is necessary; this can be performed either by visual inspection or by algorithmic means. To visually inspect a DEM it can be displayed in either raster or vector format using a GIS. As a raster image the heights are represented as grey levels, thus the simplest and quickest inspection can be carried out by density slicing. This can be used to reveal unexpected discontinuities in the DEM, discrepancies with known map data, or artefacts caused by attributes of the matching or interpolation algorithms. In vector format the DEM can be presented in many different wire frame or 'blocked in' diagrams. Many GISs will have tools to enable these to be presented in a variety of different perspectives, and even allow the viewpoint to be changed in real time, for instance, to revolve the DEM or to view it from a number of different elevations.

For a more rigorous approach it is necessary to define what exactly is meant by the 'quality' of a DEM, for example (Day and Muller 1988) the following three attributes of quality can be considered:

- *accuracy* with respect to an independently derived DEM principally through calculation of the difference between the forward geocoded height model and the reference model; RMS difference values together with histograms of the distribution of the height differences provide a range of quality metrics
- *reliability* – defined as the proportion of points with elevation errors greater in magnitude than three standard deviations
- *sampling density* – defined by the proportion of points actually matched.

Algorithms for automatic quality control rely on applying a number of geophysical constraints to the DEM. For example, if existing elevation data (e.g. from map contours) is available for the area, constraints such as the minimum and maximum heights for the area can be applied. Other constraints include maximum slope, i.e. the maximum rate that elevations can change within a prescribed horizontal distance. If the locations of water features can be determined either from digital map data or by thematic classification of the image themselves (see Section 1.5), then the constraint 'all elevations over a water feature must be equal' can be applied. Thematic information is also important for quality control of extended surface features such as buildings and trees, which will introduce additional uncertainty into the elevation determined; for example, the satellite will image the top of a canopy of trees rather than the underlying surface topography.

In the case of InSAR a number of different approaches (Small and Nüesch 1996) can be used for validation of DEMs including:

- DEM flattening using a synthetic interferogram. This has the advantage of avoiding the need for phase unwrapping but the disadvantage of not validating height directly.
- Backward geocoding of slant-range height model. This works directly on the height values but since it uses height values which were previously part of the geocoding process it is not a 'blind' test. It is, however, a useful process for validating coherence, the local incidence angle and interdependencies in the height error model.
- Forward geocoding of slant-range height model. This enables a true end-to-end validation of the InSAR height model generation process.

Local surveys of height generated from a GPS may also be used to provide ground-control for DEMs (Cross *et al.* 2000). For example, in order to validate the accuracy of an InSAR DEM of the British Isles, 15 kinematic GPS profiles, each of about 200 km long, were measured. Data for each profile was collected using two GPS receivers, one placed in a road vehicle and the other remaining static at a location close to the profile being measured. By using high-precision kinematic techniques (see Section 8.2.1) an accuracy of about 1 metre horizontal and about 3 metres in height in the calibration data is achievable.

3.5 SHUTTLE RADAR TOPOGRAPHY MISSION (SRTM)

Successfully launched onboard the NASA Space Shuttle Endeavour on 11 February 2000, the Shuttle Radar Topography Mission (SRTM) is a joint project between the United States National Imagery and Mapping Agency (NIMA) and the National Aeronautics and Space Administration (NASA). During its ten-day mission an unprecedented collection of topographic data of most of the world's land surfaces, between latitudes 60°N and 60°S was obtained. The physical configuration of SRTM (see Fig. 3.10) consists of radar electronics and antennas located in the Shuttle payload bay with receiving antennas located at the end of a 60-metre mask deployed into space. This makes it possible to conduct single-pass interferometry which

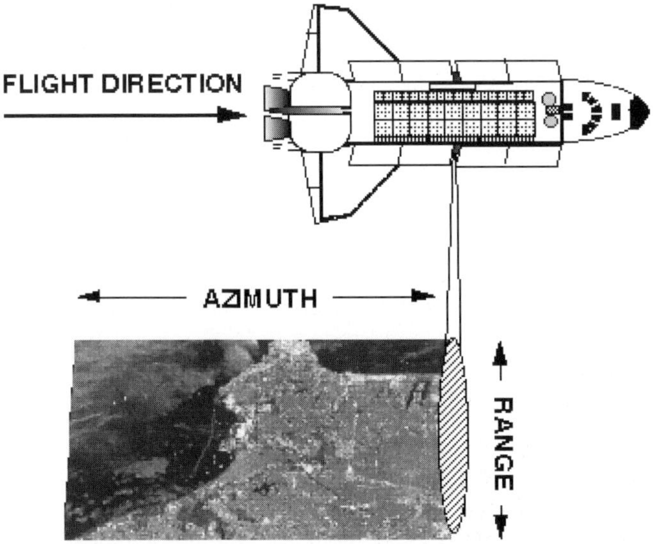

Fig. 3.10 SRTM Shuttle configuration (courtesy NASA/JPL).

removes the problems of image coherence associated with images with a large temporal separation.

The radar used provided a combination of C-band and X-band data. C-band data is distributed through the United States Geological Survey's EROS Data Centre, while the X-band data is distributed by the German Aerospace Centre, DLR. The processing system (JPL 2001) consists of three components:

- the interferometric processor, which creates the basic DEM product from the raw radar data
- the mosaic processor which combines the image data and DEMs a continent at a time
- the verification system which performs quality control.

Figure 3.11 shows the coverage of the Earth achieved by the mission. Areas in red (see rear cover) could not be mapped. The principal coverage is over land but small amounts of data were collected over the water for calibration purposes

Ultimately, the final released SRTM DEM will be at 30 metres posting for the US and at 90 metres posting for other countries. The absolute horizontal accuracy (90 per cent circular error) is expected to be 20 metres and vertical accuracy (90 per cent linear error) is 16 metres, referred to the WGS84 ellipsoid Earth model.

3.6 APPLICATIONS OF DEMS

The format of DEMs make them very amenable to manipulation to derive further information from them. An obvious example is the ability to interrogate a DEM to

Fig. 3.11 SRTM Coverage Globes (courtesy NASA/JPL).

find the height value at any grid point and, by interpolation, at intermediate points. More complex algorithms can be used to determine the two important secondary indicators of topography (slope and aspect). Another use of DEMs is for intervisibility analysis, that is, the determination of which areas can be seen from a given point, and applications of this model include the siting of communications masts and military planning. Related to this, is the calculation of shadow zones which display areas of a landscape that would be in shadow given a specified Sun azimuth and elevation. Other DEM manipulations include display of cross-sections across a terrain, the appearance of the horizon from a particular vantage point, and determination of slope values along a road using non-linear profiles. DEMs can also be used to correct for the effect of lengths and areas calculated on a digital map which assume that the surface is flat and viewed from above (Williams 1984). In areas of undulating or sloping terrain the true area will be greater than the estimated area. Slope information from a DEM can be used to correct for this using a simple trigonometric adjustment.

The local topography is an important component of many ecological models and DEMs can be used for modelling catchment areas and watersheds and providing simulation of the flow of water over the earth's surface, which can then be incorporated in hydrological models representing run-off calculations or flood prediction. A DEM will often be an important component of an environmental model within a GIS, for example in soil science, pollution studies or forestry (see Section 4.3). DEMs can also be used to calculate the volume of material needed to fill-in a terrain feature and similar approaches can be used for excavation purposes, either to estimate the volume extracted or to provide a revised model of what the surface would look like, for example, following a mining operation. Other civil engineering applications include road building/modification and the construction of reservoirs and dams. Finally, visualisation techniques, such as those described in Section 4.4.2, can also use DEMs within detailed flight simulators, etc. (Muller *et al.* 1988). Other uses for DEMs include:

- topographic information about soil types, studies of vegetation and habitat monitoring
- physical processes such as glaciation, river flooding, subsidence
- personal infomobility and outdoor recreation applications;
- military mission planning, cross-country route evaluation, flight simulation and training
- civil engineering, town planning and environmental impact assessment
- mobile phone and telecommunication industries.

Table 3.2 gives a cross-reference to the other chapters of the book where the above examples and many others are described in fuller detail.

Table 3.2 Cross-reference to DEM image applications

Application	Section reference
Orthoimage generation	1.2.1, 8.1.4
Orthoradar generation	1.3.4
Geocoding validation	1.3.5
Radiometric correction	2.1
Combining SAR and optical imagery	2.5
Synthetic Image Generation	2.5
Image classification	4.2
Modelling with GIS	4.3
Knowledge-based GIS	4.3.5
Three-dimensional visualisation	4.4.2
Virtual GIS	4.4.4
Forestry applications	5.2
Telecoms applications	5.4

REFERENCES

Barnard ST and Thompson WB (1980) Disparity analysis of images. *IEEE Transactions on Pattern Analysis and Machine Intelligence*, **PAMI-2**, **4**, 333–40.

Belward AS and Valenzuela CR (eds) (1991) *Remote Sensing and Geographical Information Systems for Resource Management in Developing Countries*. Kluwer Academic Publishers, Dordrecht, The Netherlands.

Brockelbank DC and Tam AP (1991) Stereo elevation determination techniques for SPOT imagery. *Photogrammetric Engineering and Remote Sensing*, **57** (8), 1065–73.

Cooper PR, Friedman DE and Wood SA (1987) The automatic generation of Digital Terrain Models from satellite images by stereo. *Acta Astronautica*, **15** (3), 171–80.

Crippa B and Crosetto M (1998) Optical and radar data fusion for DEM generation, *Proceedings of the ISPRS – Commission IV Symposium, Stuttgart (Germany)*; *Int. Arch*. Vol. XXXII, Part 4, pp. 128–34.

Crosetto M and Aragues FP (1999) Radargrammetry and SAR interferometry for DEM generation: validation and data fusion. *CEOS '99*.

Cross PA, Barnes J, Walker AH, Muller J-P and Morley JG (2000), GPS Validation of IfSAR Digital Elevation Models from LANDMAP. *Remote Sensing Society Conference 2000, University of Leicester*.

Day T and Muller J-P (1988) Quality assessment of Digital Elevation Models produced by automatic stereomatchers from SPOT image pairs. *Photogrammetric Record*, **72** (12), 797–808.

Durand D (2000) *DEM stereoscopic aspects of SPOT*. Committee on Earth

Observation Satellites (CEOS), *Resources in Earth Observation* (4th Edition), Centre National d'Etudes Spatiales (CNES), France.

Frisby JP (1979) Seeing – *Illusion, Brain and Mind.* Oxford University Press.

Frost VS, Styles JA, Shanmugam KS and Holzman JC (1982) A model for radar images and its application to adaptive digital filtering of multiplicative noise. *IEEE Transactions on Pattern Analysis and Machine Intelligence*, **PAMI-4**, pp. 157–66.

Greenfeld JS (1991) An operator-based matching system. *Photogrammetric Engineering and Remote Sensing*, **57** (8), 1049–55.

Gruen AW (1982) Adaptive least squares correlation: a powerful image matching technique, *South African Journal of Photogrammetry, Remote Sensing and Cartography*, **14** (3), 175–87.

Gruen AW (1986) High precision image matching for Digital Terrain Model generation. *International Archives of Photogrammetry and Remote Sensing*, **26** (3/1), 284–96.

Guindon B (1989) Development of a shape-from-shading technique for the extraction of topographic models from individual spaceborne SAR images. *Proceedings of IGARSS '89 Symposium, Vancouver, Canada*, 10–14 July 1989, pp. 597–602.

Hill J (1991) Remote sensing systems: sensors and platforms. In Belward and Valenzuela (eds), pp. 111–43.

JPL (2001) Shuttle Radar Topography Mission, web-site.
http://www.jpl.nasa.gov/srtm/index.html

Kauffman DS and Wood SA (1987) Digital Elevation Model extraction from stereo satellite images. *Proceedings of IGARSS '87 Symposium, Ann Arbor*, 18–21 May 1987, pp. 349–52.

Konecny G, Lohmann P, Engel H and Kruck E (1987) Evaluation of SPOT imagery on analytic photogrammetric instruments. *Photogrammetric Engineering and Remote Sensing*, **53** (9), 1223–30.

Lee JS (1981) Refined filtering of image noise using local statistics. *Computer Graphics and Image Processing*, **15**, 380–9.

Lin Q, Vesecky JF and Zebker HA (1992) New approaches in interferometric SAR data processing. *IEEE Transactions on Geoscience and Remote Sensing*, **30** (3), 560–7.

Marr D (1982) *Vision.* WH Freeman & Company, San Francisco, California.

Massonet D (1997) Satellite Radar Interferometry. *Scientific American*, February 1997.

Mori N, Takaoka H, Tonoike K, Komai J and Murai S (1988) Investigation of the effectiveness and applications of Japanese ERS-1 stereoscopic images. *International Archives of Photogrammetry and Remote Sensing*, **27** (Part B1/1), 109–19.

Muller J-P, Day T, Kolbusz J, Dalton M, Richards S and Pearson JC (1988) Visualization of topographic data using video animation. *International Archives of Photogrammetry and Remote Sensing*, **27** (B4), 280–8.

Muller J-P, Morley JG, Walker AH, Mitchell K, Jung-Rack K, Cross PA, Barnes J, Dowman IJ, Kitmitto K, Smith A, Chugani K and Quarmby N (2000) The LANDMAP project for the automated creation and validation of multi-resolution orthorectified satellite image products and a 1″ DEM of the British Isles from ERS tandem SAR interferometry. *Remote Sensing Society Conference 2000, University of Leicester.*

Norvelle FR (1992) Window shaping and DEM corrections. *Photogrammetric Engineering and Remote Sensing*, **58** (1), 111–15.

Otto GP and Chau TKW (1989) 'Region-growing' algorithm for matching of terrain images. *Image and Vision Computing*, **7** (2), 84–94.

Polidori L (1992) Potentialities and limitations of topographic mapping from Synthetic Aperture Radar. *Proceedings of Satellite Symposia 1 and 2: Navigation and Mobile Communications, and Image Processing, GIS and Space-assisted Mapping, from the 'International Space Year' Conference, Munich, Germany*, 30 March–4 April 1992, ESA ISY-2, pp. 225–8.

Rocca F and Prati C (1992) Innovative applications of repeated satellite SAR surveys. *Proceedings of Satellite Symposia 1 and 2 : Navigation and Mobile Communications, and Image Processing, GIS and Space-assisted Mapping, from the 'International Space Year' Conference, Munich, Germany*, 30 March–4 April 1992, ESA ISY-2, 193–7.

Rodriguez V, Gigord P, de Gaujac AC and Munier P (1988) Evaluation of the stereoscopic accuracy of the SPOT satellite. *Photogrammetric Engineering and Remote Sensing*, **54** (2), 217–21.

Simard R, Plourde F and Toutin T (1986) Digital elevation modeling with stereo SIR-B image data. *Symposium on Remote Sensing for Resources Development and Environmental Management, Enschede, Netherlands*, August 1986, pp. 161–6.

Small D and Nüesch D (1996) Validation of height models from ERS interferometry. *Proceedings of ESA-FRINGE '96 Workshop, Zürich, Switzerland*, 30 September–2 October, 1996.

Thomas J, Kober W and Leberl F (1991) Multiple image SAR shape-from-shading. *Photogrammetric Engineering and Remote Sensing*, **57** (1), 51–9.

Trinder JC and Donnelly BE (1988) SPOT software for Wild Aviolyt BC2 analytical plotter. *International Archives of Photogrammetry and Remote Sensing*, **27** (B4), 412–21.

Williams JM (1984) *REGIS – an RAE Experimental Image-based Geographic Information System*. Royal Aircraft Establishment (Farnborough, UK), Technical Report 84103.

4

Introduction to GIS for Earth Observation

One of the most significant developments in the handling of geographic information (comparable perhaps to the invention of the map itself) has been the advent of Geographic Information Systems (GISs). GIS technology is evolving rapidly and there is by no means common agreement as to what comprises a GIS, one widely accepted definition being '. . .*a powerful set of tools for collecting, storing, retrieving at will, transforming and displaying spatial data from the real world for a particular set of purposes*' (Burrough 1986).

Until relatively recently, the approach adopted with much remote-sensing research was to analyse and interpret images in isolation, that is, not making use of any external information. With most images, however, the use of external geographic information (such as digital map data) can assist greatly in the interpretation process, the results of which can, in turn be used to update some underlying 'world model' – for example, an algorithm designed to extract road networks from SPOT images could use a digital map of existing roads to provide a more focused search space. This is an example of the integrated use of GIS which is an essential development in realising the potential of satellite images as a geographic information source, and has progressed tremendously over the last few years.

In this chapter we provide an overview of GIS technology and its relationship to Earth Observation:

- defining a GIS – components of a typical GIS, geospatial data structures, including the incorportation of raster objects; the use of GIS data sources for geocoding and simple information extraction operations
- how GIS provides effective tools for the accurate classification of images
- advanced modelling techniques including knowledge-based information extraction
- presentation of GIS data including animation and three-dimensional visualisation techniques.

In the next two chapters we present some real-world applications of geographic information from space. Chapter 5 looks at land applications and Chapter 6 at oceanographic, atmospheric and cryospheric applications. Following on from this,

Chapter 7 looks at the important role this information can play in global environmental monitoring and related areas whereas Chapter 8 considers the growing field of geomatics where GIS attains a central role.

4.1 WHAT IS A GIS?

4.1.1 Components of a GIS

Geographic Information Systems have had nearly 40 years of development. For example (Hartnell 1994), the Canada Geographic Information System (CGIS), designed for overlaying polygons for land use assessment, was first proposed in 1963. In the same decade the Harvard Laboratory for Computer Graphics SYMAP system performed similar overlay operations using raster data. The 1960s also saw the start of vector-based digital mapping and the concept of linking the graphic data to a relational database for the storage of attribute data. This basic model has been continually enhanced in terms of functionality, user interaction and performance to the GIS systems available today. A further parallel development has seen the convergence of Computer Aided Design and GIS technology, particular in areas such as cadastral mapping and property terriers.

Nowadays there is a wide range of spatially enabelled functions that can be performed by a GIS including input, manipulation, display and output of point, vector and raster data; DEM handling; topological and network modelling of vector data; analysis of related data in an associated database; and high-quality cartographic output. A 'context model' of the components of a typical GIS (Zeiler 1999) can be seen in Fig. 4.1. in which the key components are as follows:

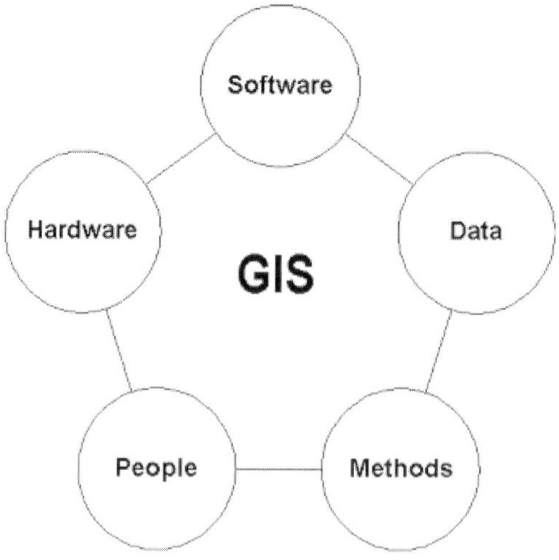

Fig. 4.1 Components of a typical GIS (copyright ESRI).

- *Hardware* – for the majority of applications mainframe systems are a thing of the past. Typical hardware configurations now encompass enterprise scale client–server installations, PC-browsers with web-based access to powerful 'map servers' and an ever-increasing range of mobile GIS hardware from palmtops and in-car computers to GIS-enabled mobile phones.
- *Software* – this provides a range to tools to manipulate the spatially related data. Most GIS packages also provide software development tools to enhance the functions available. Proprietary GIS languages are rapidly becoming supplanted by industry-standard ones such as Visual Basic for Applications (VBA) and Java.
- *Data* – this is probably the most important component of a GIS. Throughout this book we have described how satellite data can be obtained and made 'GIS ready'. Other sources of data are commercial data brokers, government departments and in-house digitisation or scanning of paper maps. An increasingly valuable source of data is to 'spatially enable' existing corporate information held in databases, for example, to 'geocode' postcode information into geographic Eastings and Northings so that customer addresses can be plotted on a digital map background.
- *GIS people* include business analysts, system designers and implementers, data modellers, digitisers and database analysts. GIS users range from scientists, business users and decision makers to the 'people in the street' who may not even realise that they are using GIS technology.
- Finally, the successful GIS will have a series of well-defined and considered methods or *business rules* which govern every aspect of its operation and ensure that the geographic information is accurate, current, consistent, and reliable.

Of course, many of the above aspects are shared by generic information and data-processing systems. What makes a GIS special can be shown by three complementary viewpoints (Maguire *et al.* 1991):

- A GIS provides geographic extensions to database management systems.
- A GIS provides a wide range of tools for spatial analysis.
- A GIS provides powerful methods of automated digital cartography.

The first aspect is described below; spatial analysis and digital cartography are described in further detail in Sections 4.3.1 and 4.4.1 respectively.

4.1.2 Geographic databases

Usually the 'geodatabase' (Zeiler 1999) itself is implemented using a commercial Database Management System (DBMS). Such a database typically provides functions such as schema definition, transaction support, data security, tabular queries, report generation, data back-up and system administration. The GIS extends the standard functionality of the relational database to provide a wide variety of spatial analysis, mapping and visualisation tools. These include storage of the geometric attributes of an object (as points, lines, polygons, surfaces, etc.), definition of topology (the spatial relationship between objects such as

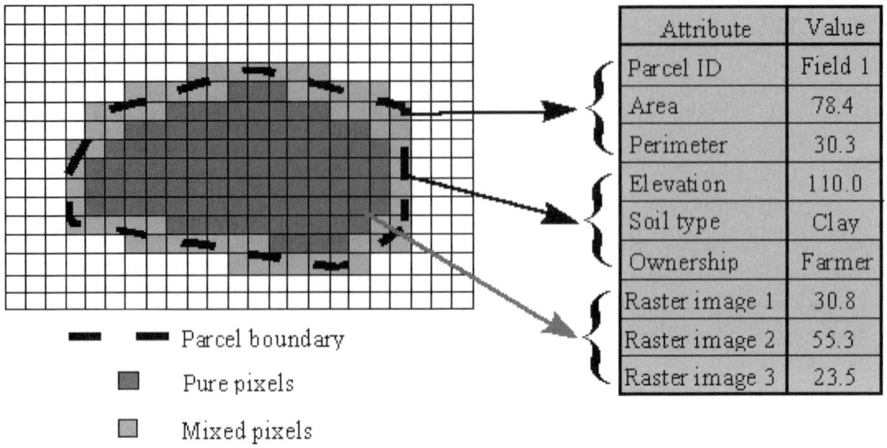

Fig. 4.2 Association of a map object with attribute data (from Hartnell 1997).

connections between pipes in a water network), rendering of map displays, spatial indexing to allow efficient searching of objects and spatial analysis tools (see Section 4.3.1). It should be noted that commercial DBMSs are becoming increasing spatially enabled and the distinction between GIS and database systems will become less-and-less important. Attribute (or tabular) data can be associated with each map object as a database record containing fields with defined values (see Fig. 4.2).

The data model (showing how the various classes of graphical or tabular objects are related) for a geographic database management system is often described in terms of an Entity-Relation Diagram (ERD). Figure 4.3 shows the ERD for the GLIMS glacier database (Kargel *et al.* 2000) which is referred to again as a cryospheric application in Chapter 6.

Each box on the diagram shows a different table with the lines between them showing how the tables are linked (for example, 'Displacement consists of vectors'). 'Crows feet' on the diagram indicate a many-to-one relationship, i.e. an image may contain one or more bands. A triangle indicates a dependent relationship – for example, for every displacement, vectors to describe that displacement must also exist. Finally, a bar indicates a mandatory relation – for example, a record in the image table must be linked to one and only one record in the instrument table.

4.1.3 Data structures

There are many possible approaches to the data model that underpin the geographic database, with a great deal of current research aimed at providing a suitable object-oriented database – for example, to permit the use of objects as attributes of other objects, such as associating scanned photographs of motorway bridges as attributes of a motorway object stored in vector format. It is still commonplace, however, for a geographic database to effectively consist of three distinct categories of data structure:

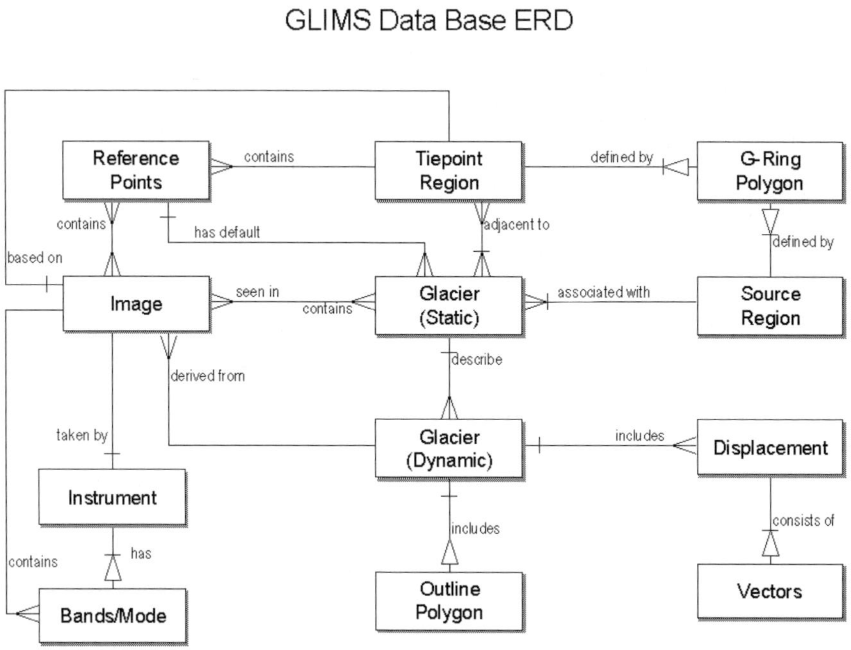

Fig. 4.3 Entity-Relation Diagram for a geographic database (from Kargel *et al.* 2000).

- *raster* – for example, images, scanned maps and DEMs
- *vector* – for example, digital maps, contours and point data
- *tabular* – for example attribute and statistical data.

Raster data

Raster data is the 'native format' for satellite images and data such as scanned map sheets, and consists of a regular array of picture elements or pixels, each of which is referenced by a two-dimensional (column, row) index. The raster format is convenient for inputting large amounts of bulk data but has two potential disadvantages:

1. It can be wasteful of space when storing scanned maps, because digital maps are characteristically more sparse than satellite images and, when transformed to raster format, will tend to have a lot of blank pixels.
2. The raster format does not contain topological information about the relative location of geographic objects.

An alternative raster format is the quadtree, which is constructed by dividing an image into quarters, and the quarters into quarters, and so on hierarchically until the entire image is divided. The end of any 'branch' on the quadtree is defined when any of the square blocks produced by the splitting process at a particular level contains

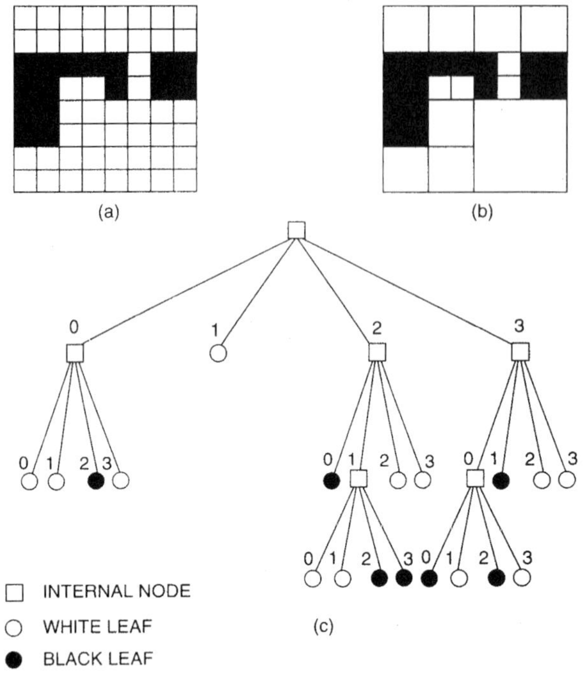

Fig. 4.4 Representation of binary raster data by a quadtree: (a) the original dataset; (b) the leaf quad-square diagram; (c) the resultant quadtree. (From Ibbs and Stevens 1988.)

pixels with all the same value. Thus the quadtree represents the values of the image by a hierarchical tree of homogeneous squares of various sizes (see Fig. 4.4).

One of the advantages (Hogg and Gahegan 1986) of using quadtrees is that they can be used in a top–down manner to provide a representation of the image at various scales, with all finer detail being hidden at lower levels of resolution. The quadtree representation is extremely efficient for performing arithmetic calculations and deriving image statistics, and is relatively easy to implement in software.

Vector data

The representation of spatial information as vectors is commonplace in most GISs, particularly for digital map data. Vector representation makes use of three simple topological objects:

- *points* – zero-dimensional objects
- *lines* – one-dimensional objects
- *areas* – two-dimensional objects.

The location of a single point can be expressed by its (X, Y) position in some coordinate system. A straight-line segment can similarly be expressed by the location of its two end-points (X_0, Y_0) and (X_1, Y_1), or expressed as vectors $\mathbf{X}(0)$ and $\mathbf{Y}(0)$ (see Fig. 4.5)

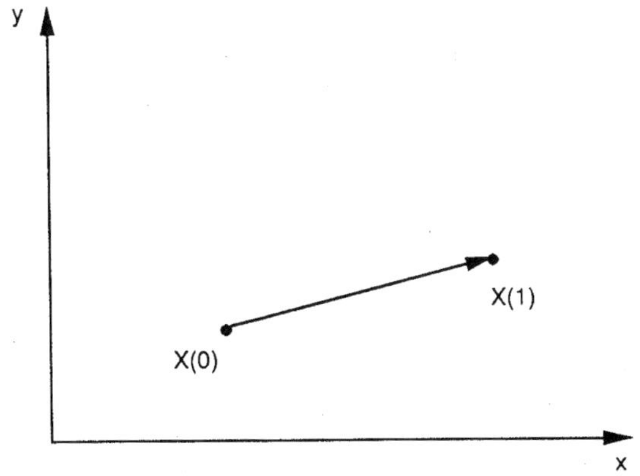

Fig. 4.5 Representation of a straight-line segment in vector format.

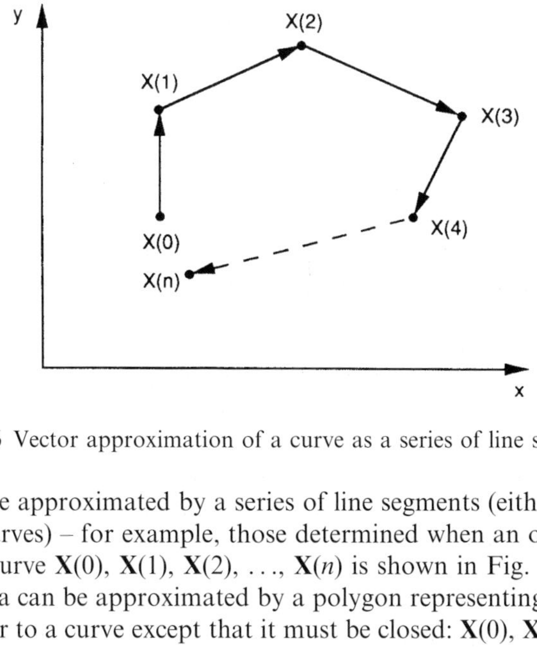

Fig. 4.6 Vector approximation of a curve as a series of line segments.

A curve can be approximated by a series of line segments (either straight lines or parameterised curves) – for example, those determined when an operator digitises a map. A typical curve **X**(0), **X**(1), **X**(2), . . ., **X**(n) is shown in Fig. 4.6.

Finally an area can be approximated by a polygon representing its boundary; the polygon is similar to a curve except that it must be closed: **X**(0), **X**(1), **X**(2), . . ., **X**(n) → **X**(0). Another variant of the vector format is the Triangulated Irregular Network (TIN) which is mainly used for the representation of DEMs (Peucker *et al.* 1978). In a TIN a network of triangles is constructed with its vertices at sample points. An example of this is shown in Fig. 4.7.

As well as the ability to store objects as points, lines or areas it is necessary to correctly represent the topology or spatial relationships between objects. For example, consider two adjacent fields which have been digitised from a map (Fig. 4.8) as two separate objects. The common boundary has been digitised twice, and inevitably there will be some discrepancy in the values recorded which can give rise to spurious sliver polygons between the two positions of the boundary. This can be

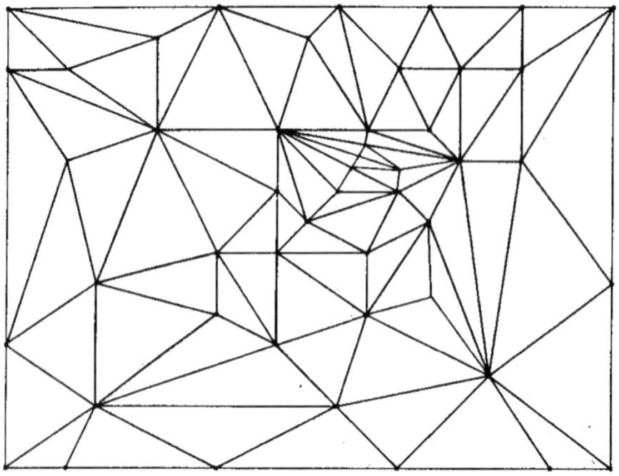

Fig. 4.7 Representation of a DEM as a Triangulated Irregular Network (TIN) (from Burrough 1986).

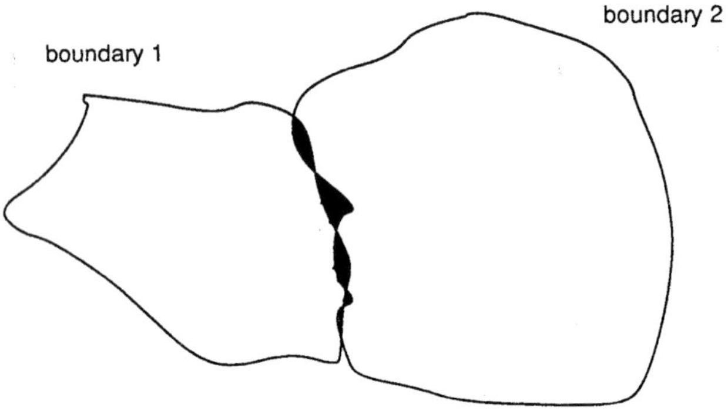

Fig. 4.8 Spurious sliver polygons may be produced by 'double digitising'.

avoided by digitising the boundary only once and indicating that it actually belongs to two separate objects. Other topological aspects that need to be handled include dealing with areas that have 'islands' and 'holes' and polygons that contain loops. Various topological models can be used (Valenzuela 1991), a typical one of which is the arc–node model, where an arc is a sequence of connected vectors which begins and ends in a node that can be the point of intersection between two or more arcs.

Tabular data

Tabular data consists of virtually any data that can be stored in a standard relational database format as a number of records each containing fields with defined values. Thus, tabular data could include the population of major cities in an area, traffic flow statistics for a road network or parameters associated with a particular species of tree.

One of the most important types of tabular data within a GIS is attribute data,

Table 4.1 Tabular representation of attribute data (from Barker 1988)

Attribute	Value
Operating area number	5
Operating area acres	327
Site index	80
Forest type	Pine
Topography	Flat
Age	38
Condition class	Mature
Volume	32.45

which is additional information linked to a geographic object to describe its properties – for example, the ownership of a field, the materials used to surface a road, or the land cover classes associated with a classified image. An example of the tabular representation of attribute data (in this case of a forest stand) is shown in Table 4.1.

Other classes of attribute data include status and object history information (often included as 'metadata' – see Section 8.3.1). The status information (Jackson 1988) stored with a geographic object provides additional background data such as area of coverage, date of compilation, geographic accuracy, and applicable standards. Additional status information may include the geographical location, map projection, scale, and feature codes. The object history provides an audit trail of the data that was used to produce the object and the processing performed, tracing back to the original source of the data; for example, a historical record for a DEM might include the identity of the original map containing the topographic information, the process used to digitise the contours and the algorithm used to interpolate contour-heights to a regular grid format. Without such information, little confidence can be placed in the accuracy and reliability of the other attribute data.

4.1.4 Integrating satellite data in a GIS

This section provides some introductory examples of how satellite data may be incorporated in a GIS. Sections 4.2, 4.3 and 4.4 provide more sophisticated examples of the synergistic use of GIS for image classification, environmental modelling and visual presentation of information. In addition, Chapters 5 to 8 provide many examples of real-world applications of these principles.

Satellite data as a GIS layer

Imagery, either from satellites or large-scale aerial photography, is widely used for updating a wide range of GIS datasets from base maps to thematic overlays. This data can be used to provide a background to help users to familarise themselves with the spatial context. Additionally, information from the image layer can be captured in vector format by 'heads-up digitisiting' where the operator interactively traces around an item of interest (such as a field boundary) and the system automatically generates the corresponding map object. This is particularly valuable where the revision cycle for map databases is too long to meet the update requirements of an application such as navigable road databases or topographic databases for telecommunication planning.

Use of GIS for image processing
Vector objects in a GIS can be used for a wide range of image-processing functions to mask areas for processing. A simple example of this would be to use the coastline to mask out the sea or land when applying special processing algorithms to the other. Linear vector geometries representing boundaries of spatial changes in spectral response can be used to define fuzzy edges for mosaicking operations. The use of vector-derived 'regions-of-interest' can also prove of use for calculating statistical values from an image only of a specific area.

Use of GIS for satellite operations
Another more esoteric use of GIS for EO applications is as a spatial indexing technology for satellite data acquisition, design of validation data acquisition and producing map layers showing the location of EO products. For example (Morley *et al.* 2000) GIS played a major part in the organisation of the LANDMAP mapping project to produce a DEM of the United Kingdom. Prior to image acquisition for ERS SAR data the European Space Agency (ESA) provides details of the map coordinates of the 'footprint' of the area to be imaged. There may be many hundreds of these footprints to be considered for a particular project and selecting the optimum combination is facilitated by defining and manipulating them in a GIS as vector polygons. These are readily processed to determine location, overlap and attributes such as whether a 'tandem pair' is available, as well as helping in the process of defining GCPs and the positioning of GPS receivers for data validation, using a digital map of the road network.

Integrating GPS in a GIS
GPS data is a 'natural' for incorporating in a GIS as it already available in a spatially referenced format (Steede-Terry 2000). GPS data can be incorporated as three main types of geographic object:

- a series of independent (x, y) points for location of assets such as fire hydrants or signals, or for use as GCPs.
- As polygons when defining such items as the boundary of a property.
- As a directed polygon, or track, when used to monitor, for example, the movement of animals such as elephants (see Section 8.4.3).

The raw GPS data may need to be processed before it can be incorporated in a GPS, for example, to interpolate between survey positions to fix inaccessible points or to smooth a boundary defined by a series of GPS locations.

4.2 CLASSIFICATION USING GIS

Another of the advantages of using an integrated GIS model is the additional capability it affords for the classification of satellite imagery, leading to an entirely new category of Geographic Information Product:

> '... *classification products produced by integrated GIS are not classified satellite imagery...but are a qualitatively different amalgam combining features of both satellite and cartographic data.*' (Davis and Simonett 1991)

A simple example of integrated classification was used for the analysis of SAR images for agricultural applications (Wooding 1985). The approach was to incorporate digital map data, in this case digitised field boundaries, into the classification process. By using the field boundaries as ground truth representative statistics such as mean backscatter can be collected for each field and related to the crop type contained therein. DEMs lend themselves particularly to integrated classification since often features of interest are known to be confined to certain elevation ranges or to have spectral characteristics which are strongly correlated with terrain parameters. Such approaches can prove of value in the application of satellite images to vegetation mapping as the topography of the underlying terrain is a major ecological factor in determining the occurrence, distribution and density of a particular species. In the integrated approach to classification (Hutchinson 1982) the sources of information can be introduced at one of three stages of the classification process:

- before classification
- during classification
- after classification.

In addition, GIS data can play a major part in the calculation of land cover areas from satellite imagery and a further section describes this.

4.2.1 Image stratification

The idea behind pre-classification stratification is to divide the input image into a series of smaller sub-images, each of which can be classified more easily. This allows, for instance, a division into sub-units, each of which is statistically more homogeneous, or the separation of features which are spectrally similar. This is likely to increase classification accuracy because the pixels in a particular class can often have a great spectral variation over an entire image. A typical use of digital data is the classification of soil types, where the areas in an image might be pre-classified using a geological map. Other criteria used for stratification (Taylor *et al.* 1997) include differences in site condition, topography, parcel structure, irrigation, number of harvests per year, and the regional agricultural practice. The statistical basis of this stratification is that it breaks down the training samples into smaller sets of samples each having a smaller variance than the original data. In many classification algorithms, such as maximum likelihood, the variance is the basis of measuring classification distance; thus, with a smaller variance and tighter training clusters greater discrimination is possible. Stratification can also be used to reduce the confusion between spectrally similar features, for example, by means of a land cover map.

Image stratification is particularly suitable for agricultural applications: a 'per field' classification is generally the most effective approach whereby a single classification is given to each field rather than to each pixel within the field (Pedley and Curran 1991). This method gives the classification greater reliability as untypical pixels – such as mixed pixels at the field edges – can be easily filtered out using buffering tools in a GIS. Another common approach is to mask out permanent non-agricultural areas such as roads, urban areas, lakes and airports to avoid mis-classification.

4.2.2 Classifier modification

The geographic information can also be used to modify the classification process itself. Typical GIS datasets that could improve the classification process include a DEM, rainfall, slope/aspect and existing land-use information. The simplest approach is to use the map-data as an additional 'band' in the classification. This is a limited technique, however, due to the differing natures of image and map data. Image data will vary greatly within a single feature class and it is the statistics of this that forms the basis of class discrimination. Digital maps, on the other hand, describe the Earth's surface in terms of a number of discrete classes such as built-up areas, roads, deciduous woodland. There is no indication of any variance within these classes, and thus no useful statistics on which a multi-spectral classification can be based. This approach can be used in a limited way, however, with non-parametric classifiers such as the box classifier. In this case, if a pixel is to lie within a particular box it must satisfy two constraints:

- it must lie within the training box in feature space
- its map class must correspond to one of the map classes present in the corresponding training sample.

A more sophisticated approach is to use the GIS data to modify the prior probabilities used, for example, in a maximum likelihood classifier. The basic classification algorithm does not make any *a priori* assumptions about the probability of a pixel belonging to any given class, it simply calculates the most likely class based on the sample statistics presented to it. Other geographic information can be used to modify probabilities: if a forestry map indicates that a pixel corresponds to an area marked as coniferous woodland, this provides additional evidence that the pixel should be classed as coniferous. Note that even if the map is many years out of date it can be used to provide some information: indication that an area was once coniferous will convey some probability that it is still coniferous.

The use of DEMs in classifier modification has certain advantages over other types of digital map data. Firstly, variables such as height and slope can be considered as continuous functions and can thus be more easily integrated into statistically based classification algorithms. Secondly, the derivation of DEMs is objective in its nature – the height and variation of the terrain is a well-defined

- dynamic models
- functional models.

An object model is a static model of an object's structure and relationships (for example, the structure of a satellite sensor in relationship to a satellite platform). This form of modelling represents the real-world object directly as a 'snapshot' of its physical form, and is used when geometry and location in relation to other objects is of most importance. A dynamic model represents how an object's state changes with time (for example, the position and orientation of a satellite platform) and is used where the timing and sequence of events (the object's history) is of most importance. Finally, functional modelling consists of the use of algorithms (either as explicit equations or using statistical techniques such as regression) which transform sets of inputs to a physical system – for example (Battad and Loh 1993) soil slope and land use to a set of outputs (in this case hydrological response). This last approach is generally of most value for modelling complicated environmental problems. The modelling approaches that we look at here are:

- spatial analysis, based on topological information available from a geographic dataset
- models based on EO data
- temporal evolution of models
- error modelling in GIS
- knowledge-based GIS
- techniques based on neural networks.

4.3.1 Spatial analysis

Spatial analysis is an important technique for combining data layers in a GIS to derive new layers and associated information based on their geometric and topological properties. Examples of spatial analysis models include:

- *Polygon overlay* – features on one data layer are overlaid onto those of other data layers and a logical combining operation is applied. For example, a weighted combination of soil, topography and drainage can provide a map of suitability for agricultural production.
- *Buffer generation* – buffers can be created around points, lines or areas at an equal distance in all directions. These can be used, for example, to define river corridors or exclusion zones for disease quarantining.
- *Feature dependent symbology* – a look-up table is applied to 'colour in' an object depending on its attributes. For example colouring a DEM as a series of height ranges. This is also referred to as thematic or chloropleth mapping.
- *Network analysis* – tracing through a network of line features such as pipes connected according to a series of rules ('Sewerage pipes must not be joined to mains supplies'). Network analysis may be used to simulate flow of gas, traffic, etc. or to find the quickest route between two points on a network, such as deriving turn-by-turn maps for in-car navigation.
- *Area and length calculations* – for line and polygon features.

- *Contouring* – to derive a continuous surface from discrete point values, for example, to derive height contours from a gridded DEM or isobars from local air pressure measurements.

As an example of an application of spatial analysis consider the situation of a planning officer who wishes to know all the locations in a region that are suitable for housing development (Williams 1984). The problem to be solved could typically be framed as:

> 'Find the location of all areas greater than 10,000 m^2, on land, below 150 metres, on a slope of less than 1 : 50 and within 3 km of a major road.'

One way of solving this problem within a GIS would consist of the following steps:

1. Assemble the data sets required to solve the problem:
   - digital map of land areas (D1)
   - Digital Elevation Model (D2)
   - digital map of roads (D3).
2. Use the Digital Elevation Model (D2) to produce a map layer of areas below 150 metres (D4).
3. Use the Digital Elevation Model (D2) to produce a map layer of areas of slope less than 1 : 50 (D5).
4. Use map of roads (D3) to produce a buffer-zone layer of areas within 3 km of a major road (D6).
5. Combine layers D1, D4, D5 and D6 to produce a map of areas on land below 150 metres on slope of less than 1:50 and within 3 km of a major road (D7).
6. 'Sweep' map D7 to remove all areas of less than 10,000 m^2.

One of the powerful aspects of a GIS is the ability to rework the problem with modified parameters (a 'what–if?' exercise). For example, the parameters of the above query can be easily modified to solve the problem:

> 'Find the location of all areas greater than 50,000 m^2, on land, below 100 m, on a slope of less than 1:25 and within 1 km of a major road.'

4.3.2 Models based on EO data

A mathematical model can be thought of as a specification for a number of related processes, each of which can be governed by a set of parameters. The model can be adapted by changing the parameters, by adding, removing or modifying processes and by changing the 'rule base' which governs the prioritisation and interaction between processes. Alternatively, instead of modifying the model, a 'frozen' model can be run on different datasets and the results compared. Of course many of the approaches to the use of GIS described previously (such as spatial analysis) can be viewed as simple models. For many environmental applications, however, a more sophisticated mathematical model is required: one capable of defining a number of different processes on varying spatial scales, and quantitatively representing their interaction and temporal evolution.

One such example of environmental modelling was a project to investigate the ecological significance of land use change in the United Kingdom (Griffiths and Wooding 1988). Studies of landscape ecology show that the abundance of a particular species is dependent not only on the presence of its preferred habitat within a landscape but also in the spatial arrangement of habitat patches throughout that landscape. Satellite images can be used to derive the land cover for a particular area, which can then be used in a more detailed model of the landscape pattern (Baskent 1993) that describes the functional interaction between the various patches that make up a landscape in terms of the flow of energy, material (especially nutrients) and species of plant and animal between them. For example, in a patch-based model of forest stands (Joy *et al.* 1993) a forest is represented as a grid of patches each 0.1 hectare in size. The interaction between patches is defined by a mathematical model which provides parameters for the shading and seed dispersal associated with each patch. The model is then used to predict the possible future evolution of forest patches and their effect on wildlife habitat, forest harvesting and fire disturbance.

Another good example of an integrated approach to an environmental application is an evaluation of the urban environment in Warsaw, Poland (Bochenek and Polawski 1992). The base image used for 1:100,000 scale mapping was acquired from Landsat TM, after geocoding a digitised urban boundary of Warsaw was overlaid. Other geographic information used included topographic and thematic maps of the Warsaw area, Cosmos space photographs and ground-truth information. The Thematic Mapper image was classified into 11 land-cover/land-use classes which were validated using the supporting geographic information. This classification formed the base layer of the Warsaw database (see Fig. 4.11), and two other thematic layers were also derived from the satellite data:

- vegetation classes derived from the Normalised Difference Vegetation Index (using bands 3 and 4)
- surface temperature map of the Warsaw urban area derived from Landsat thermal band (a good application of this under-utilised dataset).

On top of the base data digital soil-maps and maps of agricultural use were introduced. These maps included six types of soil and seven types of cropland. The final layer includes statistical information about annual dustfall and sulphur dioxide concentration. Once the GIS database was populated, the next stage was to use environmental models to derive new combinations of data. Three main analyses were performed:

- combination of land-use map and agricultural suitability map to determine best cropland categories
- identification of forests within areas of low pollution
- combination of poor agricultural areas, specific land uses and areas of low pollution to determine areas suitable for urban expansion within the Warsaw boundary.

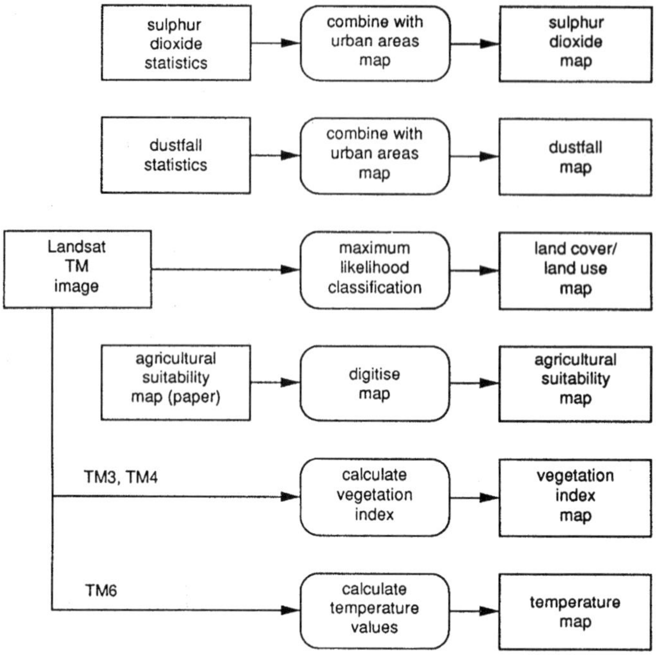

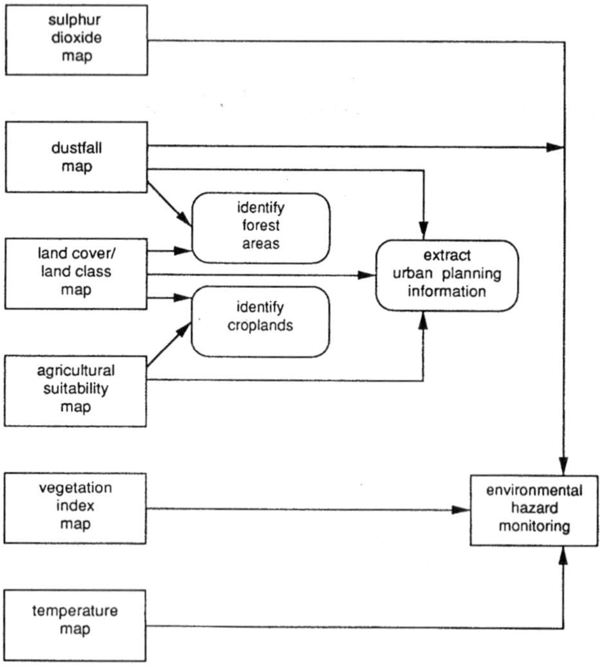

Fig. 4.11 Integrated GIS used to analyse the urban environment in Warsaw, Poland: (a) production of the geographic database; (b) details of geoprocessing. (Derived from Bochenek and Polawski 1992.)

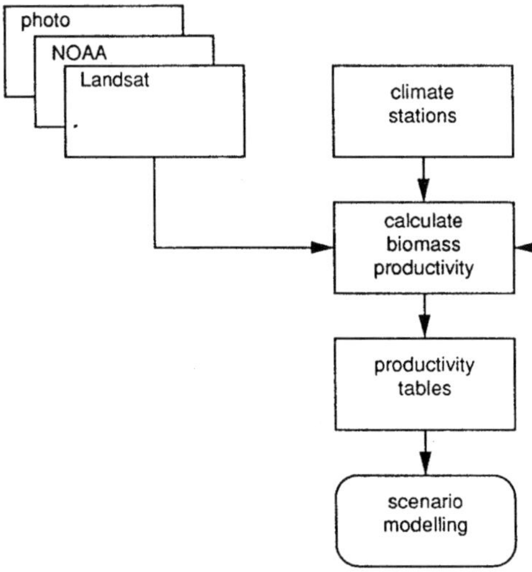

Fig. 4.12 Integrated GIS used for modelling of climatic change in Quebec, Canada (adapted from Viau *et al.* 1993).

The final stage was to use the urban model to produce a map of environmental hazard within the urban environment. This shows areas that are most suitable for living purposes as well as local 'trouble spots'.

Satellite images, in this case AVHRR, are also an important part of a climate model being used for impact modelling of climatic change in Quebec, Canada (Viau *et al.* 1993). The basic data structure of the model can be seen in Fig. 4.12 with the three main sources of information being imagery (AVHRR, Landsat and aerial photography), data from climate stations, and other terrestrial data such as forest maps and bio-geophysical records.

The AVHRR data is used to derive a 1-km grid land surface classification of the main ecological domains. This information, together with the other data sources, is fed into the ecological model to produce biomass productivity estimates, which in turn can be used to model the dynamic response of the forest to various climate change scenarios.

4.3.3 Temporal evolution of models

The other factor that must be considered in conjunction with each geographic object is its temporal nature, and this requires a model that can evolve over time (see Fig. 4.13). However, even in the (hypothetical) situation that no real-world changes had occurred it is still desirable to be able have a model which can be iterated to produce more accurate results. Such iterations could include the ability to filter out high-frequency noise and to increase confidence in the stable components of a model as well as the ability to store alternative and historical versions of 'fuzzy' data – one of the characteristics of most complicated mathematical models is that relatively small

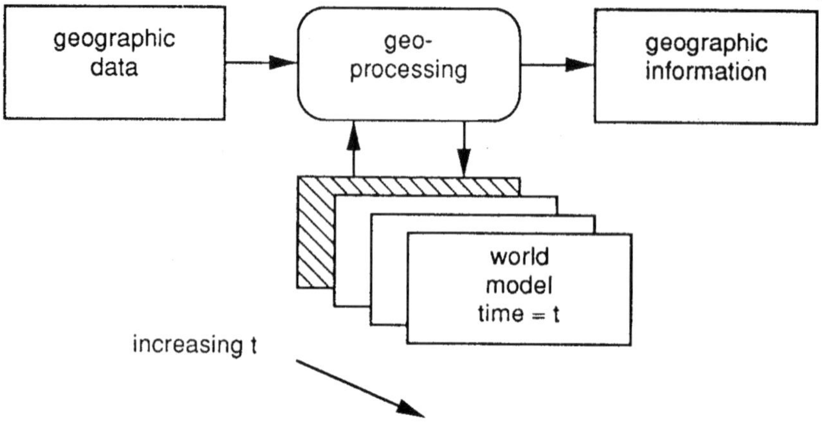

Fig. 4.13 Temporal evolution of a GIS model.

changes in the starting parameters of a model can lead to very large variations in the state of the model over time. It is important to be able to discern the implications of this for a particular model, e.g. for a long-term global warming model. Other aspects of temporal evolution include:

- modelling of the changes that have occurred over time (see Section 2.6.1)
- the ability to represent objects that are not homogeneous in time – for example, map sheets may be revised selectively over a number of years and will never present a true 'snap-shot' of an area at a particular moment in time
- the ability to 'roll-forward' the model to predict the status of geographic objects at some stage in the future – for example, the nature of global tropical deforestation in the next decade – and this operation could be performed several times with varying parameters.

Increasingly the computer power available has led to the use of animation techniques to provide the temporal aspects of a GIS model. This is described in further detail in Section 4.4.3.

4.3.4 Error modelling in GIS

Because the information within a GIS comes from a variety of disparate sources, all with their characteristic errors, it is important to be able to perform some sort of error modelling in order to determine the confidence that can be placed in the output information. The problem is not solely associated with errors in the geocoding and classification of satellite data, although it is essential that error metrics (such as residual errors and confusion matrices) should be considered as part of the model when satellite data is incorporated in an integrated GIS. There are just as likely to be sources of error in a conventional map (Hutchinson 1982), including:

- spatial variation of attributes such as soil colour or vegetation cover within an area marked as homogeneous on a map

- interpolation of point samples (such as soil type) into arbitrary boundaries between features
- variation of individual interpretation of the same sample data (for example, analysis of aerial photographs)
- generalisation, that is, errors introduced as the result of deriving maps at one scale from those at a different scale without resurveying
- temporal non-homogeneity – typically only major features will be amended when a map is revised, leaving many relatively minor changes unrecorded.

As well as these intrinsic errors additional errors may be introduced as the result of digitisation – for example, the production of sliver polygons where the common boundary between two features is digitised twice with slight variations each time (see Section 4.1.3).

There are five main categories of error in a GIS (Weir 1991):

- errors in source data such as data dropout and striping in satellite images, errors in the original source maps, interpolation of point data and temporal errors
- errors in data input including inaccuracies in GCP selection, errors in training area specification, digitising errors, and incorrect keying of attribute data
- errors in data storage – for example, accuracy lost during raster to vector conversions (and vice versa) and numerical approximations in computer arithmetic
- errors in data processing such as geocoding and classification errors in satellite images, DEM interpolation and errors propagated by dataset combination
- errors in data output – for example, the reproduction limitations of output devices.

In an ideal GIS all sources of data would include information about the statistical distribution of errors such as range, mean and standard deviation, or 95 per cent confidence limits. It is also desirable to be able to 'date stamp' error metrics and to quantify the decreasing confidence in data accuracy as the data ages. It is also desirable to have a set of rules which define the error metric associated when a new object is created from two existing ones. (The rule defined here is a very simplistic and artificial one to introduce the concepts.) Consider, for example, a straight-line segment, A, with length 100 metres. Because of inaccuracies in measurement there will be an error associated with the length (say 1 metre). Using the notation:

$$\text{length(object)} = [L, \delta L] \tag{4.3}$$

where L = estimated length
δL = error in length measurement

then length A = [100, 1].

Using this 'fuzzy' definition of length a few simple rules for defining the error metric associated with various operations on objects can be formulated. For example, the rule for estimating the length of two objects placed end-to-end is given by:

$$\text{length (object 1 @ object 2)} = [L_1 + \delta L_1, L_2 + \delta L_2] \quad (4.4)$$

where @ = geometric operation of 'place-end-to-end'
L_1 = estimated length of object 1
L_2 = estimated length of object 2
δL_1 = error in measurement of object 1
δL_2 = error in measurement of object 2

Thus if line B has estimated length 200 metres and error in length 3 metres, then

$$\begin{aligned} \text{length } B &= [200, 3] \\ \text{length } (A \text{ @ } B) &= [100+200, 1+3] \\ &= [300, 4] \end{aligned}$$

Unfortunately it is difficult to quantify many sources of error in geographic information and harder still to produce realistic metrics and valid models for error propagation within a system which, by its very nature, involves many complicated manipulations and combinations of widely differing data sources. Error modelling within integrated GIS, although of utmost importance for the widespread acceptance of the technology, is still very much in its infancy; in the meantime '*all persons using GIS technology . . . must be aware of the errors which can occur and where possible take appropriate steps to keep such errors to an acceptable minimum*' (Weir 1991).

4.3.5 Knowledge-based GIS

An integrated GIS provides the capability for a skilled operator to include many different sources of geographic data in an integrated analysis to solve a wide range of problems and to generate new geographic information (Gahegan and Flack 1996). The next emergent technology in the integrated GIS area is likely to be in the realm of knowledge-based systems which could help to automate many of the skilled tasks needed for 'image understanding'. This is particularly important as the number of new Earth observation missions planned in the time-frame to circa 2010 will provide a space segment producing quantities of data virtually impossible to analyse by conventional means (even using an integrated GIS). The development of knowledge-based GIS is, however, still very much at the research stage and this section presents two examples active areas of this research:

- extraction of roads
- knowledge-based image interpretation.

Extraction of roads

One area in which knowledge-based techniques have been successfully applied is the extraction of roads from satellite imagery. Not only is the automatic extraction of roads vital for mapping purposes (the rate of building of new roads in most countries rapidly outstripping the map revision cycle) but also for acting as pointers to other human activities (such as urban developments, agriculture and deforestation activities) and the identification of road junctions for use as GCPs. The shape of roads ('long narrow features without many bends') is straightforward to describe mathematically and there is much scope for introducing information about how roads intersect, how they join onto towns and how they appear in conjunction with other image features such as rivers – the spatial analysis tools within a GIS being ideal for manipulating this information.

The use of feature detection techniques has been discussed in the context of DEM generation (see Section 3.2.2) and similar techniques can be used to extract the primitive line segments that represent parts of roads. As well as this low-level information a knowledge-based model of how roads 'behave' is used to link the road fragments to form a complete network. Characteristics that can be used to define a road segment include:

- it is a straight line (or has a curvature subject to certain limits)
- it has a known maximum width
- it has a homogeneous and predictable intensity
- if multi-spectral imagery is used then a road will have a characteristic spectral signature
- it will (generally) have a high contrast with the surroundings
- roads do not occur above a certain height
- roads do not occur above a certain slope, or on undulating terrain (these last two points suggest the potential benefit of using a DEM).

Low-level operators can be defined, based on some of the above constraints, to extract candidate road segments from the entire image; however, they will also extract other linear features such as canals, rivers and boundaries of fields. A context-based model can be used to help determine genuine road segments. Such a model can make use of two categories of knowledge:

- generic knowledge about roads (such as their widths, curvature, the nature of junctions) which can be applied to a wide class of images
- scene-specific information such as the weather conditions and the known positions of existing roads (in terms of the integrated GIS paradigm the 'world model' is a network of roads which can be used to help extract road information from images and which can, in turn, be used to update the knowledge-base).

An example of this approach to road extraction (Ton *et al.* 1989) attempts to classify roads into one of three categories:

1. Major roads – two- to four-lane highways typically 15–45 metres in width.
2. Local roads – typically 8–20 metres wide.
3. Minor roads.

The low-level road extraction is performed using a series of road-sharpening operators which can extract magnitude and direction of candidate edge pixels. Because roads are characteristically of the order of one pixel wide, a road is often surrounded by mixels; by using a deliberately widened ('one-pixel-away') operator there is more likelihood of using pure background pixels. The result is a set of candidate road pixels which are thresholded to form a set of road seed points. A road-following algorithm uses the seed points to produce a 'primitive road file' in which each road segment is classified as level 1, 2 or 3, depending on the result of the road-following algorithm. The high-level processing consists of three stages:

1. *Segment labelling* – all level 1 road pixels which are neighbours (in the 8-connected sense) are linked, and all T- and Y-junctions are identified.
2. *Segment merging* – three parameters are calculated for each segment (length, average direction and average curvature). A series of merge rules (see Table 4.3) are applied which allow two segments to be merged.
3. *Road labelling* – a road is labelled as *major* if it is long (greater than 100 pixels) and smooth (roughly 10° per pixel), as a *local* road if it is medium (30–100 pixels) in length and smooth, and *minor* if it is longer than 10 pixels and has not been placed in one of the preceding categories.

Table 4.3 Merge rules (based on Ton *et al.* 1989)

Merge two segments IF:
segments are separated at a junction AND points in directions either side of the junction are consistent
OR
segments are disconnected by a very small gap) AND all pixels in the gap are level 2 or 3 road pixels AND all directions are consistent
OR
segments are disconnected by a small gap [3–8 pixels] AND both segments have similar directions AND length of both segments > three times the length of the gap AND more than half the gap pixels are level 2 or 3 road pixels

An alternative approach is adopted by Wang and Newkirk (1987) which uses a method similar to unsupervised classification. The pre-processing uses Laplacian filters to produce edge images (in each spectral band) which are scaled onto the range [0, 255]. Since only two classes are required ('road' and 'non-road') only two initial cluster points are used for the unsupervised classification. In the case of three image bands the initial points (0, 0, 0) and (255, 255, 255) are used. The result of the iterative process is a binary image where a value of 1 indicates a road. Figure 4.14 shows the extraction of roads from an IRS image of Sweden.

Fig. 4.14 IRS Image of Karlstad, Sweden, with roads extracted (OM&M Observation, Mapping and Monitoring AB).

Knowledge-based image interpretation

The traditional approach to multi-spectral image classification (described in Section 2.2) is to process each pixel individually and attempt to assign it to one of a defined set of classes. In reality there is no reason to suppose that the area represented by a pixel (which is a function of systematic parameters such as sensor IFOV) should correlate exactly with a 'real-world' feature. A knowledge-based approach to feature extraction uses models of features, built-up of simple context-based rules, which can then be used to relate the low-level image features (pixels) to high-level real-world objects. A simple example of the representation of knowledge in a rule-based system (Yee 1987) is:

IF	(side 1 of a road segment is adjacent to water)
AND	(side 2 of a road segment is adjacent to water)
THEN	(the road segment is a bridge)

Fundamental to the feature extraction process is the ability to segment the image into regions with related attributes. The basic concept is to use a suitable mathematical model to describe real-world objects which can then be extracted using corresponding algorithms. Many segmentation methods are based on edge-detection – objects will appear homogeneous with little difference in reflectivity but will be separated from other object by narrow edges of rapidly changing values.

An example of this is an experimental system developed in the UK for the knowledge-based segmentation of images of land (Corr *et al.* 1989). The input to the system is of two types: a multi-temporal sequence of geocoded image data, and a digitised map of the study area. The system works by building an initial model of the image classes, seeded by information from the digital map, and refining it by analysing the multi-temporal image set. After initialisation, each subsequent image in the sequence is processed using an edge-detection filter with an automatically determined threshold. The results are processed to form a network of boundary edge pixels. The system then compares the edge network with the current iteration of the model, by using the edges to assign a confidence to the boundaries currently in the model. Each confidence is a measure of the number of edge pixels within one pixel distance of the model boundary. The regions in the segmentation are then refined using the external knowledge in one of the following ways:

1. Regions may be split into smaller regions.
2. Regions may be merged into larger regions.
3. A region's class (or confidence level) may be changed.

The initial knowledge from the map data includes classes of linear features (roads, railways, rivers, paths) and areal features (such as fields, forestry, urban areas and inland water). The knowledge-base is used as the basis for a series of segmentation rules. For example one 'split' rule is provided to cater for fields which are joined by a narrow strip and further rules specify classes which are not to be split: these include urban areas, water features and wood. There are also a number of merge rules to amalgamate large homogeneous features from several smaller components, these are based on attributes such as the shape of the common boundary, contrast and length of boundary and the concavity of the region. The consistency rules are used to adjust the confidence in the class based on the available evidence. Map data is incorporated where appropriate: for example, a 'water body' rule states that the height contours must not change across a water feature. The confidence in a region is also strengthened by comparison with its neighbour; for example, compatible classes are field and field boundary, urban and road, or where two neighbouring regions have the same class.

Another example of a knowledge-based approach is that used for monitoring of urban areas in Europe (Barnsley 1993). The objective of the project is to use satellite images together with digital mapping to obtain accurate and consistent statistics about urban areas within the EU. Traditional pixel-based approaches are rarely successful on urban areas which present a complex patchwork of land cover types such as residential and industrial buildings, recreational areas, transport and water features. The methodology adopted is to infer land use statistics from the spatial arrangements of land cover classifications using a two-stage process:

- a low-level, per-pixel classification into a number of broad land cover types
- 'spatial reclassification' into landuse classes based on both the frequency and spatial arrangement of the class labels within each part of the image.

Thus, for example, residential districts may be characterised by the expected spatial arrangement of roads, building materials, grass and trees.

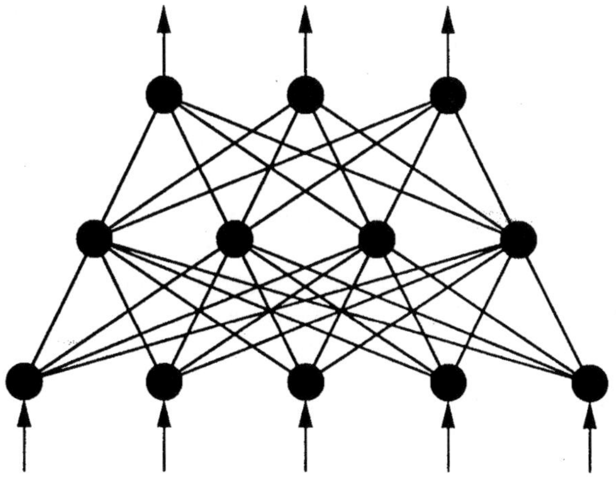

Fig. 4.15 Schematic representation of a typical neural network.

4.3.6 Techniques based on neural networks

Traditional multi-variate classification methods can become increasingly inapplicable to multi-source data due to its being 'multi-type' (Benediktsson *et al.* 1990); for example, spectral data, elevation range and non-numerical class types. One approach that is becoming increasingly adopted is the use of neural networks, a typical example of which can be seen in Fig. 4.15.

This particular network consists of a number of 'neurons', each of which can have several inputs and only one output. There are a great variety of possible architectures for neural networks and the determination of the optimum configuration for particular applications is currently a very active area of research. The output of a neuron is determined by a function which combines the values of all the inputs connected to that neuron, for example by a weighted sum:

$$F(x) = \Sigma\, w_i x_i \qquad (4.5)$$

where x_i = value of ith input
w_i = weighting to be used.

This is often used in conjunction with a threshold value such that the neuron only 'fires' if the set of input stimuli produce an output value above the threshold. Neural networks are wired so that a set of stimuli on the input layer propagates through the internal neural connections ('hidden layers') to produce a second set of signals on the output layer. Thus a neural network can be viewed as a black-box which takes a number of input values (such as grey-levels within an image window) and produces a number of output values (such as probabilities of being within a particular class). In order to train such a network it is presented with a number of samples of different image features together with the class that they represent. During the training process the network continually adjusts the weights of the functions controlling neuron firing until it 'knows' the optimum configuration for predicting the class value of a feature. It is also possible to make use of unsupervised neural networks which divide the data presented into classes based on similarity of patterns, without use of training data, in a manner analogous to that of unsupervised multi-spectral classification (see Section 2.2.2).

The neural network approach has the advantage that no prior knowledge of the statistical distribution of an image is needed. Another advantage of using neural networks is that they can take care of the weighting of evidence from different sources. However, one of the philosophical objections to the use of neural network techniques is that they operate without a specified mathematical model of the problem in hand (for example, feature extraction from a satellite image). Instead, a neural network will derive its knowledge of a problem domain from training samples in different categories of interest, which can lead to two main problems:

- large numbers of training samples are required to set up the network
- the method may not be robust to features that do not resemble the training data.

The applications of neural networks to the classification of images has possibly not lived up to its early promise (Hepner *et al.* 1990) but in some circumstance performances are comparable to that obtainable with conventional multi-spectral classifications and may in some cases be superior. Eventually the optimum approach may be to use neural networks in conjunction with a more modelled knowledge-based approach.

4.4 PRESENTATION OF GIS DATA

This section describes some of the methods of presenting GIS data graphically including digital cartography, three-dimensional visualisation, animated techniques and 'virtual GIS'. These four methods of presentation can also be thought of as a series of dimensional representation of increasing complexity (two-dimensional space, three-dimensional space, two-dimensional space plus time, three-dimensional space plus time).

4.4.1 Digital cartography

Digital (or automated) cartography is the drawing of detailed hard copy maps by computer. There is no implication that information processing has taken place but often a GIS will have been used to do the 'clever stuff' before the map is plotted. As a science, cartography has been around for many centuries – see, for example, the satellite image map of New York compared with the seventeenth-century map of 'New Amsterdam' shown in Fig. 4.16 (colour section).

Some of the standard plot options for cartographic output are:

- page size and orientation
- map grids and 'neatlines'

- scale to be plotted at, scale bars and north arrow
- font type, size and placement rules to be used for annotation
- colours and shadings to be used for polygon fills
- plot legend and other marginalia.

Digital image cartography

Basic cartographic rules should still apply to automated cartography based on image products, such as the mosaics described in Section 1.6, but a few 'space age' considerations also need to be taken into account.

Firstly, the use of wholly digital techniques permits a far wider range of products and allows experimental or prototype products to be generated with a rapid turnaround time. Cartographic features of interest that may be superimposed on a geocoded image product include urban areas and communications such as road, rail and air routes. Annotation can include the names of conurbations and major natural features such as mountains, lakes and rivers; political and administrative boundaries, and map grids to provide scale and orientation information. As with conventional map data, height information can be added in the form of spot heights or contours. Another use of map data is to use an image mask to highlight areas of interest – for example, a mask could be used to define an administrative boundary. Within the boundary the image map is enhanced to bring out features of interest; outside the boundary it is subdued to provide background and continuity information. Another alternative is to incorporate the map data as a raster by scanning the entire map into an image format which (since it will already be in a standard map projection) can be easily combined with a geocoded image.

Selecting the scale

The output cartographic product is most useful if produced at a standard mapping scale. The general rule of thumb (Colvocoresses 1986) is that the pixel size should equate to that of 0.3 mm in the final hard copy product. Thus with Landsat TM a pixel size of 30 metres would render it a suitable source for mapping at a scale of about 1:100,000 and SPOT PA with a pixel size of 10 metres somewhere between 1:25,000 and 1:50,000. A summary of suitable scales for satellite image mapping can be found in Table 4.4.

Table 4.4 Appropriate scales for image maps (based on Albertz *et al.* 1992)

Sensor	Pixel size (m)	Scale
Landsat MSS	80	1:250,000
Landsat TM	30	1:100,000
SPOT XS	20	1:100,000
SPOT PA	10	1:50,000
Ikonos	1	1:5,000

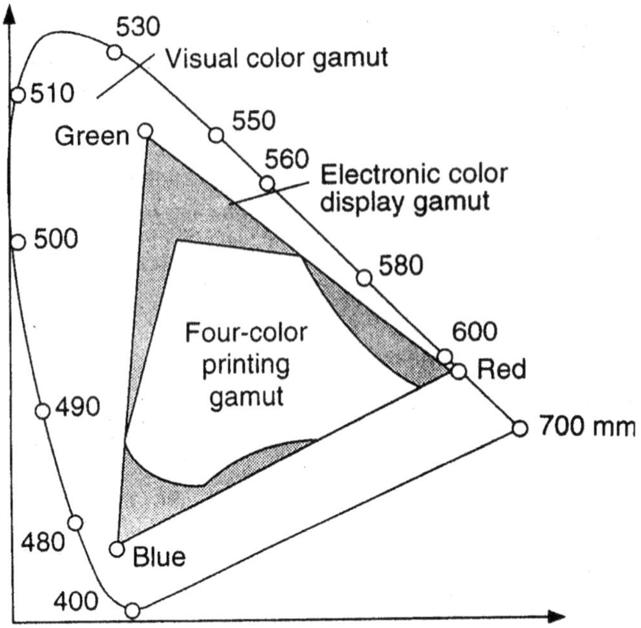

Fig. 4.17 Comparison of visual colour gamut with electronic colour display gamut and four-colour printing gamut (courtesy USGS; adapted from Williamson 1983).

Colour balance

Although, as has been demonstrated previously in this chapter, there has been a considerable body of theory underpinning the development of GIS, ultimately a considerable amount of technical and artistic skill is necessary. As Kidwell and McSweeney (1984) of the US Geological Survey describe it:

'... *image maps are both science and art in one medium*'.

One important criterion for the production of hard copy geocoded products is the range of colours available (the colour gamut). In fact three colour gamuts are of paramount importance (see Fig. 4.17):

- the visual colour gamut
- the colour visual screen display gamut of a GIS
- the colour printing gamut.

As can be seen, more colours are available in the 'real-life' gamut than on the display of a GIS, which in turn encompasses more colours than the colour-printing gamut. This should be borne in mind when the beautifully balanced colour image on the computer screen produces a disappointing hard copy product! When printing a geocoded product the subtractive primary colours (cyan, magenta and yellow) are used instead of the primary additive colours (red, green and blue respectively). Guidelines for successful visual appearance (Kidwell and McSweeney 1984) include:

1. Look for good detail and contrast on the cyan separation.
2. Make sure that the magenta and yellow separations map well onto the cyan producing clear, bright colours.
3. Check that spectral highlights and white clouds are free of colour.
4. Check that dark shadow is black or dark grey.
5. Check for good contrast in the mid-tones.

Hard copy cartographic maps are produced either by using standard printing/photo-processing methods (for example, generating a photographic negative from the digital data by filmwriting) or by direct output on a hard copy device such as an electrostatic plotter.

4.4.2 Three-dimensional visualisation

One of the most useful ways of displaying a geocoded image product is to drape it over a DEM. For example, Fig. 4.18 shows a perspective view of London generated from high-resolution Ikonos images.

GIS tools used for three-dimensional visualisation include the ability to render a DEM, rotate it, and provide height exaggeration, perspective. Sun illumination and fogging effects. As well as being able to display an image layer in its true position they will be able to add geometric models of individual real-life object, such as the buildings in Fig. 4.18, to the generalised representation of terrain relief shown in the DEM.

Another impressive use of three-dimensional visualisation is the ‘Tang Project’ (Böhler and Heinz 1997) to investigate the 18 mausoleums of the Chinese Tang Dynasty (618–907 CE) which are scattered over an area of 5,000 km in Shaanxi

Fig. 4.18 Perspective view of London generated from Ikonos (courtesy NPA Group).

province in the People's Republic of China. The size of each single mausoleum, including the surrounding wall, may reach up to 15 km. Figure 4.19 shows how a simple digital outline on a SPOT image, together with the underlying topography, can bring the archeological site to life.

The stage following visualisation is to provide the opportunity to interact – for example, to click on an object on a three-dimensional display and be provided with details of its attribution – or to define and move the viewpoint for the scene being viewed.

4.4.3 Animation

The use of animation is a powerful tool to add the third dimension of time to a two-dimensional map. Needless to say, the print medium is not a good way to demonstrate this and the reader is advised to search the Internet for some good examples of animated GIS! One simple, but highly effective example of this (Holtschlag and Aichele 2001) is for the visualisation of drifting buoy deployments on the St Clair River, which is a connecting channel of the Great Lakes between Lake Huron and Lake St Clair on the international boundary between the United States and Canada. The purpose of the project, which took place in October 2000, was to investigate flow characteristics near public water intakes. One group of buoys was released at uniform intervals in a transect across the river to measure flow patterns; a second tranche was deployed in mid-channel clusters to study turbulent dispersion characteristics. Each buoy was GPS-enabled so that its real-time track could be recorded and replayed as a graphical animation, superimposed on a standard 1:15,000 navigational chart of the river. Buoy locations were generally included at about 90-second intervals, when GPS data was available. Between measured locations, buoy positions were automatically interpolated by use of a Bezier curve. The resulting animation (which can be seen at http://smig.usgs.gov/features_0301/drifters.html.) can provide valuable insight into the impact of local inflows on the water quality at public-supply intakes and also the planning of emergency responses in the event of contaminant spills within the river.

The above example showed the use of animation to depict events in 'pseudo real time' (i.e. at roughly the same speed at which they occurred). Many of the events being modelled using satellite images take place over extended timescales of months (crop monitoring, iceberg tracking) or years (ozone depletion, urban growth). Animation at many times faster than real time can provide valuable insight into long-term trends, for example animations of vegetation indexes can vividly show the seasonal progression of the 'green wave' on a global scale. An example of this (Scipal and Wagner 2000) is the use of the ERS Scatterometer to monitor soil moisture over the African continent over the period 1992–2000. The animation can be seen at http://www.ipf.tuwien.ac.at/ww/home.htm and shows the variation over time between dry and wet conditions and areas of tropical forest and sandy deserts where retrieval of soil humidity is not possible.

4.4.4 Virtual GIS

The final refinement in GIS depiction is to devise methods of depicting three-

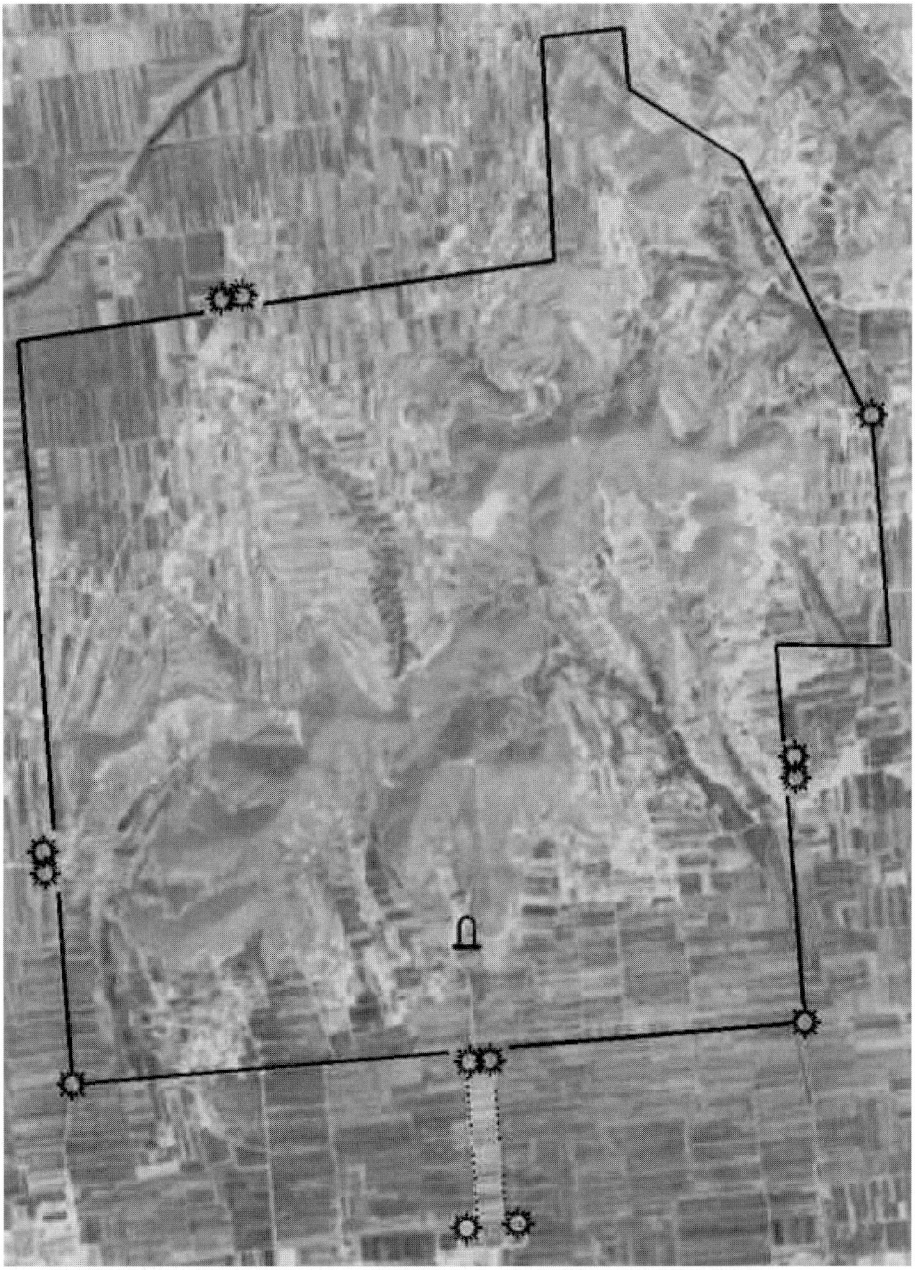

Fig. 4.19 Orthophoto and perspective view of Guangling Mausoleum. SPOT image draped over digital elevation model. Results of local surveys were included at correct locations. (Courtesy German National Aerospace Research Center (DLR), Roemisch–Germanisches Zentralmuseum in Mainz (RGZM), University of Applied Sciences in Mainz (FH Mainz) and SPOT Image.)

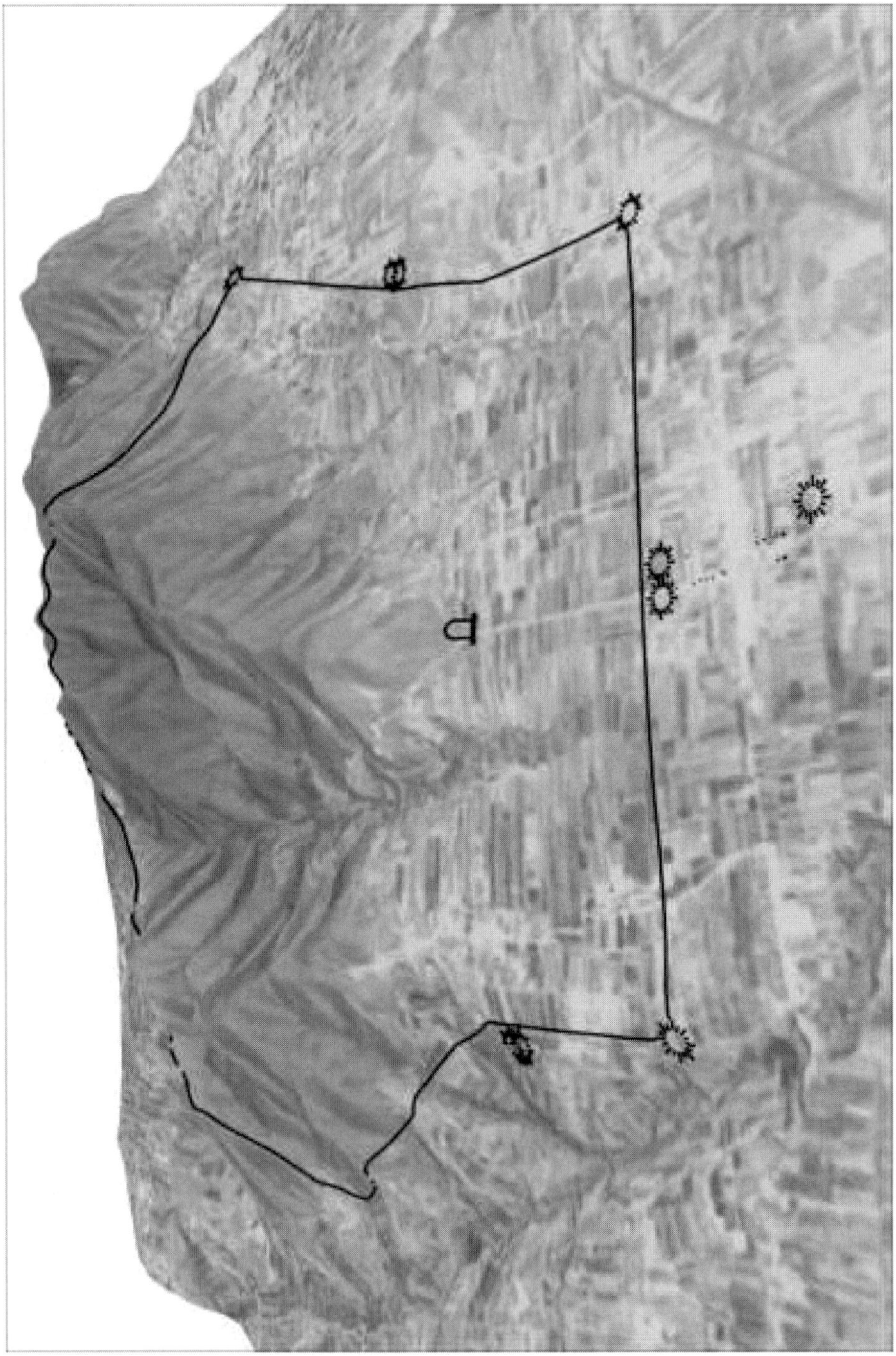

dimensional spatial information as well as temporal information in a 'Virtual GIS'. One approach to this is to animate the three-dimensional visualisations described in Section 4.4.2; this is often used for flight simulations and fly-throughs. Despite the fact that they are rendered on a two-dimensional computer screen, such animations can appear very realistic. The main reason for this is that they are depicting a surface that is not 'fully' three dimensional. For each point of the Earth's surface the height of the surface above the datum is portrayed. There may be a limited amount of vertical extension, for example to portray buildings or a tree-line, but this will be only a fraction of the vertical extension that may be required in complex three-dimensional models of, for example, the Earth's atmosphere.

For such three-dimensional models new approaches must be adopted. Increasingly the technology of virtual reality from the worlds of computer games, military simulation and remote operations are being adopted for virtual GIS.

REFERENCES

Albertz J, Lehmann H and Tauch R (1992) The production of satellite image maps – experiments at the Technical University of Berlin. *Proceedings of Satellite Symposia 1 and 2 : Navigation and Mobile Communications, and Image Processing, GIS and Space-assisted Mapping, from the 'International Space Year' Conference, Munich, Germany*, 30 March – 4 April 1992, ESA ISY-2, pp.175–80.

Aybet J and Walpole R (1994) An integrated GIS approach to the use of earth observation data – sugar beet prediction and management system. Association for Geographic Information, AGI 94, Birmingham, UK, 15-17 November 1994, pp. 15.1.1–15.1.6.

Barker GR (1988) Remote sensing: the unheralded component of Geographic Information Systems. *Photogrammetric Engineering and Remote Sensing*, **54** (2), 195–9.

Barnsley M (1993) Monitoring urban areas in the EC using satellite remote sensing. *GIS Europe*, October 1993, pp. 42–44

Baskent EZ (1993) Quantifying forest landscape structure for integrated resource planning. *GIS '93 Symposium, Vancouver, British Columbia*, February 1993.

Battad DT and Loh DK (1993) Framework for the integration of Geographic Information Systems with soil erosion simulation models. *GIS '93 Symposium, Vancouver, British Columbia*, February 1993, pp. 353–61.

Belward AS and Velenzuela CR (eds) (1991) *Remote and Geographical Information Systems for Resource Management in Developing Countries.* Kluwer Academic Publishers, Dordrecht, The Netherlands.

Benediktsson JA, Swain PH and Ersoy OK (1990) Neural network approaches versus statistical methods in classification of multisource remote sensing data. *IEEE Transactions on Geoscience and Remote Sensing*, **28** (4), 540–51.

Bochenek Z and Polawski Z (1992) Use of satellite data and GIS for evaluating urban environment. *Proceedings of Satellite Symposia 1 and 2: Navigation and Mobile*

Communications, and Image Processing, GIS and Space-assisted Mapping, from the 'International Space Year' Conference, Munich, Germany, 30 March – 4 April 1992, ESA ISY-2, pp. 377–9.

Böhler W and Heinz G (1997) Recording and visualizing topography and object geometry for archaeological documentation. *Archäologisches Korrespondenzblatt*, **27**, 1997, Heft 2, Verlag des Römisch-Germanischen Zentralmuseums Mainz, pp. 355–73.

Burrough PA (1986) *Principles of Geographical Information Systems for Land Resource Assessment*. Clarendon Press, Oxford, UK.

Colvocoresses AP (1986) Image mapping with the Thematic Mapper. *Photogrammetric Engineering and Remote Sensing*, **52**, 1499-1505.

Corr DG, Tailor AM, Cross A, Hogg DC, Lawrence DH, Mason DC and Petrou M (1989) Progress in automatic analysis of multi-temporal remotely-sensed data. *International Journal of Remote Sensing*, **10** (7), 1175–95.

Cuong PV (1992) Application of remote sensing and GIS – technology to natural resources management: selecting rice fields in Vietnam. *Proceedings of Satellite Symposia 1 and 2 : Navigation and Mobile Communications, and Image Processing, GIS and Space-assisted Mapping, from the 'International Space Year' Conference, Munich, Germany*, 30 March–4 April 1992, ESA ISY-2, pp. 363–8.

Davis FW and Simonett DS (1991) GIS and remote sensing. In Maguire DJ *et al.* (eds), pp. 191–213.

Doak SC and Lackey L (1993) Assessing change in urban natural areas with multi-date satellite imagery. *GIS '93 Symposium, Vancouver, British Columbia*, February 1993, pp.1163–8.

Gahegan M and Flack J (1996). A model to support the integration of image understanding techniques within a GIS. *Photogrammetric Engineering and Remote Sensing*, **62** (5), 483–90.

Griffiths GH and Wooding MG (1988) Pattern analysis and the ecological interpretation of satellite imagery. *Proceedings of IGARSS '88 Symposium, Edinburgh, Scotland*, 13–16 September 1988, pp. 917–22.

Hartnell T (1994) Environmental Management, through Integrating Remote Sensing. *GISGIS '94 Conference, Birmingham, UK*, May 1994.

Hepner GF, Logan T, Ritter N and Bryant N (1990) Artificial neural network classification using a minimal training set: comparison to conventional supervised classification. *Photogrammetric Engineering and Remote Sensing*, **56** (4), 469–73.

Hogg J and Gahegan M (1986) Regional analysis using Geographic Information Systems based on linear quadtrees. *Mapping from Modern Imagery, Proceedings of a Symposium held by Commission IV of the International Society for Photogrammetry and Remote Sensing, Edinburgh, Scotland*, 8–12 September 1986, pp.113–24.

Holtschlag DJ and Aichele SS (2001) Visualization of drifting buoy deployments on St. Clair river near public water intakes – October 3–5, 2000. *U.S. Geological Survey Open File Report* 01–17.

Hutchinson CF (1982) Techniques for combining Landsat and ancillary data for digital classification improvement. *Photogrammetric Engineering and Remote Sensing*, **48** (1), 123–30.

Ibbs TJ and Stevens A (1988) Quadtree storage of vector data. *International Journal of Geographical Information Systems*, **2** (1), 43–56.

Jackson M (1988) Trends in commercial GIS development. Presented *at Image Processing 1988, Blenheim Online Publications, Pinner, Middlesex, UK*, pp. 201–13.

Jensen JR, Ramsey EW, Holmes JM, Michel JE, Savitsky B and Davis BA (1990) Environmental Sensitivity Index (ESI) mapping for oil spills using remote sensing and Geographic Information System technology. *International Journal of Geographical Information Systems*, **4** (2), 181–201.

Joy M, Klinkenberg B and Cumming S (1993) An architecture for spatial modelling: linking environmental models with remote sensing and GIS. *GIS '93 Symposium, Vancouver, British Columbia*, February 1993, pp. 925–33.

Kargel JS, Kieffer HH, Raup B, MacKinnon D, Mullins K and the GLIMS Consortium (2000) First results from GLIMS. (Global Land Ice Measurements from Space) IGARSS 2000, Honolulu, Hawaii, 24–28 July, 2000.

Kattenborn G and Klaedtke H-G (1997) Classification of ERS-1 SAR data over Seville (Spain) for agricultural statistics. *The 3rd ERS Symposium (ESA), Florence (Italy)*, 18–21 March 1997.

Kidwell RD and McSweeney JA (1984) Art and science of image maps. *Proceedings ASPRS 51st Annual Meeting, Washington DC*, 10-15 March 1984, Vol. 2, pp. 771–82.

Maguire DJ, Goodchild MF and Rhind DW (eds) (1991*) Geographical Information Systems: Principles and Applications*. Longman, London.

Morley JG, Walker AH and Muller J-P (2000*) GIS techniques employed for the LANDMAP Project, RSS 2000*. University of Leicester.

Pedley MI and Curran PJ (1991) Per-field classification: an example using SPOT imagery. *International Journal of Remote Sensing*, **12** (11), 2181–92.

Peucker TK, Fowler RJ, Little JJ and Mark DM (1978) The Triangulated Irregular Network. *Proceedings of the ASP Digital Terrain Models (DTM) Symposium, American Society of Photogrammetry, Falls Church Virginia*, pp. 516–40.

Scipal K and Wagner W (2000) Large scale soil moisture monitoring in Africa using radarsatellite data. *EARSEL SIG Workshop on Remote Sensing for Developing Countries, Gent, Belgium*, 13–15 September 2000 (in press).

Sondheim M (1993) Modelling the real world. *GIS '93 Symposium, Vancouver, British Columbia*, February 1993, pp. 1099–1111.

Steede-Terry K (2000) *Integrating GIS and the Global Positioning System*. ESRI Press.

Stuttard MJ, Hayball JB, Narciso G, Suppo M and Oroda A (1994) Use of GIS to assist hydrological modelling of lake basins in the Kenyan Rift Valley. *Association for Geographic Information, AGI 94, Birmingham, UK*, 15–17 November 1994, pp. 13.4.1–13.4.8

Taylor JC, Sannier C, Delincé J, Gallego (1997*). Regional Crop Inventories in Europe Assisted by Remote Sensing: 1988–1993*. Joint Research Centre, European Commission, EUR 17319 EN.

Ton J, Jain AK, Enslin WR and Hudson WD (1989) Automatic road identification and labeling in Landsat 4 TM images. *Photogrammetria (PRS)*, **43**, 257–76.

Valenzuela CR (1991) Spatial databases. In Belward AS and Valenzuela CR (eds), pp. 311–33.

Veronese VF and Mather PM (1993) Supervised filtering: a postclassification technique based on contextual and ancillary information. *Proceedings of the 19th Annual Conference of the Remote Sensing Society, Chester College*, 16–17 September 1993, pp. 120–7.

Viau A A, Royer A, Boivin J, Ansseau C and Theriault A (1993) Integration of remote sensing data in a Geographic Information System for impact modelling of long-term climatic change in Quebec's boreal forest. *GIS '93 Symposium, Vancouver, British Columbia*, February 1993, pp. 493–7.

Vieira CAO and Mather PM (2000) An examination of the effectiveness of multitemporal crop classification, Remote Sensing Society 26th Annual Conference, Leicester, UK, 12–14 September, 2000.

Wang F and Newkirk R (1987) A knowledge based system for highway network extraction. *Proceedings of IGARSS '87 Symposium, Ann Arbor, Michigan*, 18–21 May 1987, pp. 343–7.

Weir MJC (1991) Errors in Geographic Information Systems. In Belward AS and Valenzuela CR (eds), pp. 349–55.

Williams JM (1984) *REGIS – an RAE Experimental Image–based Geographic Information System*. Royal Aircraft Establishment (Farnborough, UK), Technical Report 84103.

Williams J, Parker D, Harris R, Turner R and Baker D (2000) Area estimation of the British potato crop using earth observation and GIS. In *Aspects of Applied Biology*, **60**, *Remote Sensing in Agriculture*, pp. 229–34.

Williamson SJ (1983) *Light and Colour in Nature and Art*. New York, John Wiley & Sons.

Wooding MG (1985) SAR image segmentation using digitized field boundaries for crop mapping and monitoring applications. *Microwave Remote Sensing Applied to Vegetation*, European Space Agency, SP-227, pp. 93–8.

Wyatt BK and Fuller RM (1992) European applications of space-borne earth observation for land cover mapping. *Proceedings of the Central Symposium of the 'International Space Year' Conference, Munich, Germany*, 30 March – 4 April 1992, ESA ISY-1, July 1992, pp. 655–9.

Yee B (1987) An expert system for planimetric feature extraction. *Proceedings of IGARSS '87 Symposium, Ann Arbor*, 18–21 May 1987, pp. 321–5.

Zeiler M (1999) *Modeling our World. The ESRI Guide to Geodatabase Design*. ESRI Press.

Further reading

Brown R, Slater J and Askew D (1994) Environmental monitoring of protected landscapes – monitoring environmentally sensitive areas using satellite imagery, within GIS. *Association for Geographic Information, AGI 94, Birmingham, UK*, 15-17 November 1994, pp.15.2.1–15.2.4

CEOS (1994) Committee on Earth Observation Satellites (CEOS) Dossier A, September 1994.

Contzen J-P (1992) Earth observation programmes and plans of the Commission of the European Communities (CEC). *Proceedings of the Central Symposium of the*

'International Space Year' Conference, Munich, Germany, 30 March – 4 April 1992, ESA ISY-1, July 1992, pp. 1429–33.
van Dijk A and Galassetti-Morrey S (1993) Can GIS help feed the world? *GIS Europe*, November 1993, pp. 18–20.
Donnay J-P (1992) Remotely sensed data contributes to GIS socioeconomic analysis. *GIS Europe*, December 1992, pp. 38–41.
Jensen JR, Ramsey EW, Holmes JM, Michel JE, Savitsky B and Davis BA (1990) Environmental Sensitivity Index (ESI) mapping for oil spills using remote sensing and Geographic Information System technology. *International Journal of Geographical Information Systems*, **4** (2), 181–201.
Kempka RG, Reid FA and Altop RC (1993) Developing large regional databases for waterfowl habitat in Alaska using satellite inventory techniques: a case study of the Black River. *GIS '93 Symposium, Vancouver, British Columbia*, February 1993, pp. 581–91.
Macdonald JS (1992) Delivery of information from Earth observation satellites. *Proceedings of Satellite Symposia 1 and 2 : Navigation and Mobile Communications, and Image Processing, GIS and Space-assisted Mapping, from the 'International Space Year' Conference, Munich, Germany*, 30 March – 4 April 1992, ESA ISY-2, pp. 243–9.
Nichol J (1993) Monitoring Singapore's microclimate. *Geo Info Systems*, February 1993, pp. 51–5.
Oberg M and Andersson C (1993) Estonia: mapping environmental damage by satellite. *GIS Europe*, March 1993, pp. 50–2.
Palko S, Lowe JJ and Pokrant HT (1993) Canada's new, seamless forest cover data base. *GIS '93 Symposium, Vancouver, British Columbia*, February 1993, pp. 985–90.
Shaffer LR (1992) The Earth Observing System. *Proceedings of the Central Symposium of the 'International Space Year' Conference, Munich, Germany*, 30 March – 4 April 1992, ESA ISY-1, July 1992, pp. 945–50.
Smith G, Fuller R, Amable G, Costa C and Devereux B (1997a) Classification of environment with vector- and raster-mapping (CLEVER Mapping). *Proceedings of the GIS Research UK, 5th National Conference, University of Leeds*, 9–11 April, pp. 70–2.
Smith G, Fuller R, Amable G, Costa C and Devereux B (1997b) CLEVER Mapping: an implementation of a per-parcel classification procedure within an integrated GIS environment. *Proceedings of the Remote Sensing Society Conference, Observations and Interactions: RSS97, Remote Sensing Society, University of Nottingham*, pp. 21–6.
Wilson PM (1986) Using Landsat data within a Geographic Information System. *Mapping from Modern Imagery, Proceedings of a Symposium held by Commission IV of the International Society for Photogrammetry and Remote Sensing, Edinburgh, Scotland*, 8–12 September 1986, pp. 491–9.

5

Land applications

The successful development of a market for EO applications is strongly linked to the use of GIS. To maximise the benefit of satellite data it must be used in conjunction with digital maps and *in situ* measurements of data. This chapter describes some of the 'land applications' of EO, such as:

- agricultural monitoring and yield prediction
- forestry
- geological and civil engineering applications
- telecommunications and media
- land applications of GPS.

These applications make use of the generic techniques for image geocoding, information extraction, height determination and GIS integration described in the first four chapters. However, where specialist algorithms have evolved in particular sectors, such as geological applications, they are introduced here. Further applications are described in 'Environmental monitoring and population security' (Chapter 7) and 'Geomatics' (Chapter 8).

5.1 AGRICULTURE

With an increasing world population, effective use of renewable resources is becoming ever more vital. Satellite data is proving increasingly useful for the monitoring of these resources, particularly in agriculture and natural vegetation and forestry. This section describes two of the established agricultural applications of integrated GIS: agricultural monitoring and yield prediction. Another major area of development is precision farming which brings together images, GPS and location-enabled farm equipment in an integrated GIS. This is described as a separate Case Study in Section 8.4.2.

5.1.1 Agricultural monitoring

The classification of agriculture crops was one of the earliest applications of Landsat MSS in the early 1970s (Wiegand *et al.* 1973; Hall *et al.* 1974), the most prominent

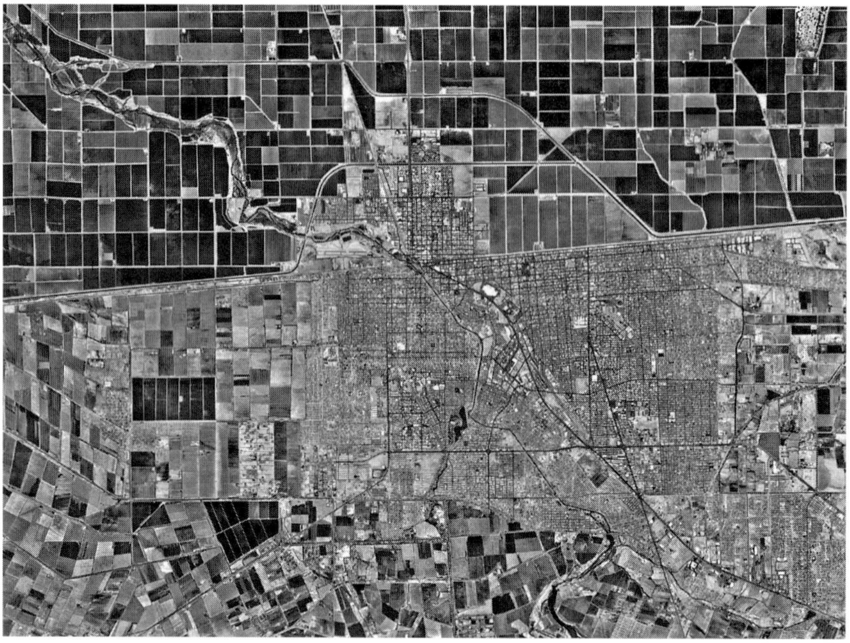

Fig. 5.1 SPOT image map of Mexicali. The US/Mexico border is clearly visible, divided by the differing agricultural land use patterns (courtesy INEGI, Mexico; copyright CNES 2001, SPOT Image Distribution).

example of which was the Large Area Crop Inventory Experiment (LACIE) carried out between the US National Oceanographic and Atmospheric Administration (NOAA), the US National Aeronautics and Space Administration (NASA) and the US Department of Agriculture. Satellite images continue to provide a synoptic overview of the agricultural practices in a region – for example, the SPOT image of the city of Mexicali in Fig. 5.1 clearly shows the difference in agricultural patterns each side of the border between the United States and Mexico.

A recent survey (Moran 2000) of the suitability of Earth observation technology for agricultural applications highlighted a wide range of conditions where EO could play a major role. These include water stress in crops, stand density, nitrogen deficiency, soil moisture, weed and insect infestations and other crop conditions. The report also concluded that relatively few of these applications had been developed operationally, due to the limitations of current EO satellites (primarily the ability to provide reliable weekly coverage during the growing season) and the ground segment (primarily to provide accurately geocoded agrithematic products within 24 hours of image acquisition).

Despite the limitations noted above, there are a number of situations where EO is used operationally for agricultural applications; for example, the use of Radarsat for monitoring rice production in Vietnam (Radarsat 2001) and operational mapping of the United Kingdom potato crop (Parker and Williams 1999; Williams *et al.* 2000) using programmed SPOT Image acquisitions and ERS radar. One of the major examples of the use of satellite imagery is to monitor the Common Agricultural Policy (CAP) of the European Union (EU). In the spring of 1992 the Council of Ministers of the EU agreed on a reform of the CAP to regulate the enormous costs that were associated with overproduction of crops in Europe. The agreement set up a system to subsidise farmers for production of main crop types and set-aside (paying farmers not to overproduce). The subsidies are based on the areas of crops produced, thus some mechanism is required to check that the areas and crop types declared by the farmers are correct. In conjunction with local field checks the EU has set up the Monitoring of Agriculture by Remote Sensing (MARS) project which carries out a number of related activities (Meyer-Roux 1993):

- *Anti-fraud measures* use satellite data as the main input to the land parcel identification GIS systems used to validate the collection of production statistics for key crops such as vineyards and olive trees.
- *Crop yield monitoring* is conducted with agro-meteorological models (see Section 5.1.2 for further details).
- *Area estimates* using high-resolution data combined with ground surveys. This method uses the area estimation technique described in Section 4.2.4 to provide accurate unbiased results.
- *Specific surveys* use an application of area frame sampling techniques to provide the rapid and specific information needed for the definition or reform of agricultural policies.

The MARS project provides regular statistical information about European crop surface estimates. This information is integrated into crop status bulletins published every month, which show the evolution at European Union level of crop areas compared to the previous year, as well as early indicators of potential yields. An example of a MARS bulletin is shown in Fig. 5.2.

5.1.2 Yield prediction

A major review of crop yield prediction (Denore and Garcia Lopez 2000) examined the roles that earth observation can play in crop yield estimation. The first of these is as a support for statistical analysis of crop yields, for instance:

- stratification for the design of field surveys and sampling schemes
- correlation of crop parameters with spectral bands or with vegetation indices integrated over time
- analysis of NDVI profiles, for example, to identify critical moments in crop development.

The second and more sophisticated approach is to combine the satellite data with a numerical 'agro-meteorological' model of crop growth, weather conditions and other factors such as soil depth. The major components of a typical agro-meteorological model used for Cereal (de Koeijer *et al.* 2000) can be seen in Fig. 5.3. The main inputs are meteorological data (radiation, temperature, rain, atmospheric humidity), soil parameters and the vegetation index derived from the satellite data.

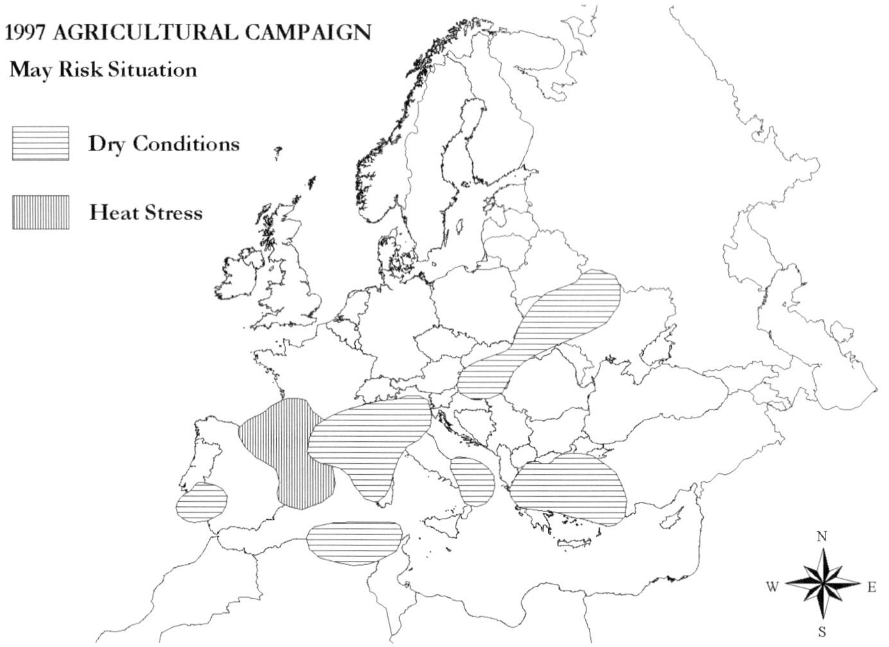

Fig. 5.2 Example of a MARS bulletin (courtesy MARS Project, European Commission).

The model also requires a list of crop-specific parameters, e.g. water use efficiency (WUE), radiation use efficiency (RUE) and light extinction coefficient (k). The model has two modes of operation depending on whether the growth is water limited or light limited. The value output is simply the minimum of these two quantities. On 'day zero' long year averages of weather data can be used to predict weather for the current season. As the season goes along this weather data set can be updated with actual data day by day. In this manner (1) the evolution of the crop can be updated on a daily basis and (2) predictions can become more accurate later in the season.

The key parameter in the growth model is the Development Stage (DVS) which is modelled as a dimensionless quantity which has a value 0 at seedling emergence, 1 at flowering and 2 at maturity. The major factor in the Development Stage is the accumulated Thermal Time (TT) defined as the summed average temperature of all days above a threshold value. An empirical function is used to relate Thermal Time to Green Area Index (GAI). A number of variants on this structure are used to reflect the difference between dryland and wetland and for the different types of crop.

The vegetation index is typically calculated from a series of three or four satellite images acquired throughout the season. As this can vary significantly throughout a region due to variations in irrigation, seed type and agricultural practice. It can be beneficial to assign different parameters for each 'stratum' in an image before activating the agro-meteorological model. An example of an operational yield

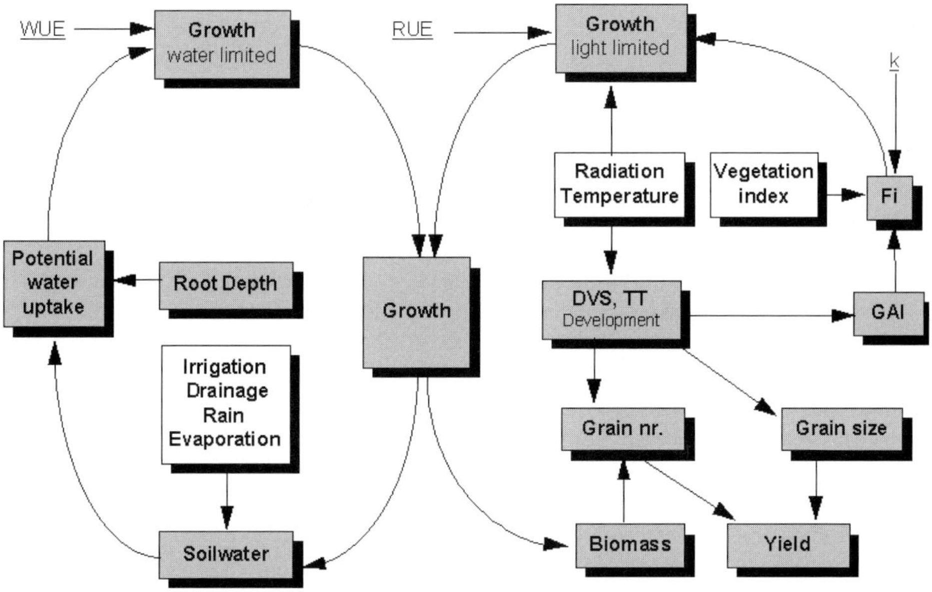

Fig. 5.3 Model used for cereal yield prediction (from Koeijer *et al.* 2000).

prediction system is provided by the 'sugar beet prediction and management system' (Aybet and Walpole 1994). Sugar beet is an economically important crop whose yield can fluctuate greatly due to certain factors, including climatic conditions, seasonal pests and diseases and crop husbandry practices. Sugar companies need accurate and timely predictions to make processing and marketing of sugar more efficient. The sugar beet prediction system utilises a conventional vector-based GIS, an image-processing system, a relational database and a numerical predictive model for yield. NDVI values derived from a multi-temporal series of SPOT and Landsat images are fed into the predictive model, along with experimentally determined sugar beet growing constants, and solar radiation derived from a standard UK Meteorological Office product. The final output is a database of estimated yield, in tonnes per hectare, for each factory area being surveyed.

5.2 FORESTRY

Forestry is also of great importance to the economic welfare of many nations and satellite images have provided valuable input to the regular and detailed forest inventories that are necessary to manage forests effectively. On a global scale, satellite data also has an important role in monitoring the effect that deforestation can have on the delicately balanced global model (this topic is described in more detail in Section 7.2.3).

One application of satellite imagery to forestry (Keil *et al.* 1992) was the use of

Landsat TM for forest classification and dead wood mapping in the largest woodland in central Europe, a conglomeration of the national park of the Bavarian forest in Germany and the national park of Sumava (the Czech Republic/Bohemia). Remote sensing can be used to monitor this environmentally sensitive region and to map 'areas of retreat' for animals and plants along the border region, which had been little mapped prior to the raising of the iron curtain.

Another forestry application was the use of SPOT images to assess storm damage in forests (Leysen *et al.* 1992). Storm damage is particularly important in a European context, for example in Flanders (Belgium) the equivalent of the annual harvest (700,000 m^3, or 1.7 per cent of the forested area in Flanders) was destroyed in just three days of storm damage in January and February 1990. Another regional forestry project (Brown and Fox 1992) using satellite data was a project to map timberland using Landsat TM in Klamath Province, California. This work was a consequence of a California state assembly bill to form a 'timberlands task force' to improve protection of wildlife resources. The aim is to develop rules to identify Wildlife Habitat Relationships (WHR) cover types over a 6 million acre study area. The classification scheme was based on a number of attributes such as vegetation classes, canopy closure and tree size classes.

Another example (Nezryt and Demargne 2001) is the use of SPOT and radar data to produce inventories of forests in Sarawak, Borneo. This enables timber companies to estimate timber reserves to ensure that they are managed sustainably. SPOT panchromatic and multi-spectral imagery was used in combination with Radarsat SAR images where cloud cover made acquisition of optical imagery impossible. Radarsat SAR stereo-pairs are also used to generate DEMs using SAR interferometry (see Section 3.3.2). In cloud-free areas a classification of forest biomass was carried out using multiple layers of SAR and SPOT multi-spectral imagery. In the cloudy areas the Radarsat interferogram was used as a substitute for the optical imagery. These two partial classifications were then merged to produce a map of forest types (see Fig. 5.4, colour section) identifying six classes of forestry:

- logging tracks, water courses and rice fields
- very-high-yield forests
- high- and medium-yield cutover
- re-entered cutover and secondary forest
- clearcuts and very-low-yield areas
- bare soils and crops.

The output product was calibrated using field evaluations of stand timber volumes. Owing to inaccessibility of the area, calibration was performed using only general mean reference values, which led to some inaccuracies in timber yield estimation. Nevertheless, a subsequent field trip showed that timber volumes were never overestimated by more than 20 per cent.

5.3 GEOLOGICAL AND CIVIL ENGINEERING APPLICATIONS

EO data can be used for a variety of geological and civil engineering applications; for example Optical/IR satellite images can be used to help detect and classify mineral resources directly from spectral signature (if the rock or soil is exposed), alternatively SAR may be used due to its ability to penetrate vegetation and surface layers and its sensitivity to surface roughness and moisture content. In this section we look at a small selection of these applications:

- the use of interferometric SAR for subsidence monitoring
- geological information extraction algorithms
- civil engineering applications.

5.3.1 Subsidence mapping

SAR interferometry is often used to detect surface changes in surface heights. Very small surface changes can be detected because radar equipment operates at extremely high frequencies, which correspond to short radio wavelengths. Thus a complete wave cycle will take place within a few centimetres (for example, 5.6 cm for ERS) and the measurement of the phase within this cycle provides a way to gauge the distance to a target with centimetre or even millimetre precision (Haynes 1999). As each 'fringe cycle' in the interferogram represents half a wavelength (2.8 cm in the ERS example), it is therefore possible to measure to fractions of a single fringe cycle.

The success of this approach does, however, rely on the reflective objects contributing to the signal remaining unchanged between image acquisitions. If changes such as leaves falling off trees, growing vegetation or vehicles moving occur, the 'random' component of phase will differ between the two images and lead to inaccuracies in the interferogram and phase unwrapping. Also, if the two images are taken from different angles the changes in geometry will introduce phase shifts. Because of this consideration, the two paths that the satellite should follow for successful interferometry must be less than 1 km apart. Fortunately, the ERS satellites, JERS-1 and Radarsat, usually comply with this requirement, although they were not designed with interferometry specifically in mind.

Figure 5.5 is an extract (Haynes 1999) from a much larger differential interferogram of an area in the north-east of England derived from two ERS images acquired 35 days apart. This shows a number of circular phase anomalies highlighted by rings, with surface displacements varying from 28 to 84 mm (i.e. one to three phase cycles). This was previously a very active area for coal mining and many of the features are known to be associated with mining activity. The figure also shows how a GIS can be used to perform spatial analyis of their proximity to major roads, railways and urban areas.

A second example from the same source (Haynes 1999) is the detection of urban subsidence in Las Vegas. Figure 5.6 (colour section) shows a colour-coded interferogram superimposed on an optical satellite image of Nevada. Groundwater abstraction on the outskirts of Las Vegas has caused subsidence of the order of 9 cm over 3 years.

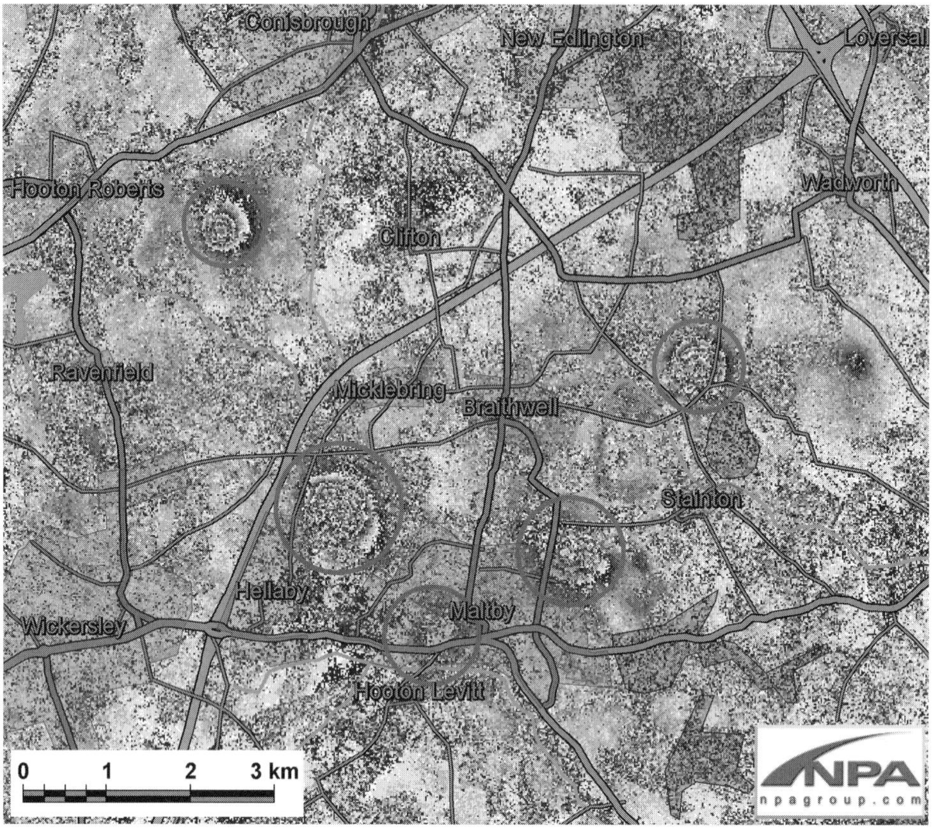

Fig. 5.5 Suspected ground displacement features (ringed) in differential interferogram extract of the north-east of England, overlaid with generalised road network. Ground displacements in the area are chiefly associated with coal mining, active at the time of radar image acquisitions. (Image copyright NPA Group 1999, ESA 1993. Mapping based upon OS Data, Crown copyright.)

5.3.2 Geological information extraction

As well as techniques of multi-spectral classification (described in Section 2.2) a number of other information extraction techniques have proved to be more specifically suited to geological applications. These include:

- Principal Components Transform
- geobotanical analysis
- band ratios
- the IHS transform
- lineament mapping.

Principal Components Transform
Principal Components Transform (PCT) has also proved of use for enhancing the value of images for geological analysis; for example, PCT was one of the techniques for drainage network analysis of Mt Olympus in Greece, using Landsat TM images (Astaras and Soulakellis 1992). In this case the first three principal components of a Landsat TM image from July 1987 were found to contain 98.6 per cent of the information (variance) of the set of six TM bands (excluding thermal band 6). Drainage lines proved to be more visible in the first principal component image (PC1) than in any of the original dataset.

Geobotany
Satellite images may also provide secondary clues such as soil moisture and vegetation changes. This latter approach, known as geobotany, is used where it is not possible to sense mineral deposits directly because of vegetation and makes use of the analysis of plant species, their distribution and health as a proxy indicator of the underlying mineral. Geobotanical anomalies that may indicate location of mineral deposits (Mather 1987) include:

- reduced ground cover or stunted growth
- unexpected species distribution
- yellowing or other pigmentation changes in leaves
- changes in the phenological cycle (such as late leafing or early senescence).

Band ratios
The motivation for using band ratios for geological purposes is similar to that for computing vegetation indexes (see Section 2.3.2); by comparing the relative value of two bands, the need for accurate absolute calibration and the need to account for the topographic effect become less paramount. Various ratios can be employed depending on the particular application; for example (Trefois and Volon 1992), a colour composite made up of Landsat TM band ratios 5/7, 5/1 and (5+3)/4 was used to emphasise hydroxyl and carbonate band absorptions in a lithologic mapping of Niger.

The IHS transform
The use of the IHS transform has already been described for visual enhancement of images. An example of the use of the IHS transform for geological mapping is a project to map hydrothermally altered rocks with Landsat data (US Geological Survey 1988). Areas of such rocks can indicate the presence of hydrothermal mineral deposits – for example, limonite anomalies can be associated with high levels of gold, silver, copper and other minerals. The Landsat sensor is particularly suitable for limonite mapping as its spectral range encompasses two characteristic limonite absorption features. The USGS project carried out in Colorado and Saudi Arabia successfully used satellite images as part of a multi-disciplinary metallic mineral assessment. The spectral difference in the images used were enhanced by image ratioing and by colour coding using an IHS transform. The results were then

subjected to further interpretation by geologists to distinguish hydrothermally altered limonite from that which may have been subject to other surface processes. The IHS transform was also used in a study of mineral exploration in the English Lake District (Forrest and Harding 1988); in this case Landsat TM data was used as a single 'band' in a multi-source GIS that also comprised digital maps of regional geochemistry and geophysics.

Lineament mapping

The mapping of lineaments is of importance geologically and a number of filters may be used to extract edges (points with rapid spatial change in grey-level), but while such methods are generally satisfactory for locating candidate edges, the knowledge-based approach described in Section 4.3.5 may be necessary to associate individual edge pixels with the linear features of interest. The Hough transform (see Section 2.3.3) is also useful for the detection of geological lineaments. It can, however, be used for any shape that can be parameterised, for example (Cross 1988) it can be used to detect circular geological features such as granite intrusions, salt domes, impact craters and volcanic features. In this case the standard equation for a circle is used:

$$(x-a)^2 + (y-b)^2 = r^2 \tag{5.1}$$

where a = x coordinate of circle centre
b = y coordinate of circle centre
r = circle radius

Because there are three parameters (a, b and r) a three-dimensional accumulator array is used.

5.3.3 Civil engineering

As we have seen in Section 5.3.1, detecting and mapping the extents and scale of subsidence activity in urban areas is of vital use to city planners and civil engineering projects. Some of the other opportunities for using EO data in civil engineering are shown in Table 5.1.

Prior to many heavy construction projects, such as road building or infrastructure development, a desk study is often invaluable. Earth Observation data can be visualised in conjunction with a DEM (which, as we have already have seen, can itself be EO-derived) as part of an Environmental Impact Assessment. Similar GIS approaches can be used to plan and visualise alternative routes and for the computation of project costs – for example, the collection of detailed geotechnical data such as soil and borehole data.

For more detailed site surveys land use maps can be derived from satellite images. The satellite data that is most suitable is usually that with the highest ground resolution which, nowadays, can be as good as 1 or 2 metres (see Section 8.1). Such mapping data can be used by planners and pipeline engineers for a number of applications, including:

Table 5.1 Uses of EO data products and services for civil engineering (based on Brucciani *et al.* 1997)

Civil engineering stage	Potential EO application
Gather data in order to select an approach to the project	Imagery for a 'first look' overview of the project
Identify the ground conditions, and develop framework models	Land use base maps (e.g. for catchment characterisation); use of digital elevation models
Select a method of analysis	Geocoded terrain characteristic/attribute data (e.g. bathymetry or landslide analysis products)
Assess/evaluate against historical data	Archived EO data and more specialised GIS processing such as spatial analysis
Apply professional judgement and experience in reports of findings	GIS presentations and three-dimensional visualisation

- correction of systematic mapping errors and filling for missing survey data
- location of changes requiring survey verification, such as differences in land parcel geometry
- extensions of surveys to new areas of interest, such as extensions of property boundary data.

EO data has also been used pre-operationally in a number of routeing and pipeline projects; for example, in Pakistan a substantial project cost is compensation paid to land owners based on the property's current use. By using panchromatic KVR images combined with existing cadastral data it was possible to demonstrate the utilility of EO data for reducing the costs associated with laying the pipeline. Note that the full pipeline routeing problem requires very accurate topographic data that cannot be provided from satellite data alone (Brucciani *et al.* 1997).

Finally EO data may be useful for generating one of the geological products described in Section 5.3.2 (for example, to map geological structure) or to provide an image backdrop onto which existing geological maps may be superimposed using a GIS.

5.4 TELECOMMUNICATIONS AND MEDIA

5.4.1 Telecommunications

There is already a niche EO market in producing Digital Elevation Models (DEMs) which are used to define transmission maps for cellular networks. As described in Chapter 3, accurate DEMs are unavailable for many regions of the world and deriving them from satellite data may be the only realistic option. Information from satellite images can also provide both a suitable classification system and sufficiently recent data for the 'clutter' information about land-use in the area.

The DEM and the clutter layers are used within a GIS by telecommunications companies to model distributions of signal strength given the location of a proposed transmitter site and parameters such as antenna position and height. The use of a model removes the need to actually visit each prospective site. This can then be used to optimise the coverage of cost, and the use of digital maps of urban areas and roads can ensure that populated regions are sufficiently well covered.

An example of this is shown in Fig. 5.7. This classification of the area around

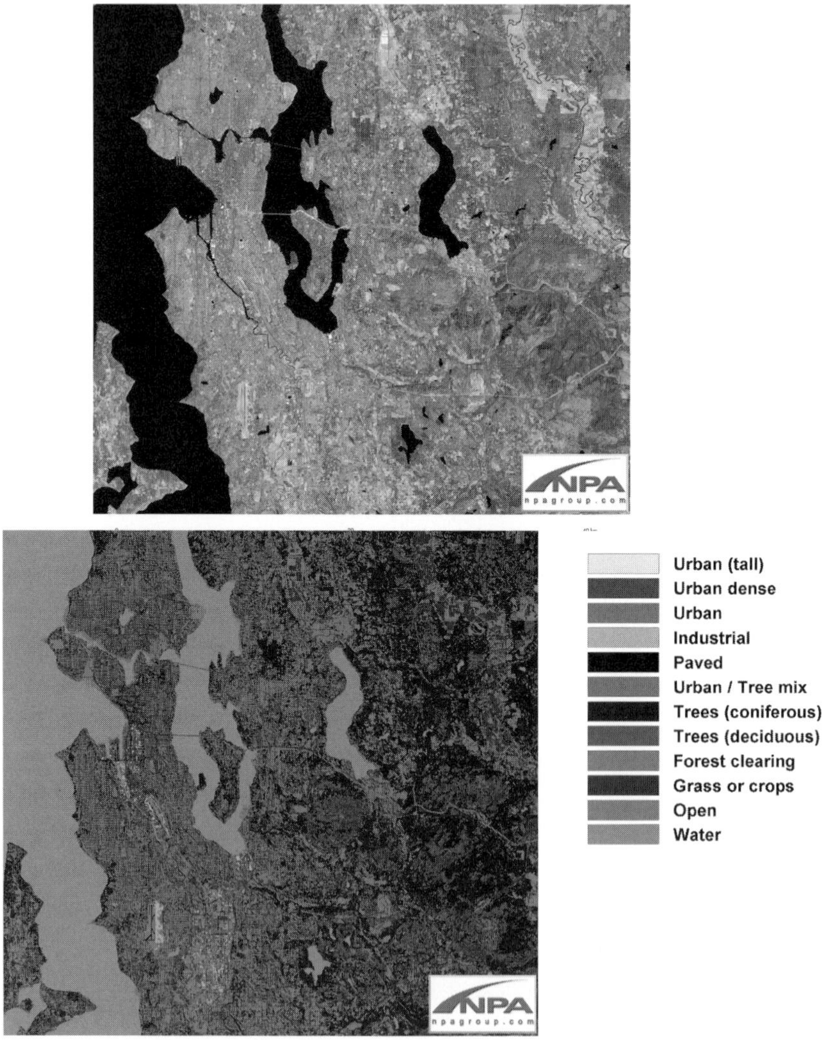

Fig. 5.7 Mobile telecoms use of EO Landsat TM image of Seattle (above); clutter map derived from image (below) (courtesy NPA Group).

Table 5.2 Earth observation contribution to macro-cell planning (based on Rosenholm *et al.* 1998)

Task	Earth observation contribution	Alternative source
Extraction of land cover data	Land cover classification from 20–30 metre multi-spectral images	Topographic maps, land cover maps/databases (e.g. CORINE)
Digital elevation models	Stereo satellite data 1–10 metre resolution	Digitised topographic maps, national or commerical height databases, airborne interferometric radar
Infrastructure information	Satellite orthoimages	Existing maps and road databases

Table 5.3 Earth observation contribution to micro-cell planning (based on Rosenholm *et al.* 1998)

Task	Earth observation contribution	Alternative source
3D city models	Extraction of 3D city models from stereo 1-metre Earth Observation data	City databases, airborne laser, stereo aerial photos, ad-hoc planning
Extraction of land cover data	Land-cover classification from 4–30 metre multispectral images	Topographic maps, city maps, colour aerial orthoimages
Digital elevation models	Stereo satellite data 1–3 metre resolution	Digitised topographic maps, national or commercial height databases
Infrastucture information	Satellite 1-metre orthoimages	Existing maps and road databases

Seattle, USA, together with elevation data, was used by the NPA Group to model the transmission of mobile phone signals. The propagation of microwave signals for communication is greatly affected by the absorption, scattering and reflection of the signal from vegetation and urban structures in addition to topography. Using such data the signal strength from a transmitter network can be modelled before it is actually established.

When planning cellular networks two categories of 'cells' are considered:

- macro-cell networks covering whole regions or counties
- micro-cell networks mainly confined to city centres.

Table 5.2 shows some of the areas where EO data can contribute to macro-cell planning, together with alternative sources of the information.

For micro-cell planning much finer resolution is needed. Until the advent of VHR imagery this was an area beyond the grasp of EO, but it is now a very large potential application particularly with the demands being made by the development of the forthcoming third generation of mobile telecoms. Table 5.3 shows some of the areas where EO could contribute to micro-cell planning.

5.4.2 News/media

With the advent of high-resolution satellite images, EO data is likely to become increasingly attractive to the print and broadcast media, both the traditional TV channels and the emerging Internet 'new media'.

Figure 5.8 shows one-metre resolution images of the United States Navy EP-3 aircraft, captured by China in 2001. These images, obtained from the Ikonos satellite, provided the world's media with a valuable source of information not obtainable elsewhere.

5.5 LAND APPLICATIONS OF GPS

Land applications of GPS-based navigation and positioning technologies are wide-ranging indeed, involving almost anything that requires positioning information. Among the main areas are:

- surveying and civil engineering
- location and management of assets for the utilities
- vehicle navigation
- intelligent transport systems
- emergency location systems
- defence and military applications
- recreational used for hikers and ramblers
- environmental mapping and monitoring
- field positioning devices for precision agriculture applications such as yield mapping and precision spraying.

The field of GPS applications is too large to be addressed in a comprehensive manner within this book. However, to give some flavour of the widespread use of GPS two representative areas, namely in-car navigation and asset management, are described here. The section is completed by a description of some of the synergistic uses of GPS and Earth Observation, and many advanced uses of GPS can be found in 'Geomatics', Chapter 8.

Fig. 5.8 Ikonos images of the US Navy EP-3 aircraft on Lingshui military airfield, Hainan Island in the south China Sea. Images acquired April 10, 2001 (above) and June 20, 2001 (below). (credit: spaceimaging.com).

5.5.1 Use of GPS for in-car navigation

In-car navigation systems are becoming increasingly available as standard equipment on a wide range of vehicles. The basic configuration for most systems is relatively straightforward. The GPS receiver can provide a near real-time location and velocity of the car, which can then be combined with digital mapping of the road network to provide clear indication of the route to the driver.

Once the basic configuration is available, a number of additional functions may be incorporated, including:

- vehicle tracking in the event of car theft
- linking to external traffic detectors – this is useful for planning optimal routes
- location-based advertising, for example, for filling stations and hotels (location-based services are described in more detail in Section 8.4.4)
- link to emergency services in the event of a breakdown.

The development of in-car navigation has driven a major effort to digitise road networks and make them available on-line. Figure 5.9 shows the availability of digital road maps in Europe as at April 2001. Road network maps omit a very large amount of contextual information and there is potential synergy between road network maps and context information provided by EO products

5.5.2 Use of GPS for asset management

The mapping and locating of fixed assets for government agencies (especially local government), utilities and infrastructure companies (electric, gas, oil, water, sewage, telecommunications, road and railway infrastructure) provide a large market for the use of hand-held/portable GPS receivers. These are used to capture locations of assets into a GIS or database. The other sector, tracking of movable assets, consists of a more varied range of applications such as tracking of freight containers, trailers etc., managing of intermodal terminal operations (e.g. sea or air ports), anti-theft devices on vehicles and other mobile assets, tracking of livestock, and tracking of weather balloons, etc.

Municipal government and utilities can use GPS to map and locate their assets, and enable them to manage the maintenance and planning of these assets. For fixed assets the main requirement is to be able to report their status – for example, condition of street lighting. The main of use of satellite-navigation is to enable field engineers to record their position when they are examining the asset. This enables information gathered in the field to be recorded directly in an asset management form. Key advantages of using a GIS with accurate locations of assets include:

- *Liability* – accurate records improve defence in lawsuits, and can reduce insurance premiums.
- *Safety* – more efficient maintenance programmes lead to safer communities.
- *Better information* – statistics of assets, for example, can aid the planning of coverage and networks of assets, decision-making and marketing activities.

There is an increasing trend for utilities and local government to equip their field engineers with mobile GIS equipment thus enabling direct entry of GPS data at the

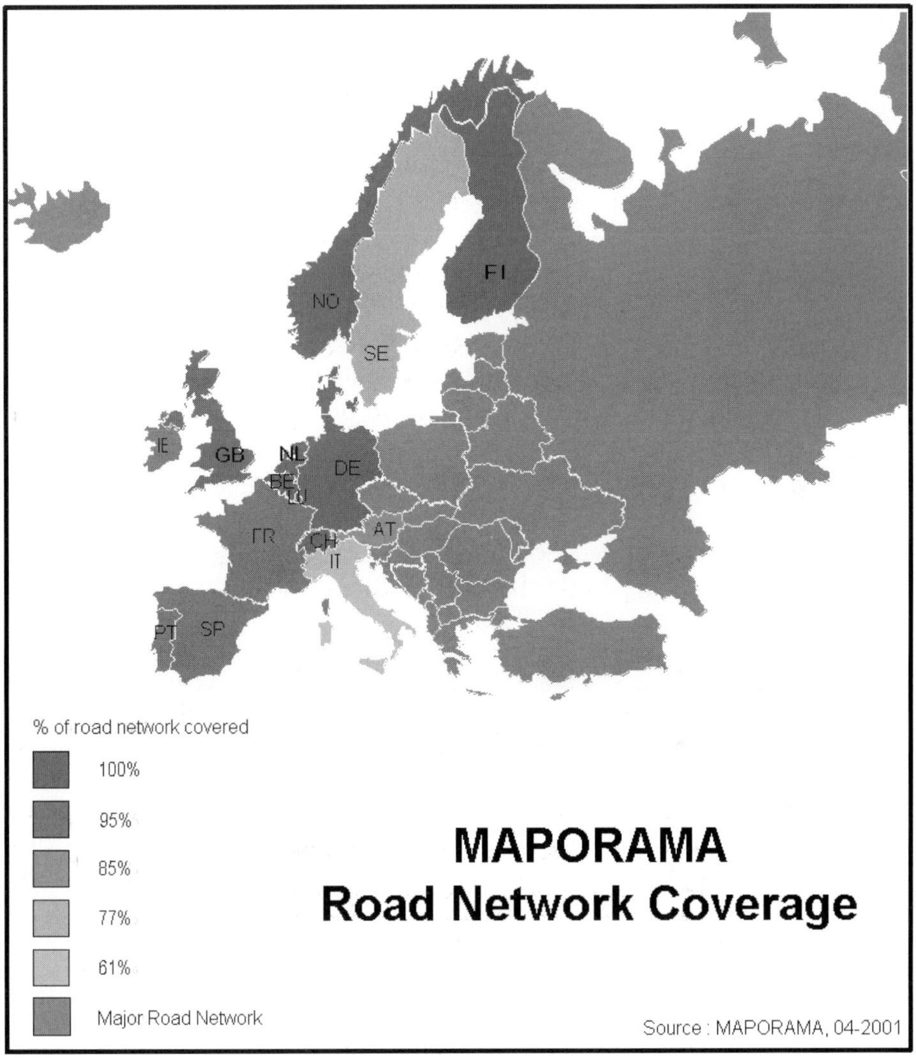

Fig. 5.9 On-line road network for the EU (courtesy Maporama).

point of entry. Examples of the sorts of assets that may be mapped by each type of organisation are shown in Table 5.4.

5.5.3 Integration with Earth Observation

Some early examples of projects that have been conducted using a combination of GPS and remote sensing (Hough 1992) that were described in the first edition of this book include:

Table 5.4 Categories of asset mappable by GPS

Electricity companies	Pylons Power lines and their surrounding electromagnetic fields. Power stations Hydroelectric, solar and wind power installations
Gas companies	Pipelines Pipeline infrastructure
Oil Companies	Pipelines, including crossings of key features such as cables, utilities, and rivers, pipeline markers Corrosion protection equipment Refineries Storage depots Oil rigs and drilling installations
Water companies	Reservoirs and wells Sewers Pipes Manholes/junctions Intakes Pump stations Control structures and valves
Telecommunications companies	Antennae/transmitters Telegraph poles Telephone lines Telephone boxes
Rail companies	Track Tunnels and bridges Hazards Signals Points Trains/wagons/carriages in sidings

- vegetation mapping of the Ituri rain forest in Zaire by the University of Utah
- production of image maps from Landsat TM for use by allied forces during the Iraqi occupation of Kuwait
- detection of gold deposits by the Geological Survey of Japan (jointly with the Research and Development Centre for Geotechnology of Indonesia) on the Indonesian island of Lomblen.

Since then use of GPS technology with EO imagery has mushroomed. GPS is also a basic 'fuel' for GIS which provides a solid baseline for extended EO processing. The synergistic use of GPS for land applications is described in several other sections of this book, for details see Table 5.5.

Table 5.5 Use of GPS for land applications: cross-references to other sections

Section	Application
3.4	DEM validation
4.1.4	Integrating satellite data in a GIS
4.4.3	Animation
7.2.4	Environmental applications of GPS
7.3.3	Earthquakes and landslides
7.4.4	Humanitarian operations and rehabilitation
8.2.1	High-precision GPS
8.2.2	The Galileo Programme
8.2.3	Personal info-mobility
8.4.2	Precision farming
8.4.3	Animal tracking
8.4.4	Location based services

REFERENCES

Aybet J and Walpole R (1994) An integrated GIS approach to the use of earth observation data – sugar beet prediction and management system. *Association for Geographic Information, AGI 94, Birmingham, UK*, 15–17 November 1994, pp. 15.1.1–15.1.6.

Astaras TA and Soulakellis NA (1992) Contribution of digital analysis techniques on Landsat 5 TM imageries for drainage network delineation. A case study from the Olympus mountain, W. Macedonia, Greece, Remote Sensing from Research to Operation. *Proceedings of the 18th Annual Conference of the Remote Sensing Society, University of Dundee, Scotland*, 15–17 September 1992, pp. 163–72.

Belward AS and Valenzuela CR (eds) (1991) *Remote Sensing and Geographical Information Systems for Resource Management in Developing Countries*. Kluwer Academic Publishers.

Brown GK and Fox L (1992) Digital classification of Landsat Thematic Mapper imagery for recognition of wildlife habitat characteristics. *Proceedings ASPRS/ ACSM Convention, American Society for Photogrammetry and Remote Sensing, Bethesda, Maryland*, pp. 251–60.

Brucciani PM, Mitchell DG and Capes R (1997) EO data utilisation: requirements of the civil engineering industry. *CEO Customer Segment Studies*, Final Report, Centre for Earth Observation (CEO), European Commission, Joint Research Centre, Ispra, Italy

Cornaert M-H and Maes J (1992) Land cover an essential component of the CORINE information system on the environment. GIS Implications. *Proceedings of the Central Symposium of the 'International Space Year' Conference, Munich, Germany*, 30 March – 4 April 1992, ESA ISY-1, pp. 473–81.

Cross AM (1988) Detection of circular geological features using the Hough transform. *International Journal of Remote Sensing*, **9** (9), 1519–28.

de Koeijer KJ, Steven MD, Colls JJ and Williams J (2000) Integration of earth observation data with a crop model for yield forecasting. *In Aspects of Applied Biology* **60**, *Remote Sensing in Agriculture*, pp. 91–8.

Denore BJ and Garcia Lopez MJ (2000) The use of earth observation for the assessment and prediction of agricultural yields: a review of crop yield estimation projects in *Aspects of Applied Biology* **60**, *Remote Sensing in Agriculture*, pp. 229–34.

Forrest MD and Harding AE (1988) Integration of Landsat TM, stream sediment geochemistry and regional geophysics for mineral exploration in the English Lake District. *Proceedings of the Sixth Thematic Conference on Remote Sensing for Exploration Geology: Applications, Technology, Economics,* 16–19 May 1988, *Houston, Texas*, pp. 469–74.

Hall FG, Bauer ME and Malila WA (1974) First results from the Crop Identification Technology Assessment for Remote Sensing (CITARS). *Proceedings of the Ninth International Symposium on Remote Sensing of Environment, Ann Arbor, Michigan*, pp. 1171–91.

Haynes M (1999) New developments in wide-area precision surveying from space. *Mapping Awareness*, November.

Hough H (1992) Satellite GPS and remote sensing. *GPS World*, February, pp. 18–24.

Keil M, Coenradie B, Rall H and Sima M (1992) The national parks of Bavarian Forest and Sumava – a Landsat TM perspective on the structure and dynamics of the largest woodland of central Europe. *Proceedings of the Central Symposium of the 'International Space Year' Conference, Munich, Germany*, 30 March – 4 April 1992, ESA ISY-1, July 1992, pp. 751–55.

Leysen MM, de Roover BP, Goossens RE and de Wulf RR (1992) Assessment of storm damage in forests using SPOT imagery. *Proceedings of the Central Symposium of the 'International Space Year' Conference, Munich, Germany*, 30 March – 4 April 1992, ESA ISY-1, pp. 773–7.

Mather PM (1987) *Computer Processing of Remotely-Sensed Imagery: An Introduction*. John Wiley & Sons, Chichester.

Meyer-Roux J (1993) *The MARS Project: Overview and Perspectives*. Publication EU 15599 EN.

Moran MS (2000) New imaging sensor technologies suitable for agricultural management. In *Aspects of Applied Biology* **60**, *Remote Sensing in Agriculture*, pp. 1–10.

Nezryt E and Demargne L (2001) Using Spot and radar data to inventory forests in Sarawak, SPOT Image web-site.
http://www.spotimage.fr/home/appli/forest/borneo/welcome.htm

Parker DS and Williams JM (1999) An assessment of the inclusion of the shortwave infrared band of SPOT 4 for detection of potatoes in East Anglia in 1998. *Proceedings of the 25th Annual Conference and Exhibition of the Remote Sensing Society, The University of Wales at Cardiff and Swansea* , 8–10 September 1999, pp. 773–80.

Radarsat (2001) Vietnam rice crop monitoring system. Project web-site.
http://www.rsi.ca/vietrice

Rosenholm D, Schumacher V and Mahlander C (1998) *Land Navigation/Digital Mapping Industry. CEO Segment Studies.* Final Report, Centre for Earth

Observation (CEO), European Commission, Joint Research Centre, ISPRA, Italy.
Trefois P and Volon C (1992) Lithologic mapping with Landsat Thematic Mapper guided by laboratory spectral measurements, Air, Niger (Africa). *Proceedings of the Central Symposium of the 'International Space Year' Conference, Munich, Germany*, 30 March – 4 April 1992, ESA ISY-1, pp. 565–9.
US Geological Survey (1988) *Mapping Hydrothermally Altered Rocks with Landsat Data.*
Wiegand CL, Gausman HW, Cuellar JA, Gerbermann AH and Richardson AJ (1973) Vegetation density as deduced from ERTS-1 MSS response. *Third Earth Resources Technical Satellite Symposium*, NASA SP-351, 1, pp. 93–116.
Williams J, Parker D, Harris R, Turner R and Baker D (2000) Area estimation of the British potato crop using earth observation and GIS. In *Aspects of Applied Biology* **60**, *Remote Sensing in Agriculture*, pp. 229–34.

Further reading

Drake N and White K (1991) Linear mixture modelling of Landsat Thematic Mapper data for mapping the distribution and abundance of gypsum in the Tunisian Southern Atlas. *Proceedings of Spatial Data 2000, Christ Church, Oxford*, 17–20 September 1991, pp. 168–77.
Quegan S, Churchill PN, Wright A, Lamont J, Rye AJ and Trevett JW (1985) SAR for agriculture and forestry. *Proceedings of a Workshop on Thematic Applications of SAR Data, ESRIN, Frascati*, 9–11 September 1985, ESA SP-257, pp.7–14.
Simonett DS, Strahler AH, Sun G and Wang Y (1987) Radar forest modelling: potentials, problems, approaches, models. *Proceedings of the Annual Conference of the Remote Sensing Society, Nottingham*, September 1987, pp. 256–70.
Taylor JC, Sannier C, Delincé J and Gallego FJ (1997) *Regional Crop Inventories in Europe Assisted by Remote Sensing*: 1988–1993. Published by the European Commission (EUR 17319 EN), 71 pp.
Wooding MG (1985) SAR image segmentation using digitized field boundaries for crop mapping and monitoring applications. *Microwave Remote Sensing Applied to Vegetation*. European Space Agency, SP-227, pp. 93–8.

6

Oceanographic, atmospheric and cryospheric applications

Chapter 6 shows some of the wide variety of oceanographic, atmospheric and cryospheric applications that can be addressed by the powerful combination of Earth Observation and GIS. The 'non-land' applications described here are:

- oceanographic applications: marine mapping, ocean colour, sea surface temperatures, ocean circulation, fishing and satellite altimetry
- meteorological applications and other atmospheric applications: chemistry, ocean–atmosphere interactions and ocean surface winds
- cryospheric applications in the 'frozen world'
- oceanographic, atmospheric and cryospheric applications of GPS.

Although some of the information extraction techniques described in Chapter 2, and use of GIS is still valid for these applications, there are a number of specialist missions, instruments and algorithms which are described here.

6.1 OCEANOGRAPHIC APPLICATIONS

The vast sizes of the Earth's oceans and their remoteness make satellite monitoring one of the most effective ways of measuring their state. This section looks at some of the areas where satellite data is used:

- marine mapping
- ocean colour
- sea surface temperature
- satellite altimetry
- fishing.

6.1.1 Marine mapping

It is of great importance to the world's naval and shipping authorities that they have access to accurate and frequently revised naval charts. The important features of such charts can be divided into two main classes: surface topographic features such as islands and coastlines and bathymetric features below the sea surface. For much of the world's sea areas such charts are unavailable, unrevised or were wrongly drawn up in the first place as the result of an inaccurate depth sounding. Thus many of the hazards to world shipping are not mapped in sufficient detail for modern needs. In ancient times, knowledge of the sea-bed was derived simply by use of rods and weighted lines and both methods were severely limited in locality and to relatively shallow areas. In the twentieth century echo-sounding equipment became widespread until it was superseded by the technically superior sonar systems now in common use. Even with such methods '... *it is still true that the world's largest vessels pass through waters last surveyed a hundred years ago or more*' (Benny and Dawson 1983).

As with many of the land-based problems described previously, remote sensing can play a major contribution to the topographic mapping, and bathymetry from satellite (although limited to shallow water applications) can be used to provide valuable input to nautical charts. Optical imagery has great advantages in three areas of nautical chart preparation:

- augmenting and updating data which already exists for an area ('putting new life into old surveys')
- as a cost-effective aid to planning hydrographic surveys
- monitoring mobile areas of sea-bed which are not currently charted due to their dynamic nature.

Shallow-water bathymetry can be carried out by considering the various routes by which light can reach the satellite sensor (see Fig. 6.1):

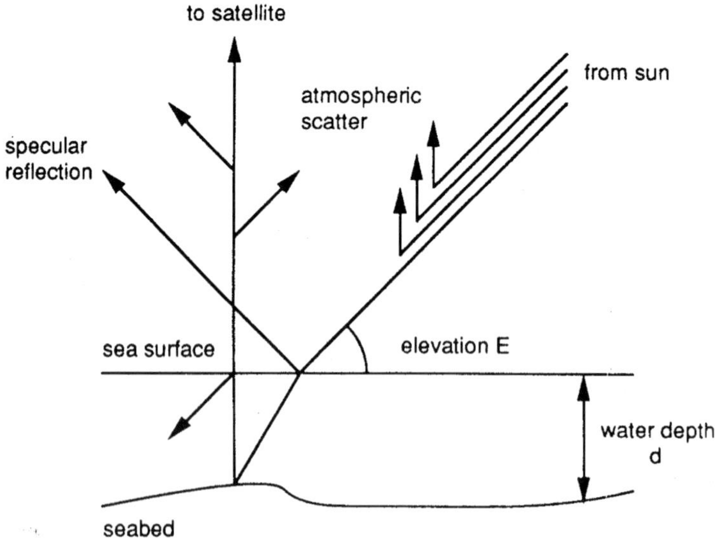

Fig. 6.1 Basis of shallow-water bathymetry.

a b

Fig. 1.23 'Cartoon' image produced by simulated annealing: (a) raw SAR image; (b) processed SAR image. (Courtesy NA Software, DERA Malvern.)

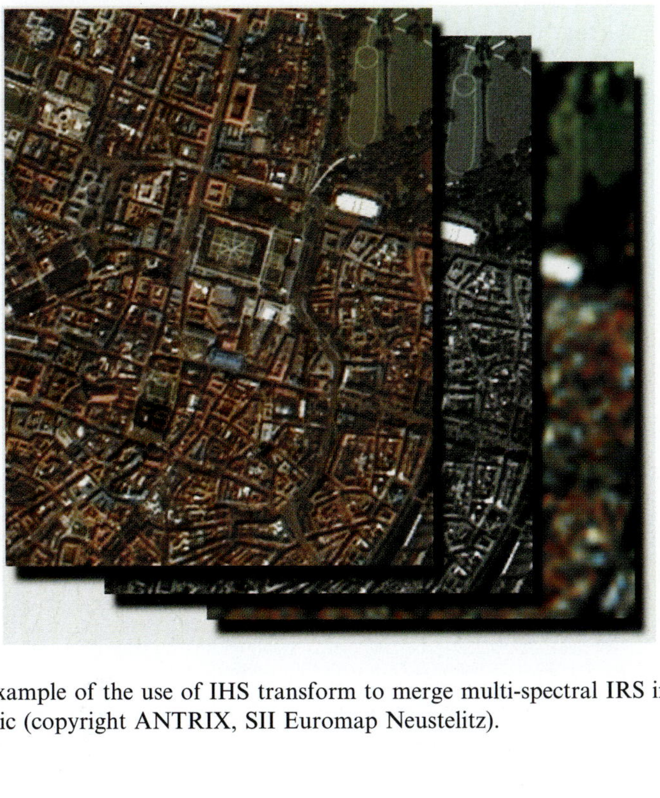

Fig. 1.26 Example of the use of IHS transform to merge multi-spectral IRS imagery with panchromatic (copyright ANTRIX, SII Euromap Neustelitz).

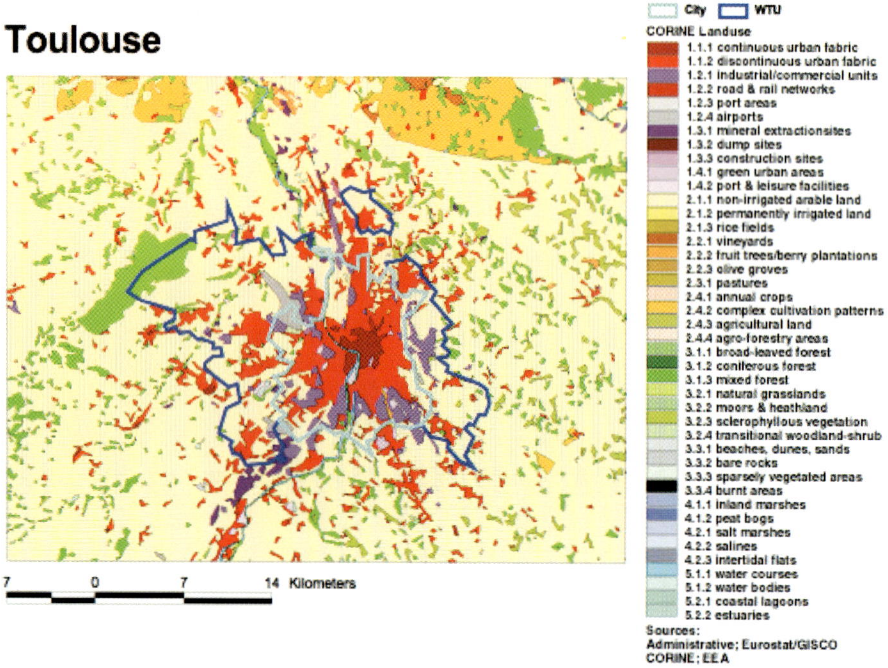

Fig. 2.13 Land cover classification of Toulouse, France, based on classification of SPOT images (courtesy Scot Conseil).

Fig. 3.9 Radar interferogram of Mount Etna (courtesy German Aerospace Research Establishment (DLR), Remote Sensing Data Centre (DFD)).

Fig. 4.16 A description of the 'Towne of Mannados or New Amsterdam' as it was in September 1661 (the 'Duke's plan' of New York, 1664; copyright 1997, British Library Board CXXI. 35) compared with a twenty-first-century image map of New York (courtesy Space Imaging).

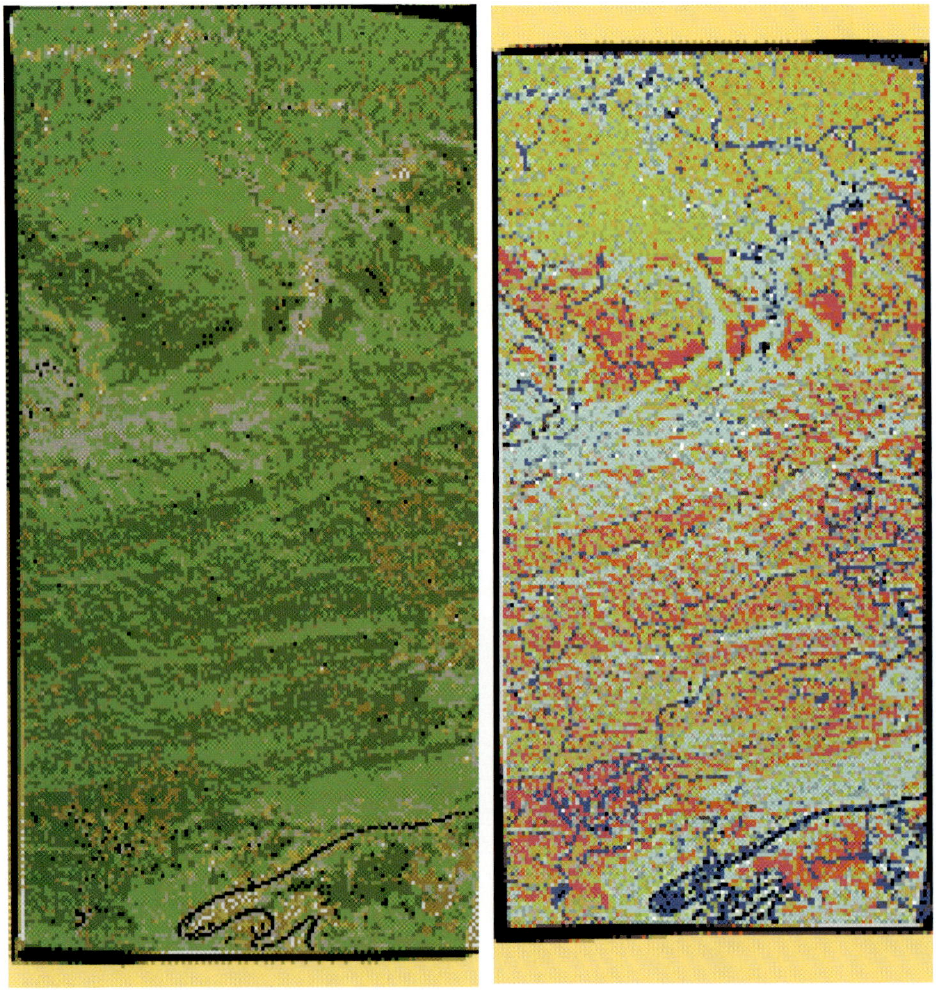

a b

Fig. 5.4 Combined SAR and SPOT forestry map of Sarawak, Borneo (courtesy SPOT Image and Radarsat International). (a) Six classes of forestry are shown logging tracks, water courses and rice fields (blue); very-high-yield forests (dark green); high and medium-yield cutover (medium green); re-entered cutover and secondary forest (light green); clearcuts and very-low-yield areas (grey); bare soils and crops (orange) (b) Timber volume maps.

Fig. 5.6 Colour-coded extract of differential interferogram over Las Vegas, Nevada, superimposed on a merged Landsat TM/SPOT PAN scene of the region. Fringes show subsidence due to groundwater abstraction amounting to 9 cm over 3 years: 6 April 1993 – 18 April 1996. (Image copyright NPA Group 1999, CNES 1993, ESA 1993/96.)

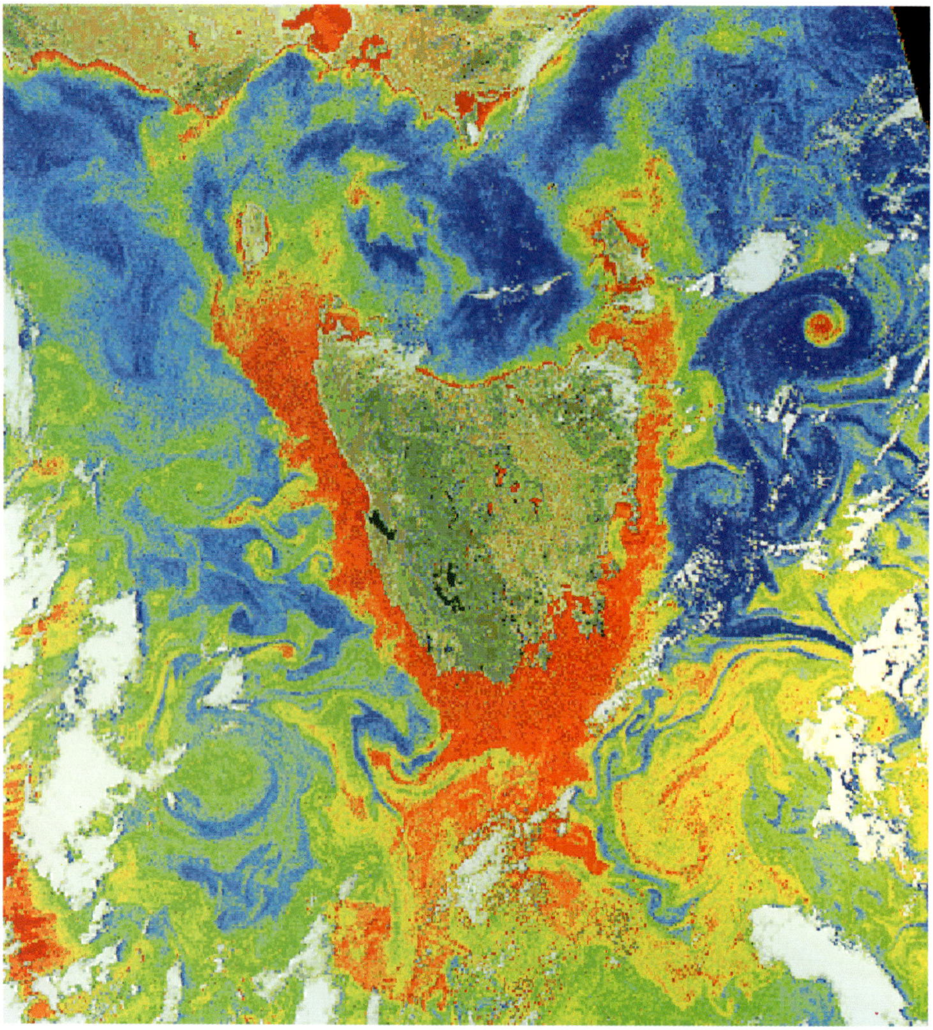

Fig. 6.2 CZCS image of Tasmania and surrounding waters, obtained on 27 November 1981. The southern coast of Australia is at the top of the image, separated from Tasmania by the Bass Strait. Yellow and red areas indicate more phytoplankton, and greens and blues less. Dark blue and purple indicate very low concentrations of phytoplankton in very clear ocean water. (Courtesy NASA.)

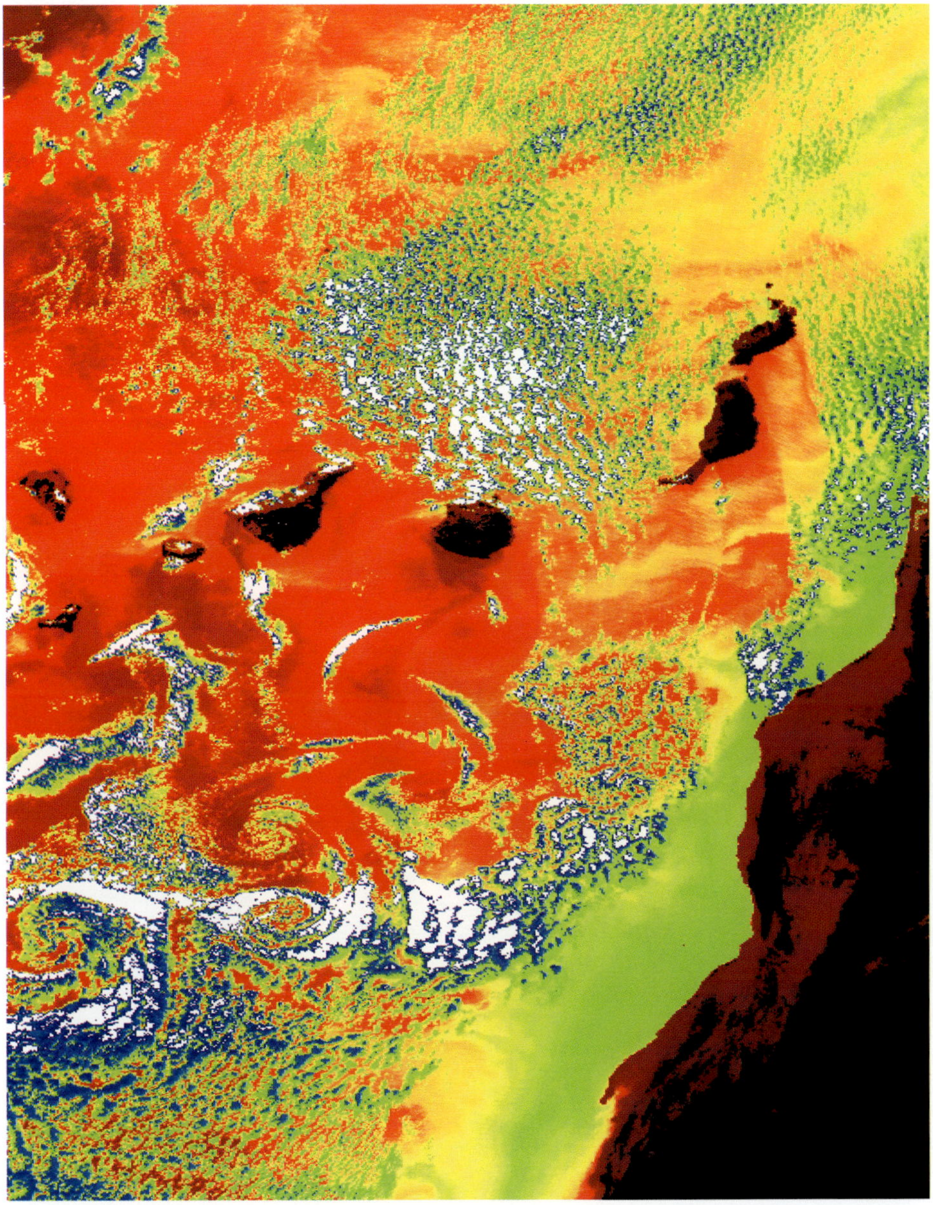

Fig. 6.3 ATSR image showing day-time sea surface temperature in the Canary Islands. The islands are shown in black and the warmest sea is red. The coast of Africa is in the bottom right. The hottest areas (shown in black) are land, and the coolest areas (shown in white and blue) are mostly cloud. The sea surface temperatures are shown using green (coolest) to red (warmest). The range of temperatures shown in the image is approximately 10 to 42 °C. (Courtesy Rutherford Appleton Laboratory.)

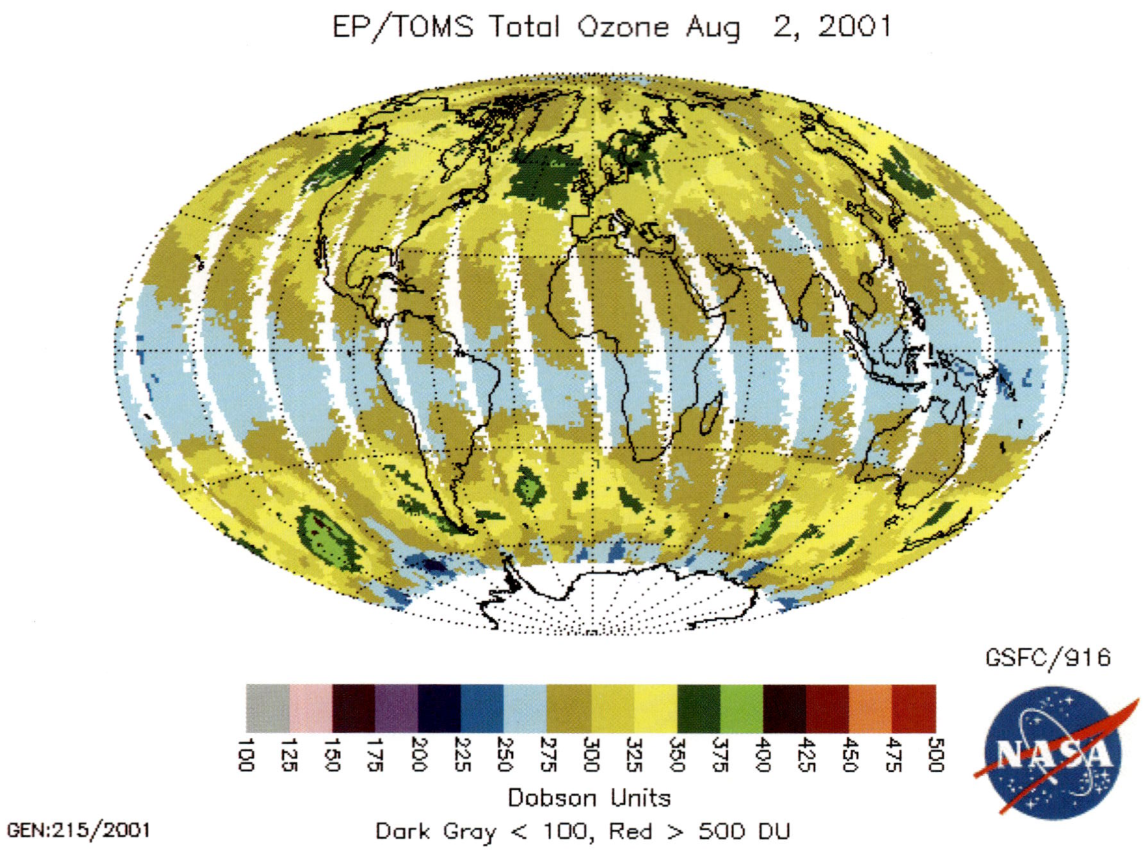

Fig. 6.7 TOMS image of ozone concentration, 2 August 2001 (courtesy NASA)

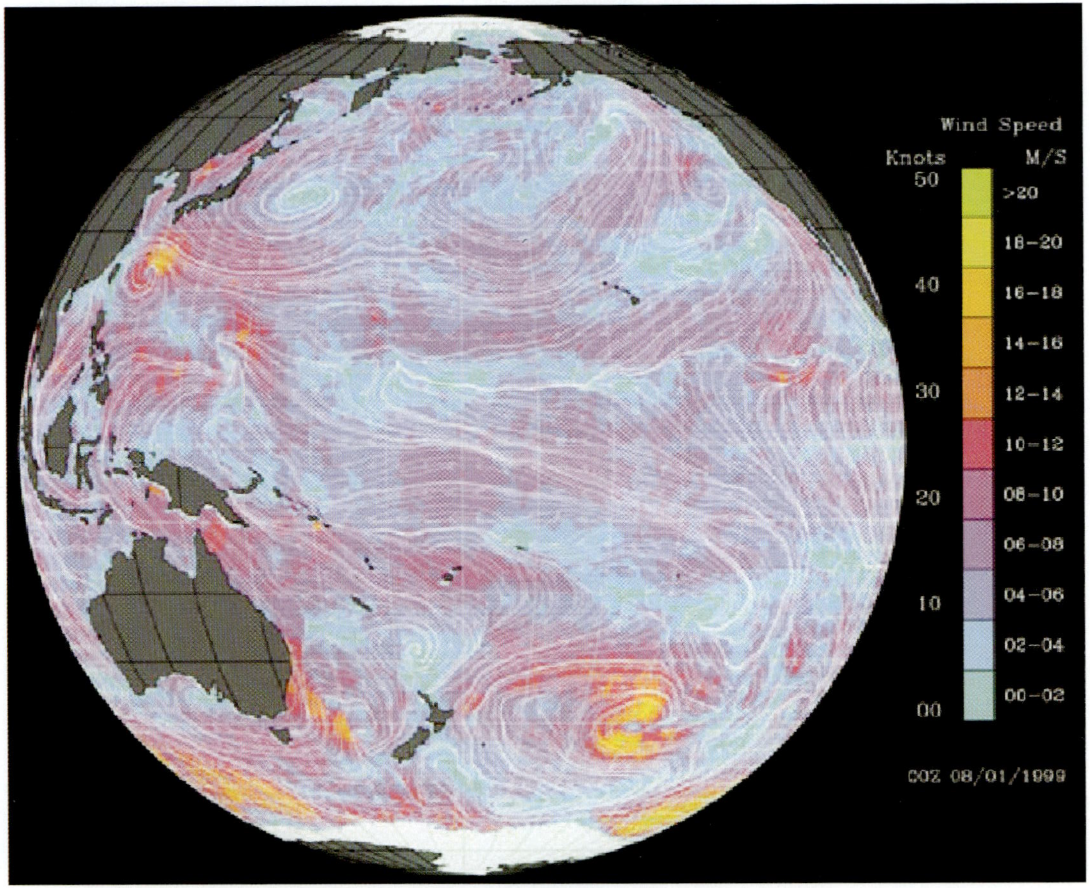

Fig. 6.11 Typhoon Olga image by Liu, Tang and Xie (NASA/JPL).

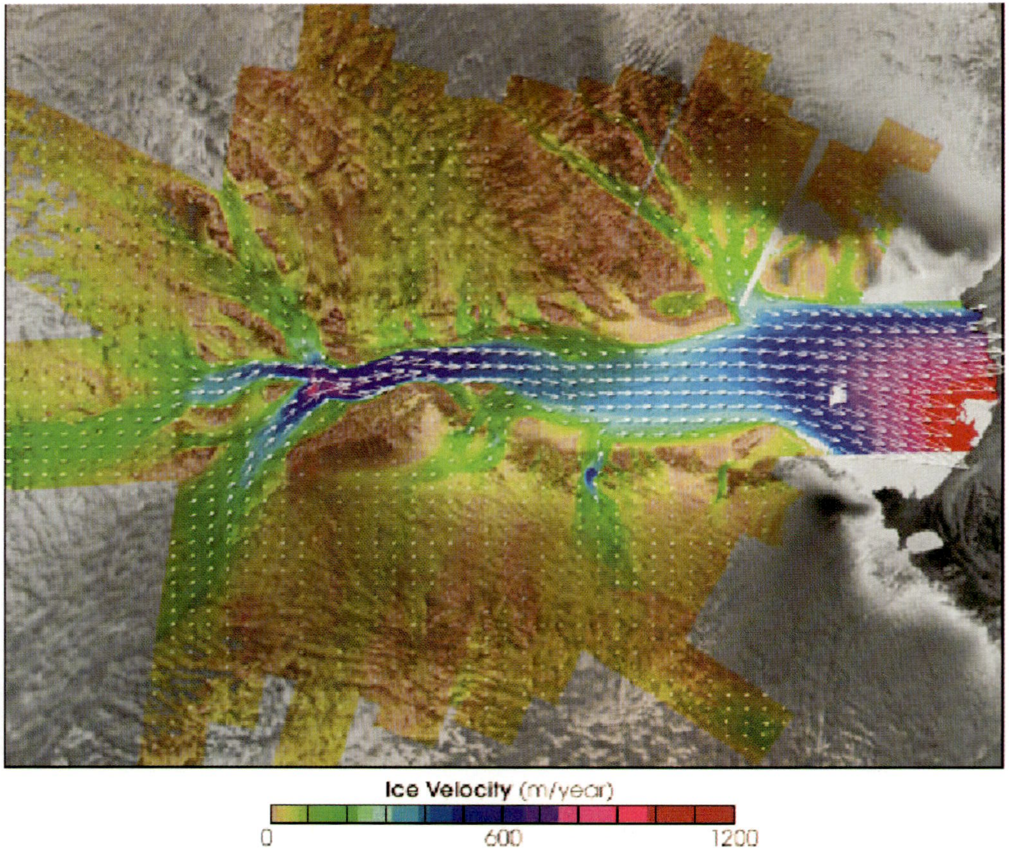

Fig. 6.15 Radar tracking of the Lambert Glacier, yellow represents the areas of no motion, which are either exposed land or stationary ice. The smaller confluent glaciers have generally low velocities, shown in green, of 100–300 metres per year, which gradually increase as they flow down the rapidly changing continental slope into the upper reaches of the faster flowing Lambert Glacier. (Image courtesy Canadian Space Agency/NASA/Ohio State University, Jet Propulsion Laboratory, Alaska SAR facility.)

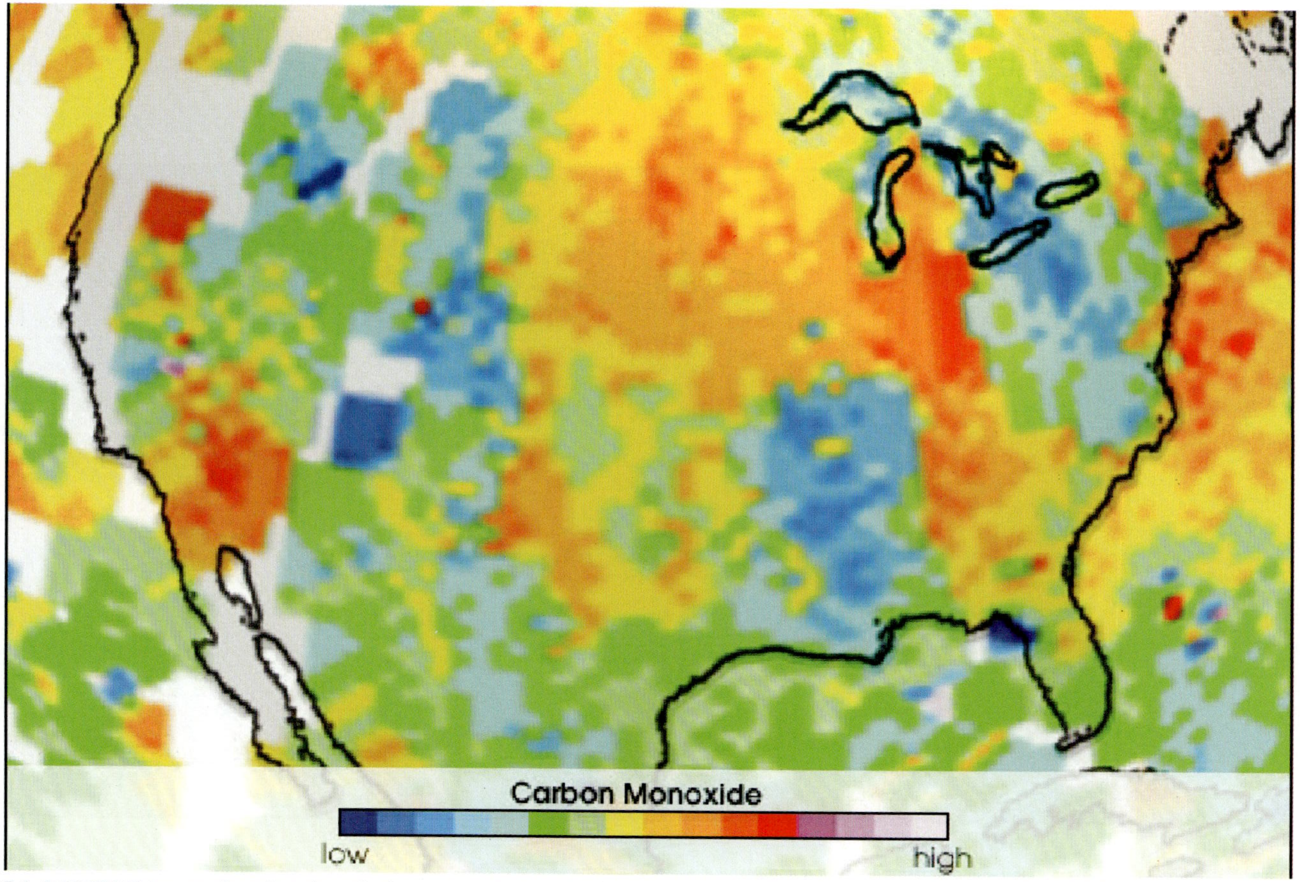

Fig. 7.1 MOPITT image showing the relative amount of carbon monoxide over North America between 5 and 7 March 2000 (courtesy NASA).

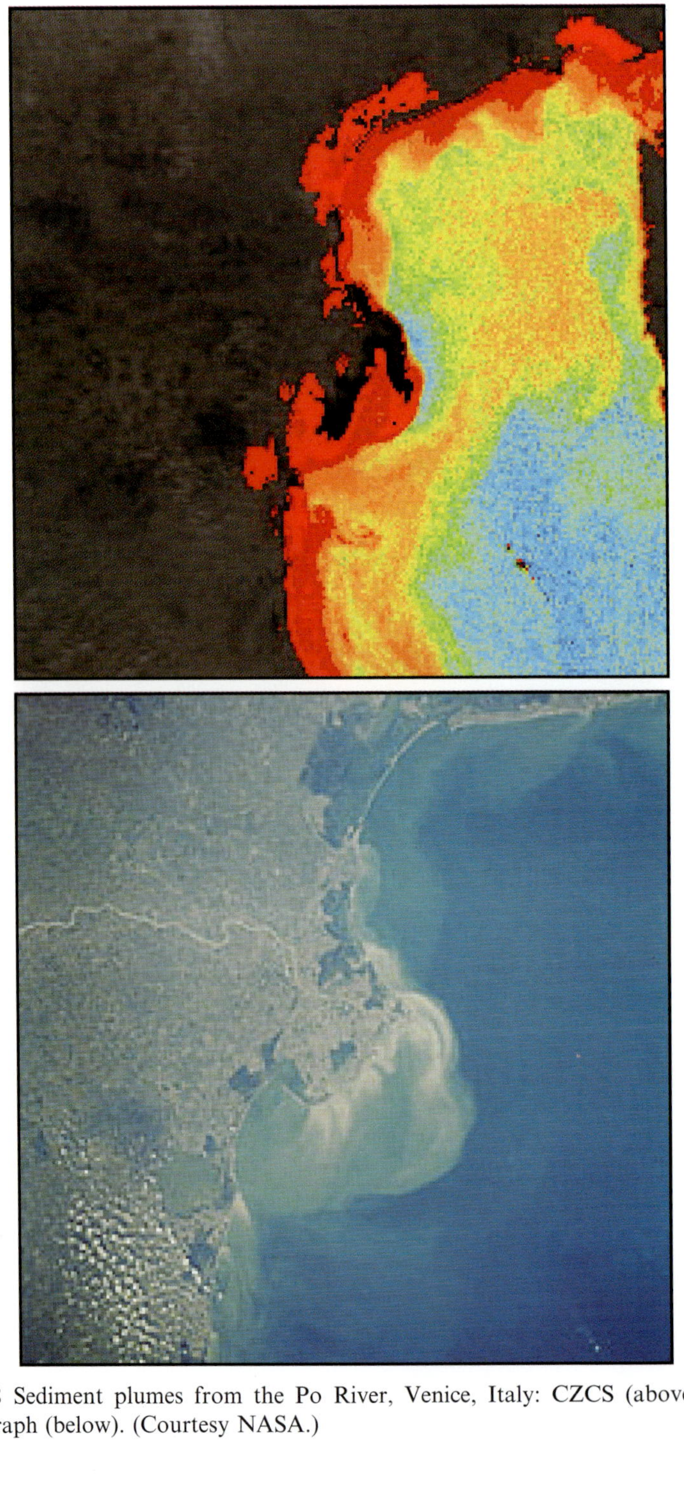

Fig. 7.8 Sediment plumes from the Po River, Venice, Italy: CZCS (above); Shuttle photograph (below). (Courtesy NASA.)

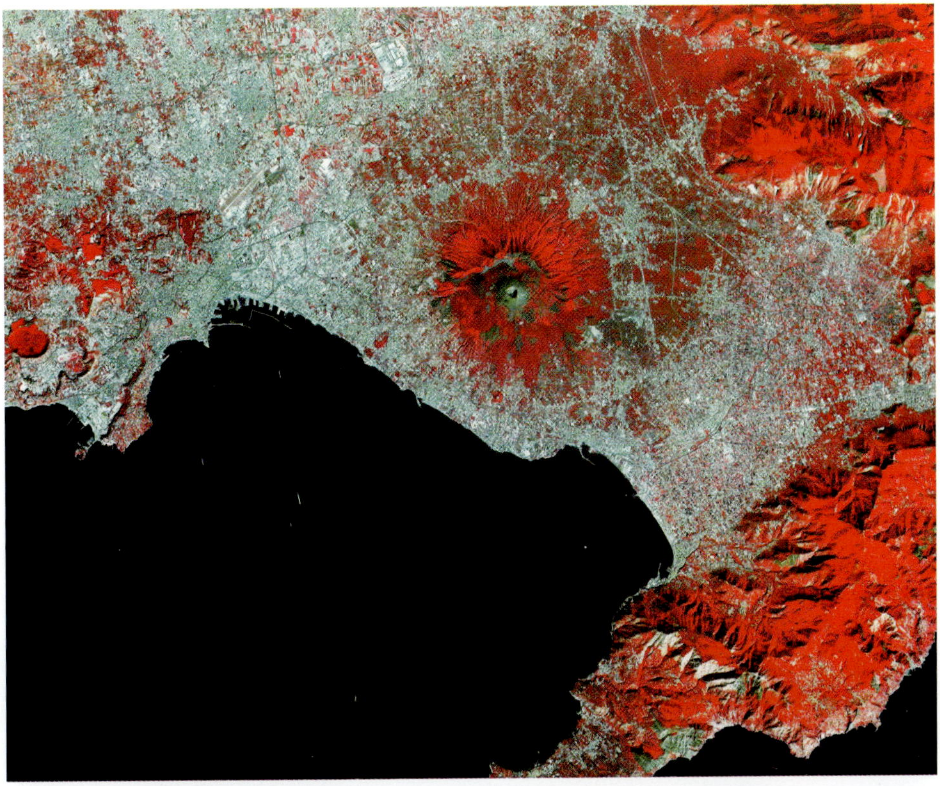

Fig. 7.13 Advanced Spaceborne Thermal Emission and Reflection Radiometer (ASTER) image of Mt Vesuvius, Italy, acquired 26 September 2000 (image courtesy NASA/GSFC/MITI/ERSDAC/JAROS, and US/Japan ASTER Science Team).

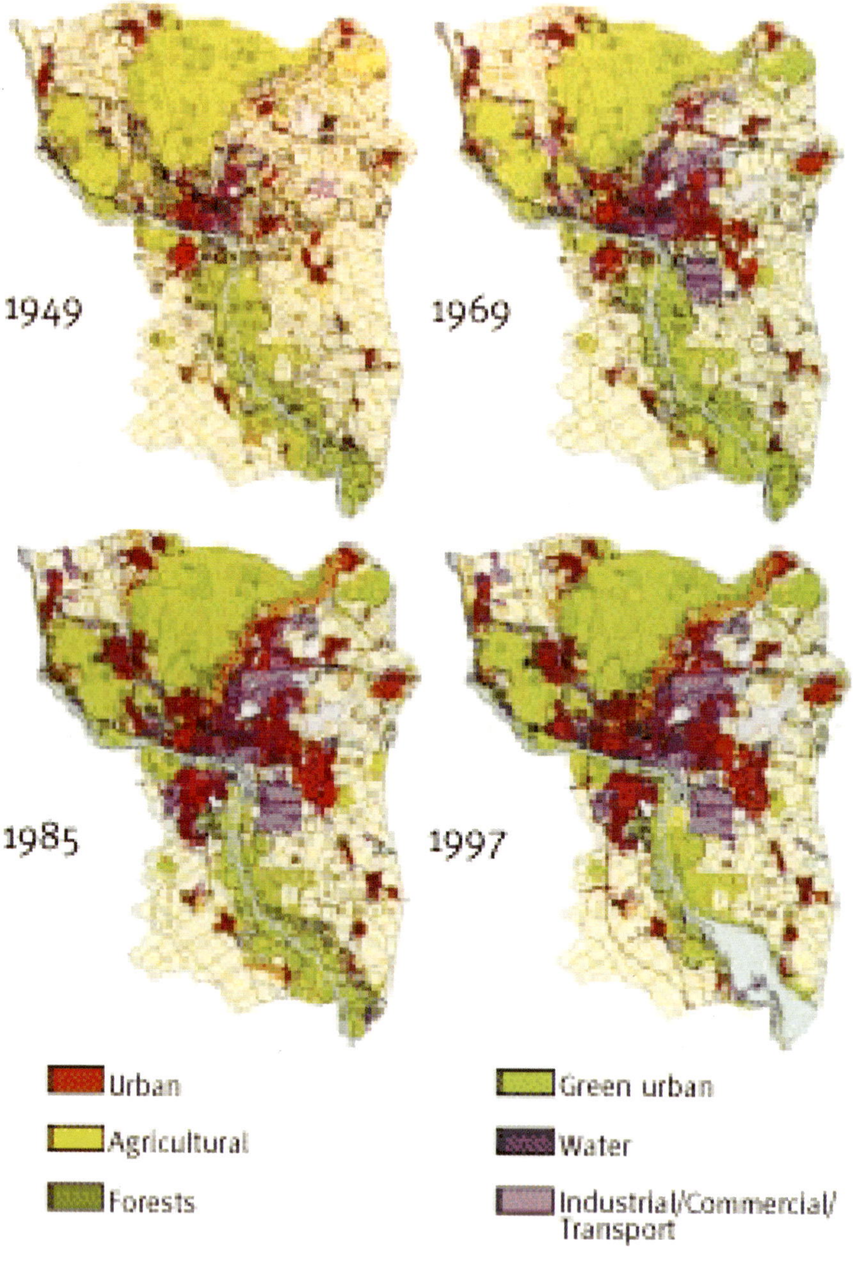

Fig. 7.17 The evolution of land use for an area for the city of Bratislava and surroundings of about 500 km^2. The four maps illustrate the land use for years 1949, 1969, 1985 and 1997. The maps are reconstructed making use of satellite imagery and of aerial photographs. (Courtesy European Commission, GMES.)

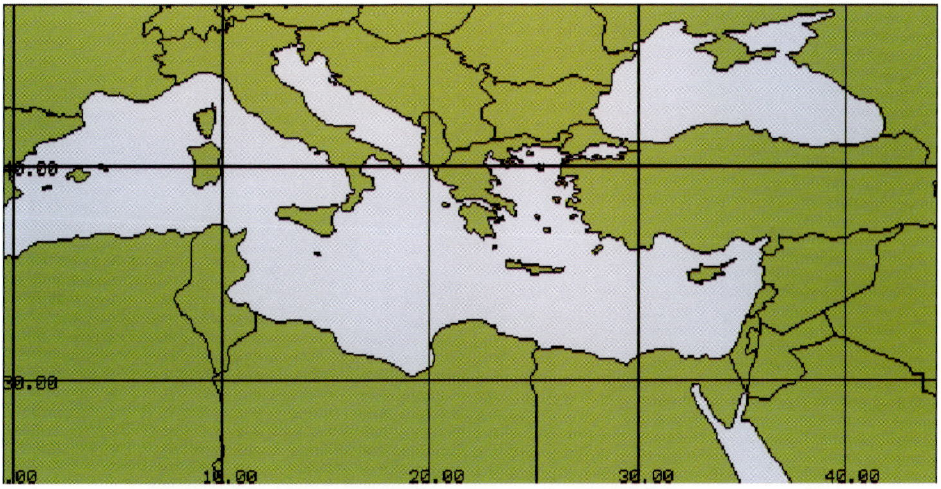

Fig. 8.7 Map-based interface to an ATSR browse catalogue (courtesy UK Natural Environment Research Council).

Fig. 8.9 Use of ARGOS for environmental data collection (copyright Service Argos Inc.).

Fig. 8.10 Big A variable rate spreader (Courtesy J & H Bunn Ltd.).

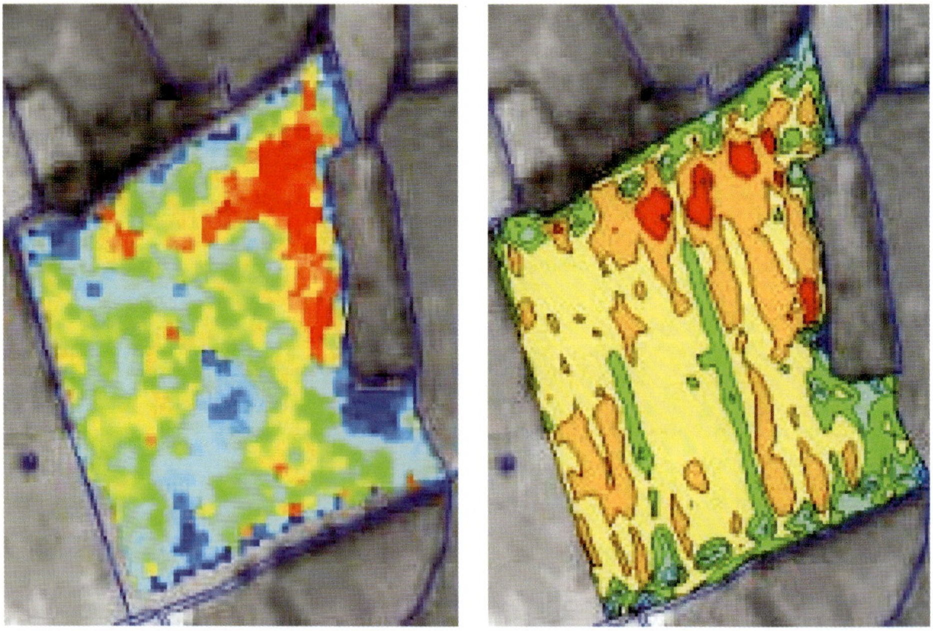

Fig. 8.11 Precision farming GIS products: (left) SPOT image enhanced to show crop vigour; (right) yield map. (Courtesy Galaxy Precision Agriculture Ltd, copyright CNES 1995, SPOT Image.)

- reflection from surface
- reflection from bottom
- reflection or scattering from particles or objects within the water.

Mirror-like (specular) reflection is also possible but only if the sea surface is almost unbroken. If the water is clear and of shallow depth the light can be returned and detected by the satellite. Light return depends on the depth of the water, the attenuation coefficient for that wavelength and the reflection coefficient of the bottom. Attenuation is characteristically low in visible bands and high in the infrared, and in general can be expressed by:

$$I = I_0 \exp(-dK) \qquad (6.1)$$

where I = amount of light remaining after attenuation by water
I_0 = strength of incident light
d = depth in metres
K = attenuation coefficient

This is the method that was used (Benny and Dawson 1983) to map a region between Ras Wadi Tiryam and Ras al Fasma in the northern Red Sea, using Landsat MSS images. The processing approach adopted was to calculate the depth maps from raw ungeocoded data and to transform the final depth contours to a map projection. There are, however, limitations to the utility of satellite images for bathymetric mapping. These are primarily:

- depth limitations (typically 20-m in clear water)
- poor spectral and spatial discrimination (for navigation purposes it is vital to be able to locate pinnacle features, no matter how narrow they are)
- inaccuracies in depth measurements.

6.1.2 Ocean colour

The launch of the Coastal Zone Colour Scanner (CZCS) onboard Nimbus 7 provided the first opportunity to make observations of ocean colour from space. Living plant cells, called phytoplankton, contain chlorophyll, the green pigment familiar in terrestrial plants. By measuring the relative proportions of reflected and absorbed sunlight, CZCS can provide accurate calculations of phytoplankton concentration. Such data helps scientists to understand the evolution of marine ecosystems and their role in El Niño and La Niña, and the role of oceanic photosynthesis and primary productivity in the Earth's carbon budget and climate (NASA 2001). Figure. 6.2 (colour section) shows a false colour image of Tasmania from CZCS, showing concentrations of phytoplankton.

As can be seen from the figure, the distribution patterns of plankton are very complex. Together with the fact that the locations are continually changing, this makes their mapping by satellite imagery far more comprehensive than oceanographic surveys from stationary ships. Such surveys rarely produce an accurate picture of the spatial distribution of phytoplankton, nor how rapidly their concentration in a region may be growing. The primary source of ocean colour data currently is the Sea-viewing Wide Field-of-view Sensor (SeaWiFs) which was launched in 1997 as a component of the NASA Earth Observing System (see Section 7.1.1) and collects global data every two days.

6.1.3 Sea surface temperature

Sea surface temperature calculations require an instrument that can measure radiation in the thermal infrared portion of the spectrum. An example of this is the Along-track Scanning Radiometer (ATSR) flown on ERS-1 and ERS-2. The ATSR measures sea surface temperatures using three infrared channels; in addition, by using two different viewing angles, it can compensate for the effects of the atmosphere on the signal received and hence produce more accurate temperature estimates. Although the ATSR is ineffective in the presence of cloud it does provide global coverage and is particularly valuable for the long-term monitoring of SST. A typical ATSR image of the Canary Islands is shown in Fig. 6.3 (colour section). A follow-up instrument, AATSR, is scheduled for launch onboard the European Envisat mission (see Section 7.1.3).

One example of the application of SST data is the NOAA CoastWatch programme (NOAA 2001) which makes satellite and associated readings from environmental buoys available to scientists and other users. For coastal areas in the Great Lakes, East Coast and Gulf of Mexico, data from the Advanced Very High Resolution Radiometer (AVHRR) are collected at Wallops Island, Virginia, and at Fairbanks, Alaska. The data is then processed using multi-channel algorithms to correct for atmospheric effects and to determine SST. The products are then converted to a GIS format (Mercator projection) and distributed daily to CoastWatch *regional* nodes in the south-east and north-east United States, Great Lakes, Gulf of Mexico and Caribbean. The end-users apply the SST data to a number of projects ranging from Gulf Stream data for offshore sailing, through tracking of sea turtles, to estimating survival times for hypothermia victims at sea.

6.1.4 Satellite altimetry

The radar altimeter is an instrument on a satellite that measures the distance from the satellite to the Earth's surface using a radar signal. The instrument transmits an electronic pulse in the microwave frequency to the Earth's surface. The microwave pulse reflects off the surface and returns to the sensor. Altitude is determined from the pulse travel time (from transmit to receive) and from the waveform of the returned pulse.

Highlights of a few of the major altimeter missions are:

- *Seasat* was the first Earth-orbiting satellite designed for remote sensing of the Earth's oceans and also featured the first spaceborne synthetic aperture radar (SAR). Seasat was launched on 28 June 1978 and operated for 105 days until 10 October 1978, when a failure in the satellite electrical system ended the mission. The Seasat radar altimeter was intended to measure spacecraft height above the ocean surface and help to lay the foundation for satellite altimetry as we know it today.

- The US Navy *GEOSAT* mission ran for 5 years and demonstrated the ability of the radar altimeter to measure the dynamic topography of the Western Boundary currents and their associated rings and eddies, to provide sea surface height data for assimilation into numerical models, and to map the progression of El Niño in the equatorial Pacific.
- *TOPEX/Poseidon* is a joint US/France altimeter mission to monitor global ocean circulation, discover the tie between the oceans and atmosphere, and improve global climate predictions.
- The successes of the ERS SAR are described in various sections of this book, but the *ERS Radar altimeter* also played a major role in the science of satellite altimetry, especially for oceanographic applications and monitoring of the Antarctic Ice Sheet.
- Launched in February 1998, and passing into operational service in November 2000, the *GEOSAT Follow-On* (GFO) programme is an operational series of radar altimeter satellites to maintain continuous ocean observation from the GEOSAT Exact Repeat Orbit.
- At the time of writing (April 2001) the *Jason 1* mission was scheduled to be launched to build on the achievements of TOPEX/Poseidon and also to monitor events such as El Niño conditions and ocean eddies.

Various algorithms are applied to the raw waveform product to account for the effect on signal propogation of the Earth's atmosphere. These include:

- sunspot and electron content information to correct for delays in the ionosphere
- integrated water vapour information to correct for delays in the wet troposphere
- total liquid water content to correct for attenuation and range errors when the signal passes through liquid water.

Satellite altimetry can play a major part in oceanographic applications, for example the British company Satellite Observation Systems have produced two data archives from radar altimetry: GRAVSAT, which calculates gravity anomalies over the oceans; and WAVSAT, which provides a significant archive of wave heights including the calculation of 50- and 100-year return values (that is, the wave height and wind speed that one would expect to occur once every 50 or 100 years). Figure 6.4 shows an example of the WAVSAT product.

6.1.5 Fishing

Fisheries are a natural for GIS, as it is of prime importance to 'know where the fish are'. In a recent study of GIS in Fisheries Management it is concluded that:

> '*Nearly all problems in fisheries are caused by the fact that different facets that affect fish populations are in disequilibrium ... Nearly all of this disequilibrium can be manifest as disparities in the spatial domain.*' (Meaden 2000)

Examples of such 'disparities in the spatial domain' include:

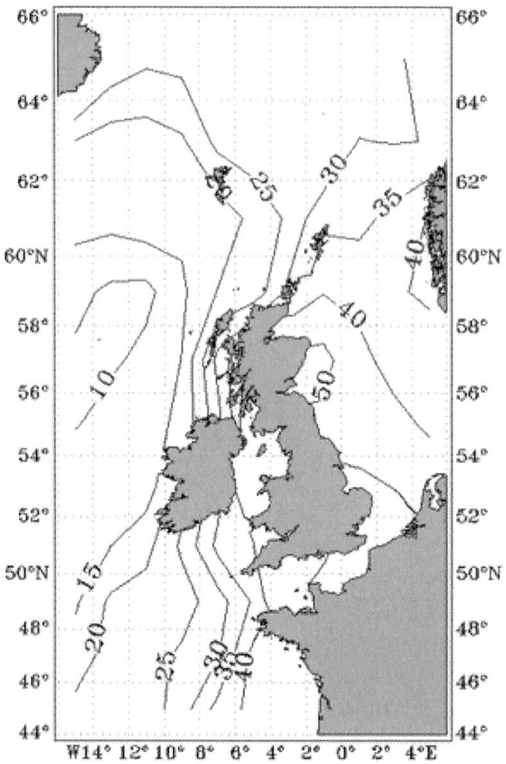

Fig. 6.4 The WAVSAT Product (courtesy Satellite Observation Systems Ltd).

- destruction of ecosystems by overfishing
- widespread marine pollution
- fishing effort being too localised
- global warming and its effect on the distribution of the marine community.

All of these problems and many related ones in fishery can potentially be addressed by GIS. For example (Meaden 2000), fish are often to be found along ocean fronts which can be detected using radar altimetry (see Section 6.1.4). In a collaborative venture, the Colorado Center for Astrodynamics Research, NASA and Scientific Fishery Systems, Inc. incorporate the altimeter products with bathymetry, SST and geo-referenced catch statistics within a GIS. Fishing personnel can also use the GIS to include tides and currents, locations of the catch, meteorological conditions, etc.

Another example of the spatial approach (Leming *et al.* 1999) is a near-real time GIS for fisheries management in the Gulf of Mexico. The satellite data layers used are SST images derived from NOAA AVHRR data and chlorophyll images derived from the SeaWiFS sensor. These are combined with digital maps of coastlines, international boundaries and bathymetric data. The GIS is available over the Internet via a Java interface.

6.2 METEOROLOGICAL AND ATMOSPHERIC APPLICATIONS

The United States launched the world's first weather satellite on 1 April 1960. Shortly afterwards meteorologists saw pictures of a mid-latitude cyclone over the north-eastern United States and the age of satellite meteorology was born. In this section we look at:

- types of meteorological platforms
- meteorological and other atmospheric applications
- examples of Meteosat image products.

6.2.1 Types of meteorological platforms

Meterorological satellite platforms are of two types: polar-orbiting or geostationary.

Polar-orbiting platforms

Polar-orbiting platforms have been pre-eminent for meteorological applications since the launch of TIROS-N by the National Oceanic and Atmospheric Administration's (NOAA) in 1978. The main sensor aboard the ensuing series of Polar Orbiting Environmental Satellites (POES) is the Advanced Very High Resolution Radiometer (AVHRR) a broad-band, four- or five-channel scanner, sensing in the visible, near-infrared, and thermal infrared portions of the electromagnetic spectrum.

Important functions of AVHRR include:

- deriving sea surface temperatures (see Section 6.1.3)
- calculating the Normalized Difference Vegetation Index (see Section 7.2.3)
- detecting atmospheric aerosols over the oceans
- monitoring volcanic eruptions and supporting an operational NOAA warning of volcanic ash in the atmosphere during eruption events (see Section 7.3.2)
- monitoring cloud patterns
- other applications requiring the high temporal resolution of global daily coverage, with moderate spectral and spatial resolution, operational sidelap stereoscopic coverage, and calibrated thermal sensors.

The main parameters of the AVHRR sensor are shown in Table 6.1.

Geostationary platforms

Geostationary satellites have an orbital period of exactly 24 hours, allowing them to 'hover' above the same place on Earth (which according to the principals of orbital dynamics must be a point above the equator). An international series of satellites provides comprehensive global coverage except in the extreme North and South. Figure 6.5 shows a mosaic generated from data gathered by these satellites.

The geostationary satellites take images over a period of a few minutes (up to 26 minutes for a GOES-8 full Earth scan) to scan the Earth line by line. The older satellites (such as GOES-7 and older, GMS, Meteosat) did so as they spun at 100

Table 6.1 AVHRR technical specification

Country of origin	United States of America
Orbit	Sun-synchronous (near polar)
Orbit period	102 minutes
Inclination	98.7° to 98.9°
Altitude	833–870 km
Coverage	Global
Repeat cycle	12 hours
Swath width	2,700 km
Pixel size	1.1 km (nadir)
Channel 1	0.58–0.68 μm
Channel 2	0.72–1.10 μm
Channel 3A (NOAA K,L,M)	1.58–1.64 μm
Channel 3	3.55–3.93 μm
Channel 4	10.3–11.3 μm
Channel 5	11.3–12.5 μm

Main applications of each spectral channel:

1	Daytime cloud/surface mapping
2	Surface water delineation, ice and snow melt
3A	Snow/ice discrimination (NOAA K,L,M)
3	Sea surface temperature, night-time cloud mapping
4	Sea surface temperature, day and night cloud mapping
5	Sea surface temperature, day and night cloud mapping

Fig. 6.5 Global composite of meterological satellite data (courtesy Space Science and Engineering Center, University of Wisconsin-Madison).

Table 6.2 Geostationary satellite technical specifications

Satellite	GOES	Meteosat	IODC (Indian Ocean Data Coverage)	GOMS	GMS
Country	USA	Europe	Europe	Russia	Japan
Operational spacecraft	GOES-East GOES-West	Meteosat 7	Meteosat 5	Elektro	GMS-5
Location	75°W (East) 135°W (West)	0°	63°E	76°E	140°E
Channels	VIS IR	VIS WV IR	VIS WV IR	VIS WV IR	VIS IR

Key to channels: VIS = visible; WV = water vapour; IR = thermal infrared

times per minute. Newer satellites (such as GOES-NEXT), as well as some older satellites (such as GOMS), are three-axis stabilised satellites which do not spin. Technical specifications of the major geostationary satellites is shown in Table 6.2.

6.2.2 Applications by wavelength

Meteorological applications can be characterised by the wavelength that the imagery has acquired, typically:

- visible (0.4–1.1 μm)
- short wave infrared (3.78–4.03 μm)
- upper level water vapour (6.47–7.02 μm)
- thermal infrared (10–12 μm)

Visible imagery

Visible (VIS) imagery (Bader *et al.* 1995) is derived from solar radiation scattered or reflected towards the satellite from the Earth–atmosphere system. The intensity of the image depends on the albedo/reflectivity of the underlying surface or cloud, thus areas of images acquired at night will appear black. For daylight images clouds will be seen as white against the darker background of the Earth. Visible images are useful for cloud detection and tracking, pollution, haze detection and identification of severe storms.

Short wave infrared

The short wave infrared window is useful for discriminating between water clouds and snow or ice crystal clouds, identifying fog at night, detecting 'hot spots' from fires and volcanoes, and for calculation of sea surface temperatures

Upper level water vapour

The upper level water vapour band can be used for tracking mid-level atmospheric motions and estimating mid-level atmospheric moisture content.

Thermal infrared

Thermal infrared (IR) images (Bader *et al.* 1995) portray the temperture of the surface below the sensor, whether it be land, sea or the tops of clouds. Warm temperatures (0–30 °C) generally correspond to the response of land or sea under clear skies. Lower temperatures mean that the clouds are getting higher and denser. Very cold temperatures are associated with very high cloud tops and associated strong convective storm activity. In visible imagery clouds will appear white against the darkened background of the Earth. In thermal imagery clouds are generally colder than the Earth's surface, therefore using a conventional thermally-related scale, clouds would appear darker. In order to provide a consistent interpretation it is often convenient to invert this scale so that the colder clouds are shown brighter. Figure 6.6 shows a thermal image of clouds from the UK Meterological Office.

6.2.3 Cloud classification

Classification of clouds (Carleton 1991) can be carried out using a combination of visible and IR imagery. There are a vast number of different methods for cloud classification based on differing attributes of the imagery, including:

- *Cloud texture* – the physical appearance of the cloud as shown by the difference in appearance of neighbouring pixels for example 'mottled' for cumulus, 'striated' for cirriform clouds.
- *Cloud brightness* – cloud reflectance will depend on thickness and phase (ice, liquid water). Deep cumulonimbus clouds tend to be the brightest. Brightness in the thermal band is a measure of cloud temperature and hence height.
- *Cloud form* – this is associated with wind speed, which can be used by tracking clouds using correlation techniques in a time series of images taken hourly or half-hourly.
- *Cloud patterns* – looking at clouds in the context of larger-scale atmospheric circulation.

A typical satellite-based cloud classification scheme based on height, albedo, directionality, shape and co-occurrence of cloud fields is shown in Table 6.3.

Table 6.3 The 'Garand' cloud classification scheme (from Carleton 1991)

1. Clear	11. Bright closed cells
2. Stratus	12. Nimbostratus
3. Scattered cumulus	13. Altocumulus
4. Broken cumulus	14. Altocumulus with cumulus
5. Scattered stratocumulus	15. Altocumulus with stratocumulus
6. Broken/overcast stratocumulus	16. Thin cirrus
7. Cloud streets	17. Multilayers with cirrus
8. Bright rolls	18. Multilayers with cumulonimbus
9. Polygonal open cells	19. Dense cirrostratus
10. Strongly convective open cells	20. Overcast cumulonimbus

Fig. 6.6 Thermal image of clouds, 19 June 2001, 18:00 hours UTC (copyright Crown, Meterological Office and EUMETSAT).

6.2.4 Meteosat: examples of meteorological products

This section provides an idea of a typical set of meteorological products (Eumetsat 2001), in this case from the European Meteosat platform which acquires images every half-hour. The history of the Meteosat programme dates back to the launch of Meteosat 1 in July 1977 by the European Space Agency. Building on early successes the Meteosat programme was transferred for operational use to the European Meteorological Satellite (Eumetsat) organisation, an intergovernmental organisation created through an international convention agreed by 17 European Member States: Austria, Belgium, Denmark, Finland, France, Germany, Greece, Ireland, Italy, the Netherlands, Norway, Portugal, Spain, Sweden, Switzerland, Turkey and the UK.

- *Cloud Motion Winds* (CMW) – The CMW product is generated by applying a correlation algorithm to sequences of three visible (half-resolution) images which can then be used to extract winds. A typical product will contain up to 750 winds per channel and is distributed for the synoptic hours of 00, 06, 12 and 18 UTC.
- *Expanded Low-resolution Winds* (ELW) – The ELW product is generated from the same data as CMW but uses a lower threshold velocity. This typically contains about 4,000 winds and is distributed every 1.5 hours.
- *High-resolution Visible Winds* (HRV) – HRV product is generated using essentially the same algorithm as the CMW product, but applied to the VIS images in full resolution. The product is generated for the synoptic hours of 06, 09, 12, 15 and 18 UTC and typically will contain up to 2,000 wind vectors.
- *Clear-sky Water Vapour Winds* (WVW) – The WVW product is similar to other wind products but generated in clear skies from the Water Vapour image. A typical product will contain about 500 wind vectors.
- *High-resolution Water Vapour Winds* (HWW) – The HWW product is generated from Water Vapour images which are divided into sub-areas of 16×16 pixels. Only segments where a cloud has been detected are processed.
- *Sea Surface Temperatures* (SST) – In the SST product a temperature value is generated for every segment where the sea surface is visible. This excludes land segments, and those segments with continuous cloud cover during the period of observation. The product is generated twice daily, for synoptic hours at 00 and 12 UTC; each product contains an accumulation of data from the preceding 12-hour period.
- *Cloud Analysis* (CLA) – The CLA product analyses cloud top temperature and cloud cover in each segment. Up to three different cloud layers can be identified.
- *Upper Tropospheric Humidity* (UTH) – The UTH product contains an estimate of the mean relative humidity for an atmospheric layer on the upper troposphere. The product is produced hourly.
- *Clear Sky Radiances* (CSR) – The CSR product contains an estimate of the mean Water Vapour channel brightness temperature from regions containing no or only low-level clouds. The product is generated every hour.
- *Cloud Top Height* (CTH) – The CTH product estimates the height of the cloud

tops specified in eight vertical intervals with a resolution of 1,500 metres. The CTH values are based on 'superpixels' of 3 × 3 IR pixels corresponding to a horizontal resolution of 15 km at the equator and about 22 km in southern Europe. The product is primarily used for aviation meteorology.

6.2.5 Atmospheric chemistry

The majority of ozone in the atmosphere can be found in the stratosphere (between 10 and 40 km above the Earth's surface). This layer of ozone prevents quantities of solar ultraviolet radiation from penetrating the atmosphere and contributing to harmful conditions in humans, including skin cancers, cataracts and immune system problems. It should be noted that the effect of ozone on human existence is highly dependent on where it is – for example, concentrations of ozone in the troposphere (up to about 10 km) have a pollutant effect on lung tissue and plant life.

One of the most important instruments for ozone detection is the Total Ozone Mapping Spectrometer (TOMS) instrument, first flown on NASA's Nimbus 7 satellite in October 1978. TOMS takes world-wide observations of the total amount of ozone in the atmosphere every day. The ozone concentration can be calculated by measuring the amount of ultraviolet radiation that is reflected at various wavelengths. The more ozone that is present, the smaller the quantity of UV that is reflected. Figure 6.7 (colour section) shows a TOMS image of global ozone concentration from August 2001.

6.2.6 Ocean–atmosphere interaction

The ocean exchanges vast amounts of gases and aerosol particles with the atmosphere as well as heat and energy. Because these air–sea interactions exert such a profound influence on the Earth's weather and climate patterns, scientists dubbed the ocean 'the global heat engine' (Herring 2001a).

The first mechanism of importance to the global heat engine is its physical coupling with the atmosphere (see Fig. 6.8). This works as follows:

1. Incoming sunlight is absorbed by the ocean, warming its surface.
2. Some of this heat is released to the atmosphere as thermal energy and water vapour, which, in turn, creates wind and rain clouds.
3. Surface wind create currents, which in turn control the distribution of cold and warm water in the ocean.
4. Where there are cool patches on the surface local air temperatures tend to be cool because of surface winds.
5. Where the ocean temperatures are warm these in turn tend to heat up the local atmosphere.

Thus '*... ocean and atmosphere are intertwined in a complex and perpetual dance – with each following the other's lead*' (Herring 2001a).

The second way that ocean and atmosphere are linked is chemical. This is a two-way process: the ocean is a source of greenhouse gases, most significantly evaporated water. This will have a heating effect on global temperature but also a cooling one as it will produce clouds which help to shade the surface; however, it is not clear yet

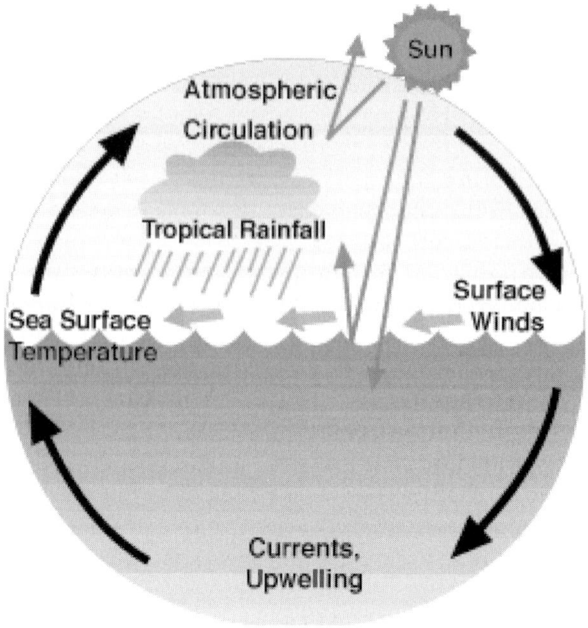

Fig. 6.8 Ocean physically coupling with the atmosphere (courtesy NASA).

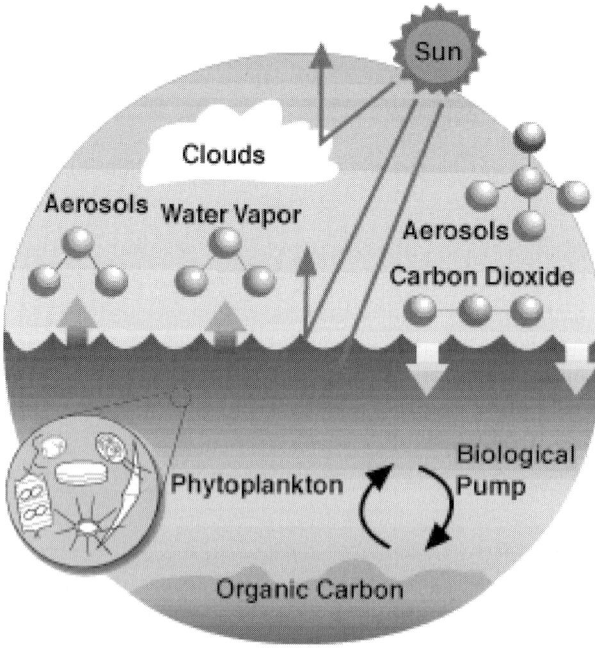

Fig. 6.9 Ocean biochemically coupling with the atmosphere (courtesy NASA).

which mechanism will dominate in the long term. In addition the ocean is a sink for greenhouse gases most importantly from the human perspective carbon dioxide. Since the Industrial Revolution, atmospheric carbon dioxide has risen by 30 per cent, while average global temperatures have climbed about 0.5 °C. On average, carbon dioxide resides in the atmosphere about 100 years before it settles into the ocean, or is taken out of the atmosphere by plants. The oceanic removal of carbon dioxide from the atmosphere will have a cooling effect on global temperatures. This chemical ocean–atmosphere interaction is shown in Fig. 6.9.

Because of the global and dynamic nature of these processes it is invaluable to be able to measure them from space. For example, some of the most well-known manifestations of the ocean–atmosphere interaction are (a) the frequent and often catastrophic El Niño and La Niña events (Herring 2001b) and (b) the 'strong and stable' Pacific Decadal Oscillation (PDO) – a long-term ocean temperature fluctuation of the Pacific Ocean that waxes and wanes approximately every 10 to 20 years. All these phenomena are clearly visible from satellite. For example, Fig. 6.10 shows imagery from TOPEX/Poseidon with a PDO pattern from January 2001 indicating above-normal sea surface heights and warmer ocean temperatures blanketing the far-western tropical Pacific and much of the north (and south) mid-Pacific.

Finally, Fig. 6.11 (colour section) provides an impressive image of Typhoon Olga which hit the China Sea in 1999, delivering high winds and torrential rains to South Korea, North Korea and other coastal communities of south Asia. The wind vectors were derived from the NASA viewing radar instrument, SeaWinds.

6.3 CRYOSPHERIC APPLICATIONS

This section looks at two of the main cryospheric applications of satellite data:

- sea ice monitoring
- snow and land ice monitoring.

6.3.1 Sea ice monitoring

Satellite Earth observation data can be used for a wide variety of sea ice applications, including the physics of ice and interaction with the ocean surface, mapping of ice conditions and as inputs to sophisticated models such as the boundary conditions for meteorological forecasting. Satellite data used for sea ice monitoring includes:

- Synthetic Aperture Radar (SAR), primarily from ERS and Radarsat for its all-weather coverage
- passive microwave radiometry such as the SSM/I sensor onboard the US DMSP programme which has particularly good sea ice/water discrimination
- optical imagery, either low resolution such as NOAA AVHRR for wide area repetitive coverage or high resolution such as Landsat or SPOT for more detailed local studies.

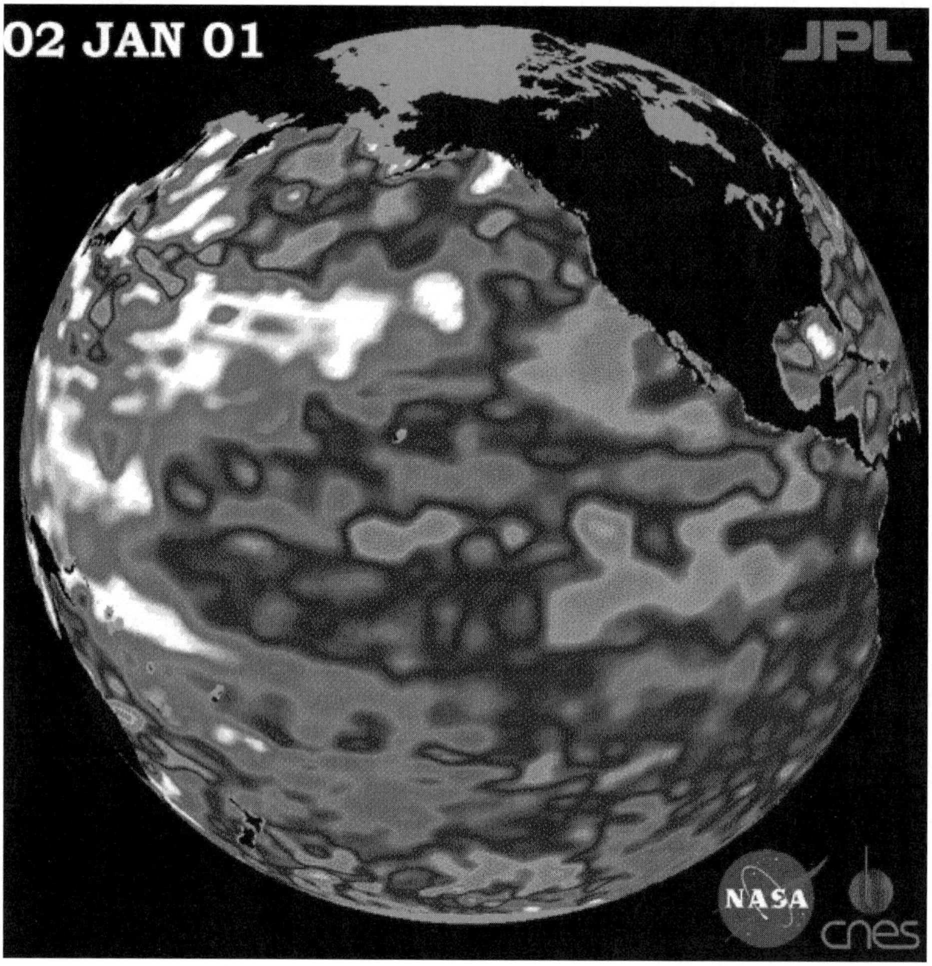

Fig. 6.10 TOPEX/Poseidon data, taken during a 10-day collection cycle ending 2 January 2001, shows above-normal sea surface heights and warmer ocean temperatures blanketing the far-western tropical Pacific and much of the north (and south) mid-Pacific. (Courtesy NASA/JPL, CNES.)

SAR is particularly useful as different sea ice types have distinctive radar backscattering properties which can be used to define ice classifications. Sea ice state is provided as an operational EO service by ice centres covering Baltic, North Atlantic and Canadian waters; indeed this was a primary design objective of the Radarsat mission. An example of sea ice monitoring is the ice classification algorithm employed at the Alaska SAR facility (Kwok *et al.* 1992) which uses a clustering technique to determine the dominant radar backscatter intensity cluster centroids within an image. The centroids are then compared to values for specific ice types placed in a look-up table, which assigns ice types to each cluster. The identified

Fig. 6.12 Alaska SAR facility sea ice product (courtesy Alaska SAR Facility, copyright ESA 1992).

ice types are grouped as multi-year ice, deformed first-year ice, undeformed first-year ice, and new ice/smooth open water. The algorithm is used for the classification of ice in ERS-1 data in the Beaufort Sea, north of 73° N, between days 1–120 and 270–365. An example of the resulting sea ice product is shown in Fig. 6.12.

6.3.2 Snow and land ice monitoring

Satellite imagery can be used for a variety of land-based cryospheric applications – for example (Vikhamar and Solberg 2000), optical images have proved suitable for the mapping of snow-covered forests. Figure 6.13 shows a time series of images acquired by the recently launched Moderate-resolution Imaging Spectroradiometer (MODIS) sensor (see Section 7.1.2) which shows the total area in North America covered by snow compared to the 35-year average snowline.

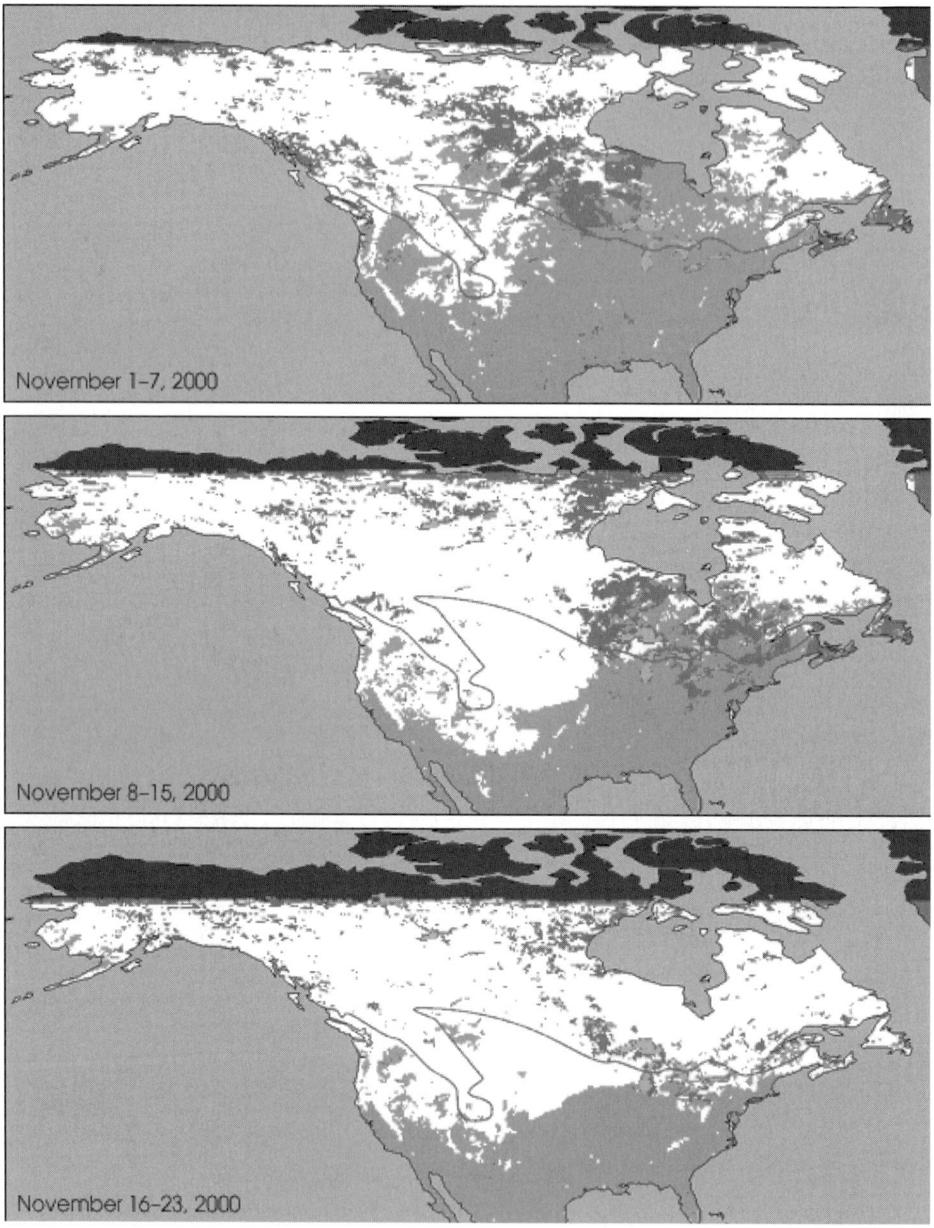

Fig. 6.13 North American snow cover November 2000 from MODIS. Each image represents an 8-day composite for the periods 1–7, 8–18 and 16–23 November 2001. Snow cover is shown in white; the dark grey northern regions are where no data was collected. (Courtesy NASA.)

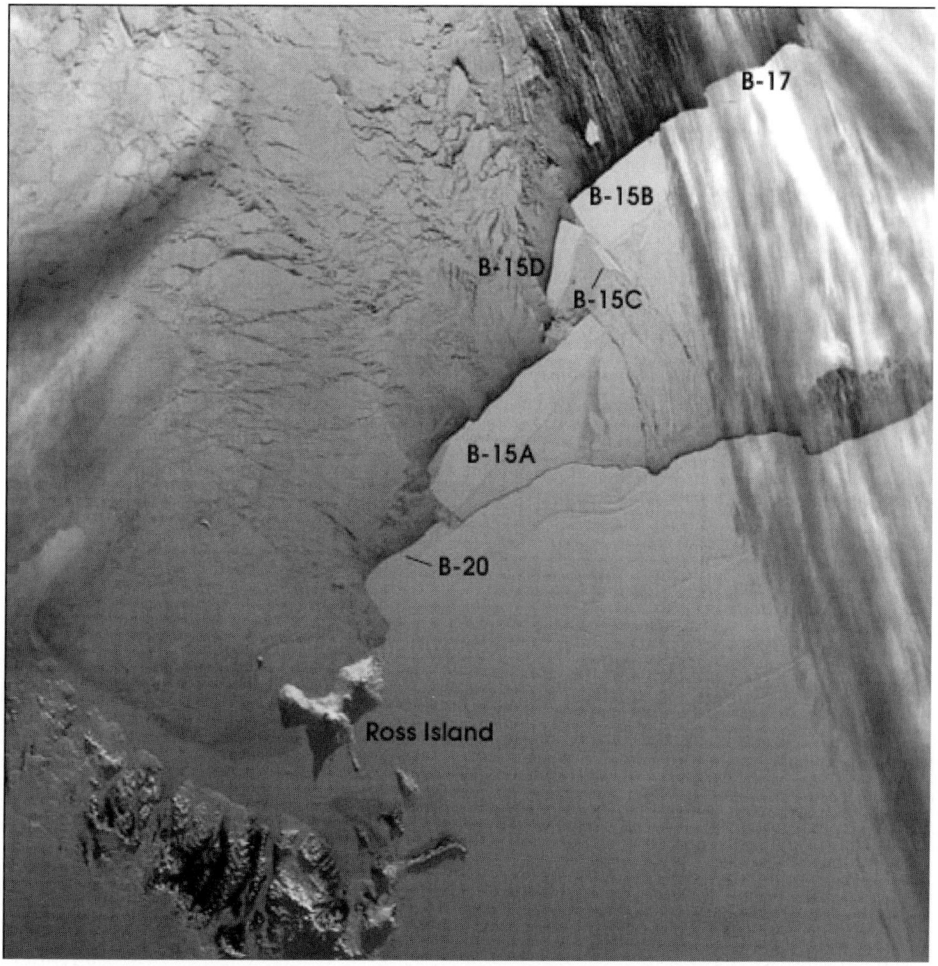

Fig. 6.14 Iceberg Production imaged by MODIS 21 September 2000. The B-15 fragments are remnants of the huge iceberg which broke away from the Antarctic shelf in late March 2000. Slightly visible is the line where iceberg B-20 broke away from the shelf in the last week of September. (Image by Brian Montgomery, NASA GSFC; data courtesy MODIS Science Team.)

Because of their synoptic coverage and frequent revisit patterns, satellite imagery is particularly suited to the tracking of icebergs and glaciers. Figure 6.14 shows another MODIS image, this time of the iceberg 'nursery' on the Ross Ice Shelf near Ross Island. The image was taken on 21 September 2000 and shows fragments of a huge iceberg which broke away from the ice shelf in March 2000. Such images of cracks in the Antarctic ice shelf are of great use in studies of global warming.

The Antarctic Mapping Mission

The Antarctic Mapping Mission (AMM) is a mission aimed at producing high-resolution maps of Antarctica using Radarsat imagery (Jezek 2001). The Radarsat satellite was rotated 180° in yaw to allow the radar to image to the left of the satellite track instead of to the right, and to steer the radar beam up to cover the South Pole. Figure 6.15 (colour section) shows motion of the Lambert Glacier on the Amery Ice Shelf as part of the Year 2000 Antarctic Mapping Mission. Prior to this mission *in situ* measurements of the ice velocity had been extremely sparse. AMM makes use of interferometric pairs of Radarsat images captured 24 days apart to provide accurate measurements of glacier velocity as well as local topography.

6.4 OCEANOGRAPHIC, ATMOSPHERIC AND CRYOSPHERIC APPLICATIONS OF GPS

As with the applications described in Chapter 5, non-land applications of GPS are widespread. Among the main areas are:

- marine navigation
- coastal habitat mapping
- tracking of oceanographic buoys
- GPS-aided weather forecasting
- monitoring glaciers.

The field of GPS applications is much too large to address in a comprehensive manner within this book. However, to give some flavour of the widespread use of GPS a few examples of oceanographic, atmospheric and cryospheric applications are given.

6.4.1 Oceanographic applications of GPS

Use of GPS for marine drifter buoys

For ocean circulation patterns the main requirement is to be able to track the position of a buoy as it floats over a number of months. This task requires a platform that can survive in the marine environment and a communication capability to report on the motion of the buoy. For surveys of relatively small areas (a few kilometres) buoys will be used in conjunction with a support ship (Viellard *et al.* 2000). For more remote locations the currently favoured technology (*Space News* 2000) is to make use of the ARGOS satellite data-collection system. This can be used to acquire data in near-real time from thousands of drifting buoys. Information from the *Global Drifter Center* in Miami (Bushnell 2000) describes a typical drifter logging hourly GPS fixes and storing them, transmitting 16 hourly positions in four interleaved data 'pages'.

Altimeter calibration

GPS receivers can be used to estimate satellite height bias and also calibration based on mean sea surface and significant wave heights derived from the buoys' measurements. For example, several GPS buoy campaigns (Martinez-Garcia *et al.* 2000) have been conducted in the Spanish north-western Mediterranean in order to contribute to the calibration of the TOPEX ALT-B in that region (CATALA campaigns in March 1999 and July 2000).

Tide gauges
The monitoring of tide gauges requires very high positional accuracy (sub-centimetre) and extensive post-processing of the GPS information, or ideally a much more accurate direct service in the first place. GPS-based tide gauges are more useful than conventional 'ruler' types because they provided height against an international geodetic reference frame rather than a local reference. This is also a growth area because of concerns about global warming; for example, the Global Sea Level Change (GSLC) project (IEESG 2000) uses GPS-enabled tide gauges to contribute to an understanding of global sea level changes, and thereby to a part of the global climate system, through studies of sea level variations on timescales from hours (i.e. tides and surges) to centuries. The sea level records obtained from tide gauges are contaminated by vertical land movement at the tide gauge site due, for example, to post-glacial rebound. This movement can be equal to, or greater than, the rate of mean sea level rise and must be removed from the tide gauge records before global mean sea level changes can be fully investigated. The location of tide gauges compared with the distribution of drifting buoys used for the World Ocean Circulation Experiment (WOCE 1997) is shown in Fig. 6.16.

Marine conservation
Marine conservation, including monitoring of fish stocks, is another example of GPS in action. For many applications, such as coral reef mapping, subsurface monitoring is necessary and may be attained by a suitable combination of the use of sonar with reference points triangulated by GPS, sometimes referred to as 'Underwater GPS' (Thomas 2000). In coastal zones extensive use is made of thermal and hyperspectral scanners such as CASI, flown on aircraft. In such cases an onboard GPS system can be used in conjunction with a GPS to provide the necessary control to georectify the imagery to a standard map projection.

6.4.2 Atmospheric applications of GPS

Support to atmospheric chemistry studies
Radiosondes used for meteorology, climatology or atmospheric chemistry studies can use GPS to report their location. Ground stations can be used for the measurements of atmospheric chemistry, and a typical example of this is a mobile sampling vehicle used by CSIRO (Australia) to measure a number of chemical species including sulphur dioxide, carbon monoxide, carbon dioxide, hydrocarbons, total sulphur and airborne particles (Sawford and Luhar). Using this system transects were made of data logged once per second.

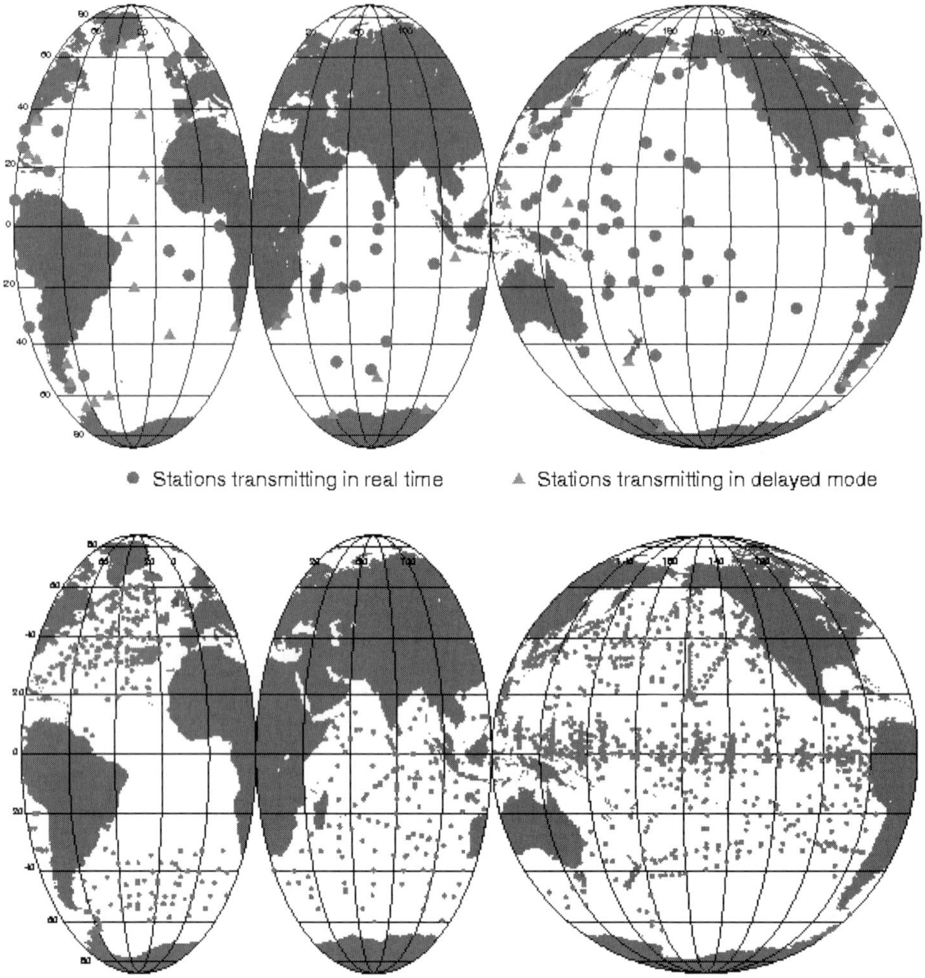

Fig. 6.16 Location of monitoring equipment (source: WOCE International Project Office). Tide gauges (above); surface drifting buoys (below).

Metreorological use of GPS

Recently developed methods (Baker *et al.* 1999) for sensing precipitable water vapour (PWV) with GPS promise improvement in short-term forecasting. Topospheric propagation delay can be modelled using a Kalman filter, and, by incorporating measurements of pressure, the wet delay can be subsequently determined, and transformed to estimates of Integrated Precipitable Water Vapour (IPWV). GPS offers advantages over traditional meteorological sensors because it is a portable, all-weather system that can provide continuous measurements with little or no need for calibration checks.

6.4.3 Cryospheric applications of GPS

Tracking of icebergs and glaciers is often facilitated by GPS, for example (SPRI 2000) digital Landsat imagery and GPS were used to map ice margins and ice divides on Severnaya Zemlya, in the Russian High Arctic. Antarctic ice can be tracked by a ruggedised tracking buoy equipped with a GPS-positioning system, an Argos transmitter, and carries an automatic weather station for the measurement of wind speed and direction, air temperature and relative humidity. Use of GPS in conjunction with satellite images is an important source of ground truth and calibration information. GPS data is also of value when used with radar altimetry for measuring baseline height and movement of ice sheets.

REFERENCES

Bader MJ, Forbes JR, Grant JR, Lilley RB and Waters AJ (1995) *Images in Weather forecasting: Practical Guide for Interpreting Satellite and Radar Data.* University Press, Cambridge.

Baker HC, Dodson AH, Penna NT, Higgins M and Offiler D (1999) Ground-based GPS water vapour estimation: potential for meteorological forecasting, *Proceedings of Symposium JSG28, IUGG General Assembly, Birmingham*, July 1999.

Benny AH and Dawson GJ (1983) Satellite imagery as an aid to bathymetric charting in the Red Sea. *The Cartographic Journal*, **20** (1), 5-16.

Bushnell M (2000) Preliminary results from global lagrangian drifters using GPS receivers, NOAA/AOML-Global Drifter Center. http://www.aoml.noaa.gov/phod/dac/GPS-rep.html

Carleton A (1991) Satellite remote sensing in climatology. *Studies in Climatology Series*, Belhaven Press, CRC Press.

Eumetsat (2001), European Organisation for the Exploitation of Meteorological Satellites, Website. www.eumetsat.de

Herring D (2001a) Ocean and Climate NASA Earth Observatory Website. http://earthobservatory.nasa.gov/Library

Herring D (2001b) What is el niño? NASA Earth Observatory Website. http://earthobservatory.nasa.gov/Library/ElNino/elnino.html

Jesek KC (2001) Radarsat-1 Antarctic Mapping Project, web-site. http://www-bprc.mps.ohio-state.edu/radarsat/

Kwok R, Rignot E, Holt B and Onstott R (1992) Identification of sea ice types in spaceborne synthetic aperture radar data, *Journal of Geophysical Research*, **97** (C2), 2391–2402.

Leming TD, May LN and Jones P (1999) A Geographic Information System for near real-time use of remote sensing in fisheries management in the Gulf of Mexico. http://coastwatch.noaa.gov/COASTWATCH/PUB/nrtgis_final_report2.html

Martinez-Garcia M, Martinez-Benjamin JJ and Ortiz MA (2000) Analysis and strategies applied to the GPS buoys data for the TOPEX Alt-B absolute calibration in the NW-Mediterranean. *Joint TOPEX/Poseidon and Jason-1 Science Working Team Meeting, Miami Beach, Florida*, 15–17 November.

Meaden G (2000) GIS in fisheries management. *GeoCoast*, **1** (1), 82-101, October.
NASA (2001) Classic CZCS scenes NASA Goddard Space Flight Centre. http://daac.gsfc.nasa.gov/CAMPAIGN_DOCS/OCDST/classic_scenes/00_classics_index.html
SPRI – Scott Polar Research Institute (2000) Form and flow of the ice caps on Severnaya Zemlya, Russian High Arctic. http://www.spri.cam.ac.uk/aemc/pi.htm
Space News (2000) Industry eyes marriage of imagery, GPS data. *Space News*, 10 July.
Thomas H (2000), Using GPS underwater. http://www.underwater-gps.com/
Viellard C *et al.* (2000) The use of GPS buoys for geo-referencing mosaic charts and environmental data collection. *Oceanology International 2000, Brighton UK*, 7–10 March 2000.
Vikhamar D and Solberg R (2000) A method for snow-cover mapping of forests by optical remote sensing. *Proceedings of EARSeL Workshop on Remote Sensing of Land Ice and Snow, Dresden, Germany*, June 2000 (in press).
WOCE – World Ocean Circulation Experiment (1997) *WOCE Data Guide 1997*. http://www.soc.soton.ac.uk/OTHERS/woceipo/dguide97/

Further reading

Barth S, Hansen E and Leben B (1999) Monitoring marine mammals from Colorado. http://www-ccar.Colorado.EDU/˜altimetry/applications/whales/
Evans RH and Gordon HR (1994) CZCS 'system calibration': a retrospective examination. *Journal of Geophysical Research*, **99** (C4), 7293-7307.
IEESG (2000) Monitoring land movement at UK tide gauge sites using GPS. IEESG Nottingham University. http://www.nottingham.ac.uk/iessg/isgres19.htm/
McClain CR, Feldman GC and Esaias WE (1993), Biological oceanic productivity. In Gurney R, Foster JL and Parkinson CL (eds) *The Atlas of Satellite Observations related to Global Change*. Cambridge University Press, New York, pp. 251–63.
McClain CR, Cleave ML, Feldman GC, Gregg W W, Hooker SB and Kuring N (1998) Science quality SeaWiFS data for global biosphere research. *Sea Technology*, September, pp. 10–16.
NOAA (2001) Coastwatch web-site. http://coastwatch.noaa.gov/
Sawford BL and Luhar AK, *The Kwinana Coastal Fumigation Study*. CSIRO Division of Atmospheric Research.
Sun Y (1996) Automatic ice motion analysis from ERS-1 SAR images using the optical flow method. *International Journal of Remote Sensing*, **17** (11), 2059–87.

7

Environmental monitoring and population security

Understanding the current state of the environment is an increasingly important activity throughout the world. Humanity's impact on the environment is growing and complex and it is difficult to disentangle human activity from natural effects on the landscape. Traditionally environmental measurements have been made *in situ*; however, this approach is often inadequate due to the limitations of spatial cover and the difficulty in keeping the information updated. Aerial photography and aircraft scanners are also used but they are limited by the cost of commissioning specialist environmental surveys. Owing to these causes coverage of environmental information in many countries may be sparse and inadequate. There are also problems with integrating data from different states due to differences in survey methods, sampling density and temporal update. Earth Observation has a role to play in solving these problems either by introducing new techniques or as a complement to existing practices. In many cases, the broader picture offered by Earth Observation data can assist their environmental protection activities.

This chapter looks at the use of EO for environmental applications at a wide variety of scales comprising:

- environmental EO missions
- environmental applications such as habitat monitoring and global vegetation mapping
- hazards and natural disasters
- population dynamics and security.

7.1 ENVIRONMENTAL EO MISSIONS

Although many EO missions have proved of value to environmental projects, here we look at some that were specifically designed for this purpose:

- The Earth Observing System
- Terra
- Envisat
- ADEOS-II

- SPOT Vegetation
- Indian Remote Sensing missions
- 'Smallsat' missions

7.1.1 The Earth Observing System

The Earth Observing System (EOS) is a series of polar-orbiting satellites which is a collaborative venture between the NASA Earth Science Enterprise (ESE) and other organisations such as the European Space Agency (ESA) and the Japanese National Space Development Agency (NASDA). EOS consists of a science component and a data archive system to enable long-term global environmental observations of the land and ocean surface, solid Earth, atmosphere and 'biosphere' (the part that we all live in). At the time of writing (January 2001) there are six EOS satellites in orbit and 14 more are planned in the time frame to approximately 2004. Table 7.1 presents a summary of the major EOS missions in chronological order and cross-references to sections where further information can be found.

Table 7.1 Earth Observing System (EOS) missions

Mission and launch date	Description
Sea-viewing Wide Field-of-view Sensor (SeaWiFS) 1 August 1997	SeaWiFS provides quantitative data on global ocean bio-optical properties. Changes in ocean colour signify various types and quantities of marine phytoplankton which have many scientific and practical applications such as fisheries monitoring (see Section 6.1.2)
Tropical Rainfall Measuring Mission (TRMM) 27 November 1997	TRMM is a joint mission between NASA and the National Space Development Agency (NASDA) of Japan. It was designed to study tropical rainfall. The energy release from tropical rainfall is an important input to global atmospheric circulation which impacts on both weather and climate studies.
Landsat 7 15 April 1999	The first Landsat was launched in 1972 making the programme the longest running civilian Earth Observation enterprise. Landsat has been used for a huge variety of environmental and other land applications. See Section 1.7.1 for an introduction to Landsat, and many other sections (particularly in Chapter 5) for details of applications.
QuikScat 19 June 1999	The SeaWinds instrument on the QuikScat mission was designed as replacement for the data failure when the NASA Scatterometer (NSCAT) lost power in June 1997. SeaWinds is a microwave radar which measures wind speed and direction near the ocean surface.
Terra 18 December 1999	The Terra satellite is the 'flagship' of EOS. It provides global data on the state of the atmosphere, land, and oceans and their interactions with solar radiation. The Terra mission is described in more detail in Section 7.1.2.

Acrimsat 20 December 1999	A series of Active Cavity Radiometer Irradiance Monitors (ACRIMs) provide long-term accurate measurement about incident solar energy falling on the ocean, atmosphere and Earth's surface.
METEOR 3M-1/SAGE III Due March 2001	The SAGE III mission on the Russian Meteor 3M-1 spacecraft will make high-latitude long-term measurements of the vertical structure of aerosols, ozone, water vapour, and other trace gases in the upper troposphere and stratosphere.
Jason 1 Due March 2001	Jason is a joint USA–France mission designed to monitor global ocean circulation, investigation ocean–atmosphere interactions and to monitor events such as El Niño (see Section 6.3.2).
QuikTOMS Due April 2001	The Total Ozone Mapping Spectrometer (TOMS) will provide long-term mapping of the global distribution of the Earth's atmospheric ozone as well as measurement of sulphur dioxide released in volcanic eruptions.
Aqua Due July 2001	Aqua (originally EOS-PM) will provide a multi-disciplinary study of the Earth's interrelated processes (atmosphere, oceans, and land surface) and their relationship to Earth system changes.
Gravity Recovery and Climate Experiment (GRACE) Due November 2001	GRACE will use a satellite-to-satellite microwave tracking system between two spacecraft to measure the Earth's gravity field over five years. Accurate measurement of long-term gravitational variations is important as they are coupled to long-wavelength ocean circulation processes and the transport of ocean heat to the Earth's poles.
ICESat Due 2001	ICESat is a Smallsat mission flying the Geoscience Laser Altimeter System (GLAS) which will measure the elevation of the Earth's ice sheets, clouds, and land to a great accuracy. See Section 7.1.7 for details of other Smallsat missions.
ADEOS II (SeaWinds) Due February 2002	The Advanced Earth Observing Satellite II (ADEOS-II), is a joint mission between NASA and the National Space Development Agency (NASDA) of Japan. The mission will take an active part in the research of global climate changes and their effect on weather phenomena.
Triana Due April 2002	Triana is a mission to investigate the relationship between solar radiation and climate.
Vegetation Canopy Lidar (VCL) Launch TBD	VCL will be used to produce the first global inventory of the vertical structure of forests across Earth. The mission will use a multibeam laser-ranging device to enable direct measurement of tree heights, forest canopy structure and global biomass with at least ten times better accuracy than existing assessments.

Solar Radiation and Climate Experiment (SORCE) Due July 2002	SORCE will include the SOLar STellar Irradiance Comparison Experiment (SOLSTICE) and Total Irradiance Monitor (TIM) to provide long-term, accurate measurements of solar ultraviolet, far ultraviolet and total irradiance.
CloudSat Due May 2003	CloudSat's primary goal is to provide data for global cloud prediction models and to understand the role of clouds in climate change.
Aura Due June 2003	The Aura satellite (formerly EOS Chemistry 1) will focus on measurements of atmospheric trace gases and their transformations.
PICASSO–CENA Due November 2003	The Pathfinder Instruments for Cloud and Aerosol Spaceborne Observations–Climatologie Etendue des Nuages et des Aerosols (PICASSO–CENA) mission is a joint mission between NASA and the Institut Pierre Simon Laplace, Paris. PICASSO–CENA will provide global observations of aerosol and cloud properties, radiative fluxes, and atmospheric state.
International Space Station (SAGE III) Due November 2003	The SAGE III mission on the Space Station will provide high latitude long-term measurements of the vertical structure of aerosols, ozone, water vapour, and other important trace gases in the upper troposphere and stratosphere.

7.1.2 Terra

The Terra satellite (King and Herring 2000) was successfully launched on 18 December 1999 and on 24 February 2000 began collecting what will ultimately become a new, 15-year global dataset on which to base scientific investigations about the Earth. Terra has been placed in an identical orbit to Landsat 7 to obtain repetitive data coverage of such variables as plant physiology under nearly identical conditions. The instruments flown on Terra are described below.

ASTER

The Advanced Spaceborne Thermal Emission and Reflection Radiometer (ASTER) obtains high-resolution images of the Earth in 14 different spectral bands ranging from visible to thermal infrared light. All three ASTER telescopes: Visible and Near Infrared (VNIR), Short-Wave Infrared (SWIR) and Thermal Infrared (TIR) are pointable in the cross-track direction. Given its high resolution and its ability to change viewing angles, one of the main uses of ASTER will be to produce stereoscopic images and detailed terrain height models.

CERES

There are two identical Clouds and the Earth's Radiant Energy System (CERES) instruments aboard Terra that measure the Earth's total radiation budget and provide cloud property estimates. One CERES instrument will operate in a cross-track scan mode and the other in a biaxial scan mode. The cross-track mode continues the measurements of the Tropical Rainfall Measuring Mission (TRMM), while the biaxial scan mode will provide new angular flux information to improve models of the Earth's radiation balance.

MISR

Multi-angle Imaging Spectro-Radiometer (MISR) is a new type of instrument designed to view the Earth with cameras pointed at nine different angles. One camera points towards nadir, and the others provide forward and aft view angles at the Earth of 26.1°, 45.6°, 60.0° and 70.5°. As the satellite orbits, each region of the Earth's surface is successively imaged by all nine cameras in each of four wavelengths (blue, green, red and near-infrared). MISR can distinguish between different types of clouds, aerosol particles, and surfaces and will be used to monitor the monthly, seasonal, and long-term trends in:

- the amount and type of atmospheric aerosol particles
- the amount, types and heights of clouds
- the distribution of land surface cover, including vegetation canopy structure.

MODIS

With a 2,330-km viewing swath the Moderate-resolution Imaging Spectrometer (MODIS) images every point on Earth every 1–2 days in 36 discrete spectral bands and a 2,330-km wide viewing swath. The wide coverage will enable MODIS, together with MISR and CERES, to determine the impact of clouds and aerosols on the Earth's energy budget. MODIS also has a dedicated channel (centred at 1.375 μm) for detection of wispy cirrus clouds – believed to contribute to global warming by trapping heat emitted from the surface. Other environmental uses of MODIS include:

- Measuring photosynthetic activity of land and marine plants (phytoplankton) to estimate how much of the greenhouse gas is being absorbed and used in plant productivity (also for observing the impact of the El Niño/La Niña 'siblings').
- Estimating areas covered by ice and snow and their seasonal trends.
- Observation of the 'green' wave that moves across continents every spring as biomass increases.
- Monitoring of hazards and disasters including fire, drought, storms and volcanic eruptions.

MOPITT

Measurements of Pollution in the Troposphere (MOPITT) is the first satellite sensor to use gas correlation spectroscopy. As reflected light enters the sensor, it passes along two different paths through onboard containers of carbon monoxide and methane. Each path absorbs different amounts of energy which leads to small differences in the resulting signals that can be used to measure the relative presence of each gas in the atmosphere. This information can add to our knowledge of the interaction between the lower atmosphere and the land and ocean biospheres. MOPITT can also measure the concentrations of carbon monoxide in 5-km layers down a vertical column of atmosphere. Figure 7.1 (colour section) shows how MOPITT can provide continental-scale carbon monoxide datasets.

7.1.3 Envisat

In autumn 2001, the European Space Agency is scheduled to launch Envisat, an advanced polar-orbiting Earth observation satellite which will provide measurements of the atmosphere, ocean, land and ice over a five-year period. It is expected that Envisat (Rast and Readings 1992) will have widespread environmental and other applications such as:

- hydrology
- soil erosion
- flood monitoring
- agriculture
- atmosphere, water vapour and clouds
- aerosol measurement
- climate research.

One of the most important aspects of Envisat (RSAC 2000) will be the ability to perform 'synergistic' analysis of the data from different instruments (for example ASAR, MERIS and AATSR). Three classes of synergistic use have been identified:

- simultaneous acquisition of data of the same location during the same orbit
- multi-sensor and multi-temporal acquisition where data are acquired from different orbits
- data fusion, combining data from individual sensors resulting in a new data product of higher information content.

Simultaneous acquisition of data is subject to operational constraints which place limits on the number and type of products which can be acquired. The second approach will provide better opportunities for data acquisition and while these may not be simultaneous, use of sophisticated modelling techniques will help to offset these limitations. Potential synergistic uses of ASAR, MERIS and AATSR have include agriculture, forestry, semi-arid zone monitoring, monitoring volcanic eruptions, various hydrological applications, climate studies, urban mapping, geology and production of DEMs. The instruments on board Envisat are described below.

AATSR

The Advanced-Along Track Scanning Radiometer (AATSR) is a follow-up to the ERS ATSR-1 and ATSR-2 datasets of precise sea surface temperature (SST) ensuring continuity of the dataset for climate research.

ASAR

An Advanced Synthetic Aperture Radar (ASAR), operating at C-band, ASAR provides continuity with the image mode (SAR) and the wave mode of the ERS-1/2 Active Microwave Instrument (AMI). It features enhanced capability in terms of coverage, range of incidence angles, polarisation, and modes of operation which should in turn lead to better land–surface discrimination.

DORIS

Doppler Orbitography and Radiopositioning Integrated by Satellite (DORIS) provides range-rate measurements of signals from a dense network of ground-based beacons. In addition to enabling orbit determination, DORIS can provide information about solid Earth dynamics, modelling of the gravity field and the ionosphere and for monitoring glaciers, landslides and volcanoes.

GOMOS

The Global Ozone Monitoring by Occultation of Stars (GOMOS) Instrument provides altitude-resolved global ozone mapping and trend monitoring with very high accuracy The primary GOMOS mission objectives are measurement of profiles of ozone, NO_2, NO_3, OClO, temperature, and water vapour needed as input to atmospheric chemistry models.

MIPAS

The Michelson Interferometer for Passive Atmospheric Sounding (MIPAS) is a Fourier transform operating in the near- to mid-infrared where many of the atmospheric trace-gases playing a major role in atmospheric chemistry have important emission features. The objectives of MIPAS are simultaneous and global measurements of geophysical parameters in the middle atmosphere; stratospheric chemistry (O_3, H_2O, CH_4, N_2O and HNO_3) and climatology.

MERIS

The MEdium Resolution Imaging Specrometer (MERIS) instrument is a 68.5° field-of-view pushbroom imaging spectrometer that measures the solar radiation reflected by the Earth, at a ground spatial resolution of 300 metres, in 15 visible and near-infrared bands. The primary mission of MERIS is the measurement of sea colour which can be converted into a measurement of chlorophyll pigment concentration, suspended sediment concentration and of aerosol loads over the marine domain. MERIS allows global coverage of the Earth in 3 days.

MWR

The main objective of the microwave radiometer (MWR) is the measurement of the integrated atmospheric water vapour column and cloud liquid water content. This is an important component of the correction algorithm used for the altimeter (RA-2). MWR data are also useful for measurement of soil moisture, ice classification and for surface energy budget studies.

RA-2

The Radar Altimeter 2 (RA-2) is derived from the ERS-1 and ERS-2 Radar Altimeters, providing improved measurement performance and new capabilities. As with the previous instruments RA-2 will be used to measure ocean topography, sea ice, polar ice sheets, and most land surfaces as well as the determination of wind speed and significant wave height at sea.

SCIAMACHY

The SCanning Imaging Absorption SpectroMeter for Atmospheric CHartographY (SCIAMACHY) will perform global measurements of trace gases in the troposphere and in the stratosphere.

7.1.4 ADEOS-II

The Japanese Advanced Earth Observing Satellite-II (ADEOS-II) is the successor to Advanced Earth Observing Satellite (ADEOS) mission. In common with Envisat, Terra and the other EOS missions ADEOS-II is targeted at the monitoring of global environmental changes such as chlorophyll, vegetation, global water energy and carbon circulation. The main ADEOS-II instruments are described below.

AMSR

The Advanced Microwave Scanning Radiometer (AMSR) will observe various physical parameters connected to water by means of weak microwave signals naturally radiated from the Earth's surface. AMSR will be able to contribute to measurement of, water vapour content, precipitation, sea surface temperature, sea surface wind, and sea ice.

GLI

The Global Imager (GLI) is an optical sensor that will observe the reflected solar radiation from the Earth's surface including land, ocean and clouds. GLI also uses an infrared channel to measure phenomena such as chlorophyll, dissolved organic matter, surface temperature, vegetation distribution, vegetation biomass, distribution of snow and ice, and albedo of snow and ice.

ILAS-II

Developed by the Environment Agency of Japan the Improved Limb Atmospheric Spectrometer-II (ILAS-II) is a spectrometer that will monitor the high-latitude stratospheric ozone using a solar occulation technique. ILAS-II will study changes in the stratosphere which are triggered by emissions of chlorofluorocarbons (CFCs).

POLDER

Polarisation and Directionality of the Earth's Reflectances (POLDER) is a pushbroom, wide-field-of-view, multi-band imaging radiometer/polarimeter developed by the French Space Agency, CNES. It will observe the polarisation, directional and spectral characteristics of the incident solar radiation reflected by aerosols, clouds, oceans and land surfaces.

SeaWinds

Developed by NASA/JPL, the SeaWinds Scatterometer will provide high-accuracy wind speed and direction measurements every two days over at least 90 per cent of the ice-free global oceans. SeaWinds will provide long-term wind data for studies of ocean circulation, climate, air–sea interaction and weather forecasting. SeaWinds is based on the NASA Scatterometer (NSCAT) which flew on ADEOS and will provide similar measurements of ocean surface winds in all weather and cloud conditions.

7.1.5 SPOT Vegetation

The SPOT 'Vegetation' instrument is a joint programme of France, Belgium, the European Commission, Italy and Sweden. The Vegetation programme supplements the locally targeted high-resolution imagery available from the SPOT Image HRV sensors by providing global daily monitoring of terrestrial vegetation, a key dataset for environmental monitoring. The Vegetation instrument, which was first flown on SPOT 4 and has been confirmed for SPOT 5, has had an operational ground segment since 1998. Figure 7.2 shows a Vegetation image of the Indian sub-continent (see front cover for colour version).

Because of the shared platform it is possible to adopt a multiscale approach to the classification of land cover using simultaneous measurements acquired through the Vegetation instrument and the HRV instruments. Environmental models to which the Vegetation instrument can contribute include:

- Surface parameter mapping (albedo, surface roughness, resistances to heat exchanges) for General Circulation Models or forecasting models. Scales addressed in GCM or forecasting models (typically about 100 km) require that land cover and its variability must be determined with a sampling of about 8 to 10 km: the basic spatial resolution needed for identification of land cover and its variability is 1 km, which is the pixel size available from Vegetation.
- Agricultural, pastoral and forest production: including the evaluation of possible global impacts of deforestation and the need for information related to political or social policies and decisions.
- Terrestrial biosphere monitoring and model development.

Table 7.2 provides a comparison of Vegetation to the SPOT HRV instrument.

Table 7.2 Comparison of Vegetation to the SPOT high-resolution instrument

Spectral band	HRVIR		Vegetation	
	Resolution	Swath	Resolution	Swath
B0 (blue)			1 km	2,250 km
B1 (green)	20 m	60 km		
M panchromatic (red)	10 m	60 km		
B2 (red)	20 m	60 km	1 km	2,250 km
B3 (near IR)	20 m	60 km	1 km	2,250 km
SWIR	20 m	60 km	1 km	2,250 km
Global coverage	26 days		1 day	

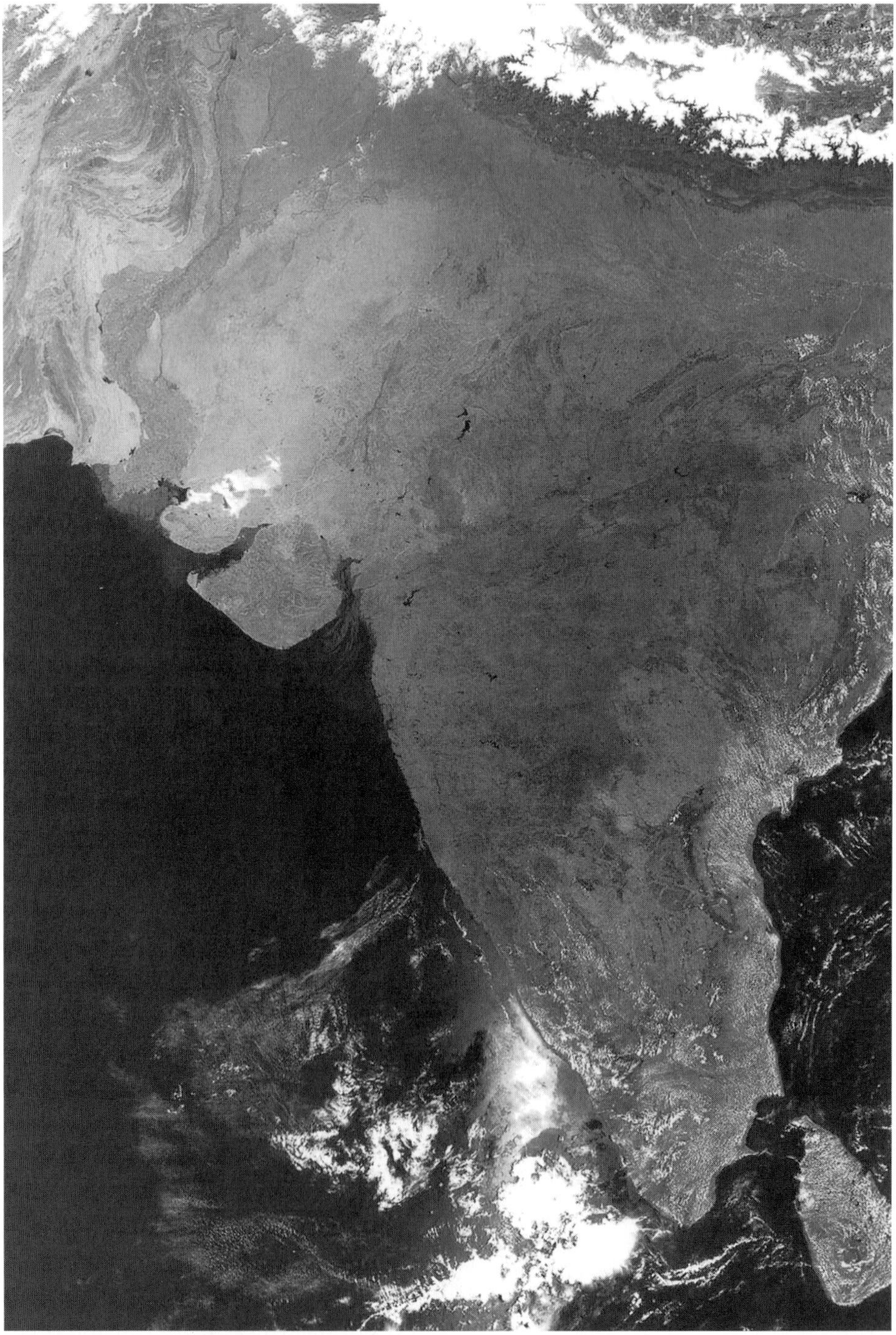

Fig. 7.2 Vegetation image of India, 2 April 1998 (© SPOT Image, CNES 1998).

7.1.6 Indian Remote-Sensing missions

India suffers from many of the problems associated with under-developed countries; it is densely populated and its 700-plus million inhabitants are greatly dependent on agriculture for a living (Rao 1991). The increased agricultural productivity has resulted in large-scale irrigation of over 30 per cent of the sub-continent, compared with a typical figure of less than 10 per cent in many 'developed' nations. However large-scale irrigation, coupled with extensive use of chemical fertilisers, poor drainage and inappropriate agricultural practices, has led to undesirable increase in soil salinity, reducing previously fertile land to an impoverished and unproductive state. India is still inadequately mapped and therefore there is a great need for the collection of information on the country's natural resources and their subsequent effective management. A key part of this process has been the continued role played by the development of India's own national remote-sensing programme, culminating in the launch of Indian Remote-sensing Satellite-1 (IRS-1) in March 1988 and forming the central component in an integrated Indian national natural resources management system encompassing not only agriculture but general land use, geology, hydrology, flood monitoring, snow and ice monitoring and marine studies. The other satellites in the series are: IRS-1B launched in August 1991, IRS-1C launched in December 1995 and IRS-1D launched on 29 September 1997.

The objective of the IRS-1 series is to provide an Indian 'end-to-end' system with India designing, building and operating the satellite, receiving and processing the data and applying the results to the pressing needs of national development. The first step taken towards this objective was the 'Bhaskara' experimental programme carried out from 1976 to1982. The Bhaskara platform featured two types of remote-sensing instruments: a slow-scan vidicon camera operating in the visible and near-infrared with a resolution of 1 km, and various microwave radiometers. Data was used to provide a land cover and geolithological mapping of India at 1:2,000,000 which was relatively coarse but still much improved on previously available information. Based on the experience of the Bhaskara experimental programme it was decided to embark on a higher-resolution pre-operational system. The objectives of the IRS-1 programme were:

- To design and develop an indigenous remote-sensing satellite.
- To develop operational support in the ground segment.
- To use data in a complementary fashion with that from more conventional sources.

Since the launch of IRS-1A the series has provided data for many successful applications in renewable resources as well as other applications such as geology, wasteland mapping and flood damage assessment. One of its most important applications is the search for ground water. India, in common with much of the developing world, needs adequate sources of water for agricultural development and industry as well as drinking and domestic use. Originally, Landsat data was used (Sahai *et al.* 1985) to help to detect the hydrological and geological features that are indicative of ground water occurrence. IRS-1 data was subsequently used (Sahai *et al.* 1991) to prepare hydro-geomorphological maps, from which potential groundwater sites around 'problem' villages can be located. Work carried out in various areas resulted in a success rate of over 90 per cent in most cases (compared with typical results of 60 per cent in cases where satellite data has not been used).

7.1.7 'Smallsat' missions

Many of the EOS and other missions described in previous sections have very large, complex and expensive infrastructures to support them. There is a growing school of thought that believes that 'smaller, faster, cheaper' missions, referred to as 'smallsats', are a more appropriate response to the needs of environmental monitoring.

Low-cost missions

The development of high-density two-dimensional semi-conductor charge-coupled device optical detectors, coupled with low-power consumption yet computationally powerful microprocessors, presents a new opportunity for remote sensing using inexpensive small satellites (SSTL 2001). An example of the new generation of low-cost satellites is UoSAT 12 developed by the UK company Surrey Satellite Technology Limited (see Fig. 7.3). As well as multi-spectral and panchromatic Earth imaging, UoSAT carries payloads for experimental S-band/L-band communications, and operational VHF/UHF store-and-forward messaging and was launched on 21 April 1999.

Another example of a proposed low-cost mission is the Laser Altimetry mission, also known as ICESAT for Ice, Cloud and land Elevation SATellite, which will accurately measure the elevations of the Earth's ice sheets, clouds, and land. ICESAT also will measure the heights of clouds for studies of Earth's temperature balance and will measure land topography for a variety of scientific and potential commercial applications. The total cost of the mission is set at under $200 million, including the launch vehicle and three years of science and data analysis.

Fig. 7.3 UoSAT 12 multi-spectral imager (courtesy Surrey Satellite Technology Limited).

Fig. 7.4 San Francisco Bay, California, taken by TMSAT (courtesy SSTL and Thai Microsatellite Company).

Another smallsat mission is the BIRD mission for the development of a new generation of imaging infrared sensors for Earth remote-sensing objectives, which can be also used for planetary exploration (DLR 2000). The primary applications for BIRD will be detection and identification of hot spots, investigation of biomass burning and processing of thematic fire maps. Figure 7.4 shows a typical image from an Earth Observation smallsat, in this case the San Francisco Bay, California.

Low-cost ground segments

The other element of EO missions where 'small is beautiful' is the ground segment. Direct satellite reception is rapidly becoming a powerful, cost-effective, and easily accessible source of data either as a conjunct to the smallsats described above or as a method of gaining access to the data from one of the larger environmental missions without having to build a complex and expensive ground segment.

Improvements in the cost and performance of computing components mean that the potential for low-cost, high-resolution data capture can now be realised with PC-

Fig. 7.5 RAPIDS installation (courtesy NRI, Greenwich UK).

based systems such as RAPIDS (Real-time Acquisition and Processing Integrated Data System; Downey *et al.* 1998). Typical 'groundstation in a box' systems use high-elevation antennae which can be built for a fraction of the cost of heavier horizon-to-horizon tracking equipment. Such systems are also easy to transport and install in areas where a rapid response is necessary – for example, in response to a natural disaster. This makes near real-time environmental information available to resource managers, planners and decision makers promptly (on demand), reliably and as inexpensively as possible. A RAPIDS installation can be seen in Fig. 7.5.

7.2 ENVIRONMENTAL APPLICATIONS

Environmental concerns remain at the top of the political agenda and there is still much controversy of the degree of threat and possible timescales involved posed by the greenhouse effect on global climatic change. What is clear, however, is that rapid action is necessary to be able to model and quantify changes taking place in the atmosphere–ocean–land system of the Earth. The problem has been formulated by Macdonald (1992):

> '... *we must be able to measure the effect we are having on the planetary environment in a way that allows us to distinguish between the natural evolution of planetary systems and changes which are caused by our activity. To do this requires that we be able to make measurements of the system as a whole, which can only be done efficiently from the vantage point of outer space* ...'

Areas where rapid developments are imperative include:

- establishing a detailed model of the Earth as a system
- increasing the scope and accuracy of terrestrial and atmospheric observations from space
- integrating space and surface observations
- making comparison with historical resources to assess evolution of processes on various timescales.

This section looks in more detail at some of the uses of satellite data in conjunction with GIS, for environmental monitoring at scales from local to global, featuring:

- habitat monitoring
- coastal ecology and marine pollution
- global vegetation mapping
- environmental applications of GPS.

7.2.1 Habitat monitoring

Satellite imagery is particularly suited for habitat monitoring playing a part, for example, in the 'UK land cover mapping project' (Wyatt and Fuller 1992), two of the motivating factors for the wide-scale use of satellite images in this project being:

- continuing improvements in the content, accuracy and availability of satellite image data
- a user community with clearly defined needs for land surface data at national and larger scales.

The project was intended to be operational in the sense that it is largely automated, sufficiently accurate for end-user requirements and produces verifiable results. The flowlines used for the project can be seen in Fig. 7.6.

The primary purpose of the project was to produce '... *a national environmental GIS which will allow investigation of interactions between land use, climate and environmental impacts as a basis for a future modelling capability*' (Wyatt and Fuller 1992). The project was intended to produce a baseline inventory for the UK – not necessarily to provide the capability for repetitive monitoring. Suggested applications of the GIS produced include:

- impact of changed land use using regional and global ecological models
- hydrological modelling such as forecast of river-flows and pollutant loading within aquifers
- mapping of ecological habitats and analysis of the relationship between biodiversity and landscape structure
- use in conjunction with DEMs and digital maps to plan cell-phone networks by modelling attenuation of radio signals by different land cover types (see Section 5.4.1).

Some of the drawbacks found in the original project are the effects of mixels,

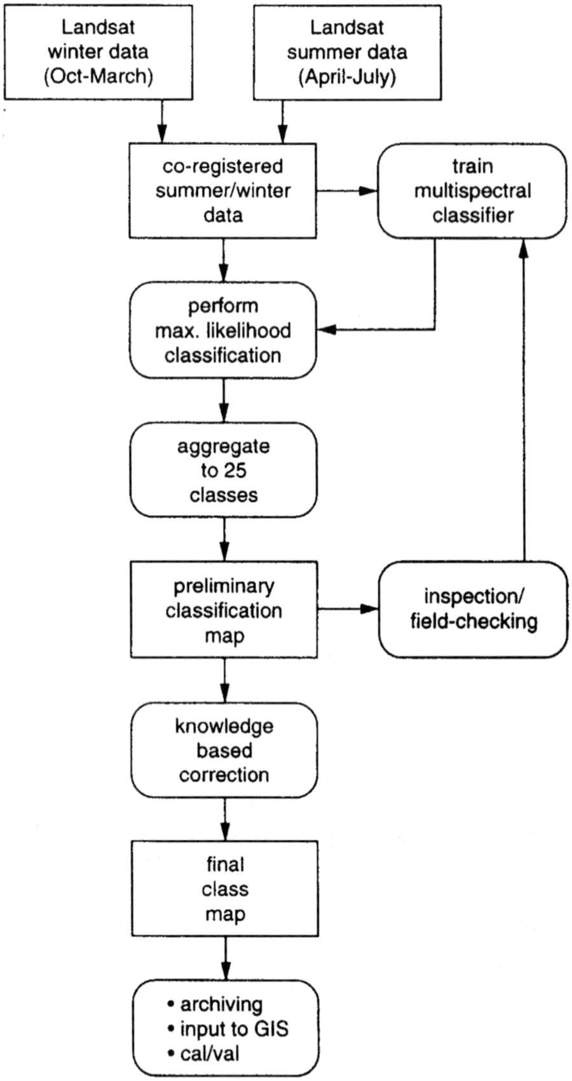

Fig. 7.6 Flowlines used to compile land cover maps of the UK (adapted from Wyatt and Fuller 1992)

misregistration problems, and the differing nomenclatures used for field surveys and image classes. Single season imagery provided inaccurate classifications, therefore classification was performed using both summer and winter imagery of an area. Two validation methods are used: quality control in the processing stage and accuracy assessment in post-processing. A field survey was conducted at 512 sites, using a stratified random selection technique; this showed a classification accuracy of between 85 and 90 per cent for each land parcel. Based on both the successes and limitations of the original project it was decided to update the mapping (Haines-Young *et al.* 2000). Four key refinements are incorporated into the production of the updated map:

- improved accuracy of classification
- added thematic detail
- compatibility with other systems of environmental survey and evaluation
- closer integration between field and satellite data using GIS technology.

The most important development is parcel-based analyses of remotely sensed data rather than using conventional pixel-based approaches. Features in the landscape, such as fields can be analysed and classified as a single entity in their true spatial context. This work was undertaken within the CLEVER Mapping (Classification of Environment with Vector and Raster Mapping) project and developed jointly with Cambridge University Department of Geography, Laser-Scan Limited and the Ordnance Survey under the British National Space Centre's LINK programme. Figure 7.7 shows a comparison of methodologies for a site at Allt a'Mharcaidh, Cairngorms (Scotland). The map from 1990 shows a number of mixed and isolated regions; whereas the map from 2000 shows a much clearer and more accurate division into distinct areas of land cover.

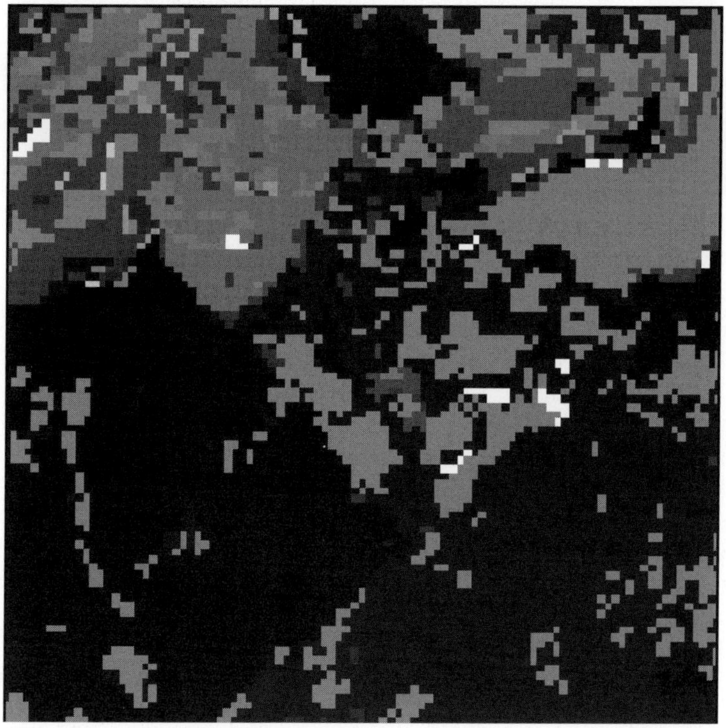

Fig. 7.7 UK Land Cover Mapping Project: comparison of the 1990 Methodology (above) with the 2000 Methodology (next page), at Allt a'Mharcaidh, Cairngorms, Scotland (courtesy DETR/DEFRA).

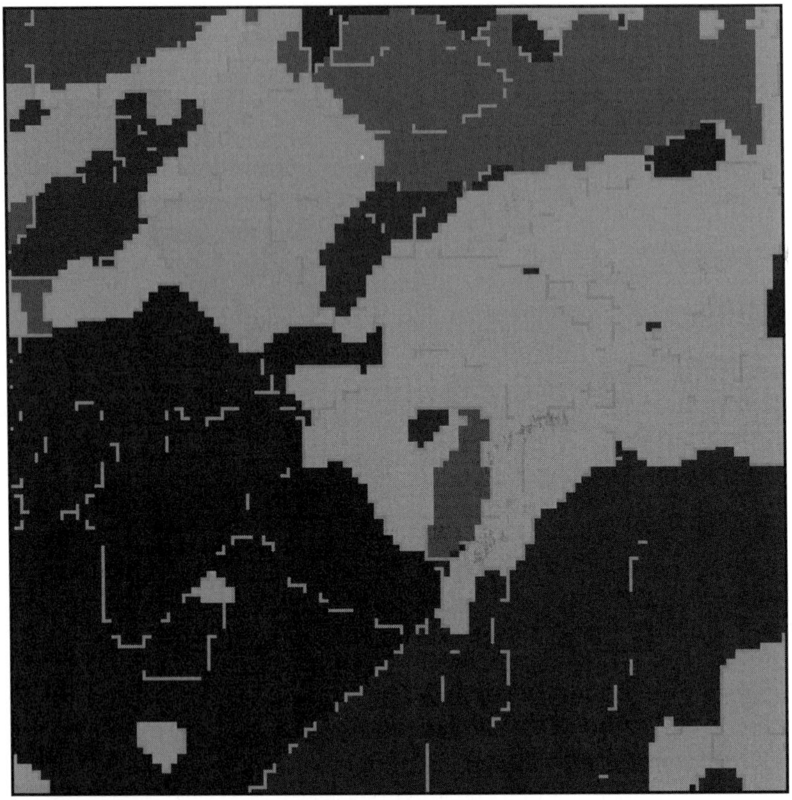

Fig. 7.7 *cont.*

7.2.2 Coastal ecology and marine pollution

Another application for which satellite data can successfully be used is coastal ecology, for example (Kuchler *et al.* 1986) the study of coral reef geography, form, surface cover and vegetation. In particular, remote-sensing techniques have been applied to the management of the Great Barrier Reef (GBR) which lies on the north-east continental shelf of Australia. The reef, designated a world heritage site, is one of the most diverse ecosystems on Earth comprising a series of about 2,500 reefs with a total length of about 2,000 km. The area of the Great Barrier Reef Marine Park is approximately 345,000 km^2, thus providing a very large object for local survey. Interest in using satellite data as a possible alternative to conventional mapping (Harrison *et al.* 1989) led to the initial development of the 'Barrier Reef Image Analysis' (BRIAN) package in 1983. This was used to produce geocoded image maps for the whole of the barrier reef at a scale of 1:250,000 using standard multi-spectral classification techniques, as well as 'depth of penetration' algorithms using Landsat TM data.

A wide range of sensors can be used for coastal applications from airborne sensors such as CASI to 'general purpose' satellites such as Landsat, SPOT and ERS and specialist missions, pre-eminently the Coastal Zone Colour Scanner (CZCS) carried on the NIMBUS 7 satellite. Since much of the CZCS instrument's observing time was dedicated to the coastal zone, some of the most visible features are the interface where fresh water encounters salt water. Figure 7.8 (colour section) shows a CZCS image acquired on 9 October 1984 highlighting a plume of sediment discharging into the Adriatic Sea from the mouth of the Po River, just south of Venice, Italy. For comparison the right-hand image is a photograph of the region taken by astronauts on board the Space Shuttle Challenger on the same day. The island city of Venice is visible in the lagoon located at the upper centre of the shuttle photograph.

The Clean Seas project (SOS 1999) was a three-year investigation, part funded by the European Commission, to investigate marine pollution monitoring using satellite borne instruments. The project produced a number of important results, for example within a 2-hour period on the morning of 15 July 1997 images were acquired over an extensive algal bloom in the Baltic Sea. For the first time, colour, temperature and SAR observations of the same bloom were available, allowing aspects of the biochemistry and imaging physics to be examined in detail. Another development is the production of maps of fronts and eddies from a variety of spatially referenced data (in particular by the use of thermal sensors) provide input to dynamic GIS models which allow the movement and behaviour of polluted components of sea water to be predicted.

7.2.3 Global vegetation mapping

At the United Nations Conference on Environment and Development held in Brazil in 1992, Agenda 21, an action programme for addressing global environmental challenges, while continuing to support sustainable economic development, was resolved. Agenda 21 clearly makes the case that baseline data, particularly spatially related information, on key environmental parameters is of great importance. For many parts of the developing world such as Africa and Asia it is vital to be able to able to maintain records of land cover, especially vegetation, over wide regions and with a rapid update rate in order to follow the dynamics of vegetation during the growing season. Before the advent of satellites the limited resources available for ground and aerial surveys in such countries often proved inadequate to contribute to the planned and effective use of the available natural resources necessary for a sustained social and economic growth. A second imperative for the reliable classification of land cover at continental and even global scales is the impact of land cover dynamics on the global energy balance, the hydrological cycle and the production of carbon dioxide. Data sources previously available have proved inaccurate, inconsistent and difficult to keep up to date – for example, global estimates of tropical rainforest areas compiled in 1979 differed from each other by up to 50 per cent. Since the availability of Landsat images in 1972 it has at least been possible to conceive of a practical earth resources monitoring programme using satellite data, but the transfer of remote-sensing technology from the research sector to operational use has proved difficult in practice. One obvious factor is the sheer volume of data involved – for example, the continental-scale coverage of Africa would require of the order of 1,000 Landsat scenes. The amount of processing required to obtain suitable images and process them on a one-off basis would be phenomenal, and to do so on a repetitive monitoring basis would be almost impossible with current technology.

A realistic alternative is the use of AVHRR-derived vegetation indexes (see Section 2.3.2). Among the topics in vegetation mapping that may be addressed by these methods are:

- detection and monitoring
- classification
- evaluation of vegetation vigour
- estimates of biomass.

In the longer term, such information could prove vital as input to models which relate the vegetation cover of the Earth to geochemical, hydrological and climatic cycles. Probably the most used vegetation index is the NDVI. Typical NDVI values determined from examining the global area coverage (GAC) of land cover categories such as oceans, lakes, deserts, grassland, and forests can be seen in Table 7.3.

An early example of the utility of AVHRR for large-scale vegetation monitoring (Malingreau 1986) was the use of weekly data to analyse a wide variety of vegetation formations ranging from the desert of Central Asia to the tropical forests of Borneo. Development curves of the NDVI index over the period 1982–85 demonstrated the large amount of useful information that can be derived. AVHRR data also formed the basis of NDVI calculations for the whole of Africa sampled at 8 km resolution (Townshend and Justice 1986). In this case temporal compositing of the images was used to reduce the effects of cloud cover and atmosphere. Areas of different land cover classes are distinguishable by multi-temporal analysis of the changing NDVI over a year. For example, deserts have a very low NDVI all the year round but tropical rainforests have a much higher NDVI as well as marked seasonal peaks (see Fig. 7.9).

Figure 7.10 shows the first 1-km resolution land cover map of all South America (Stone *et al.* 1994). South America has been divided into 39 different land covers

Table 7.3 Typical NDVI values (based on Holben 1986)

Land cover category		Red	Infrared	NDVI
Green-leaf vegetation:	dense	0.050	0.150	0.500
	medium	0.080	0.110	0.140
	light	0.100	0.120	0.090
Bare soil		0.269	0.283	0.025
Clouds		0.227	0.228	0.002
Snow/ice		0.375	0.342	–0.046
Water		0.022	0.013	–0.257

Note: Values were derived from NOAA-7 data.

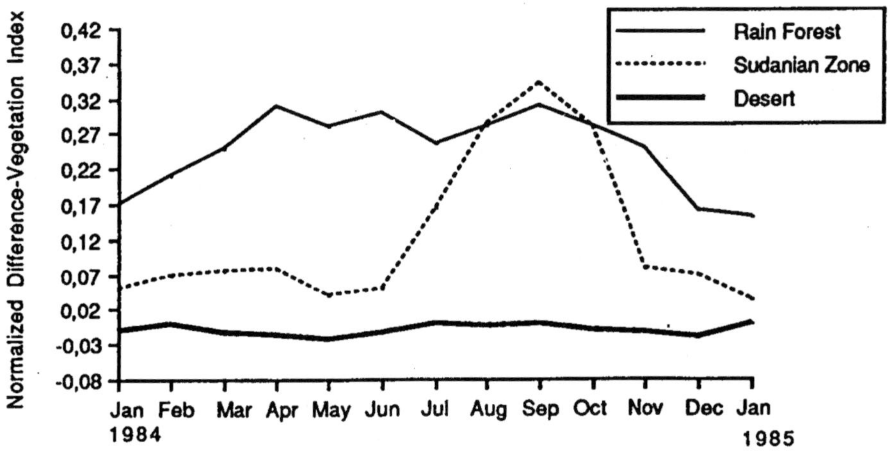

Fig. 7.9 NDVI seasonal development curves for three African land cover types in 1984 (from Belward and Valenzuela 1991).

including recent deforestation. The map uses 1 to 15 km resolution phenological (vegetation or greenness index) data from a time series of satellite data as a starting point. The work has also led into improved techniques for observing selective logging and associated fires in Brazil. For example, a single fire scar covering 1,000 km^2, discovered in TM imagery and verified in the field, is equivalent to 4 per cent of the total area estimated deforested per year.

7.2.4 Environmental applications of GPS

Typically environmentalists will use information from Earth Observation (EO) data and Geographic Information System (GIS) technology to create maps, analyse sampling data, and develop dynamic models to predict, for example, what will happen to a water well near a hazardous waste site (Fisher 1993). The reliability of any analyses or predictions depends to a large extent on the accuracy of the map features developed by the GIS. GPS can be used to provide accurate base points that the GIS uses when making its maps and data layers (for example, the geo-rectification of satellite imagery). Without GPS those map projections can be inaccurate and, therefore, predictions based on modelling can be called into question. There is a very wide range of environmental applications of GPS (US Department of Commerce 1998) including:

- *Ground mapping of ecosystems* – GPS is used to map all manner of ecosystems, from eel grasses in the Puget Sound to coastlines in Louisiana.
- *Overviews of environmental phenomena* – Comprehensive views of deforestation, as well as environmental phenomena in lakes, rivers and estuaries, are analysed through overviews of a variety of spatial features that are referenced to GPS-derived coordinates.

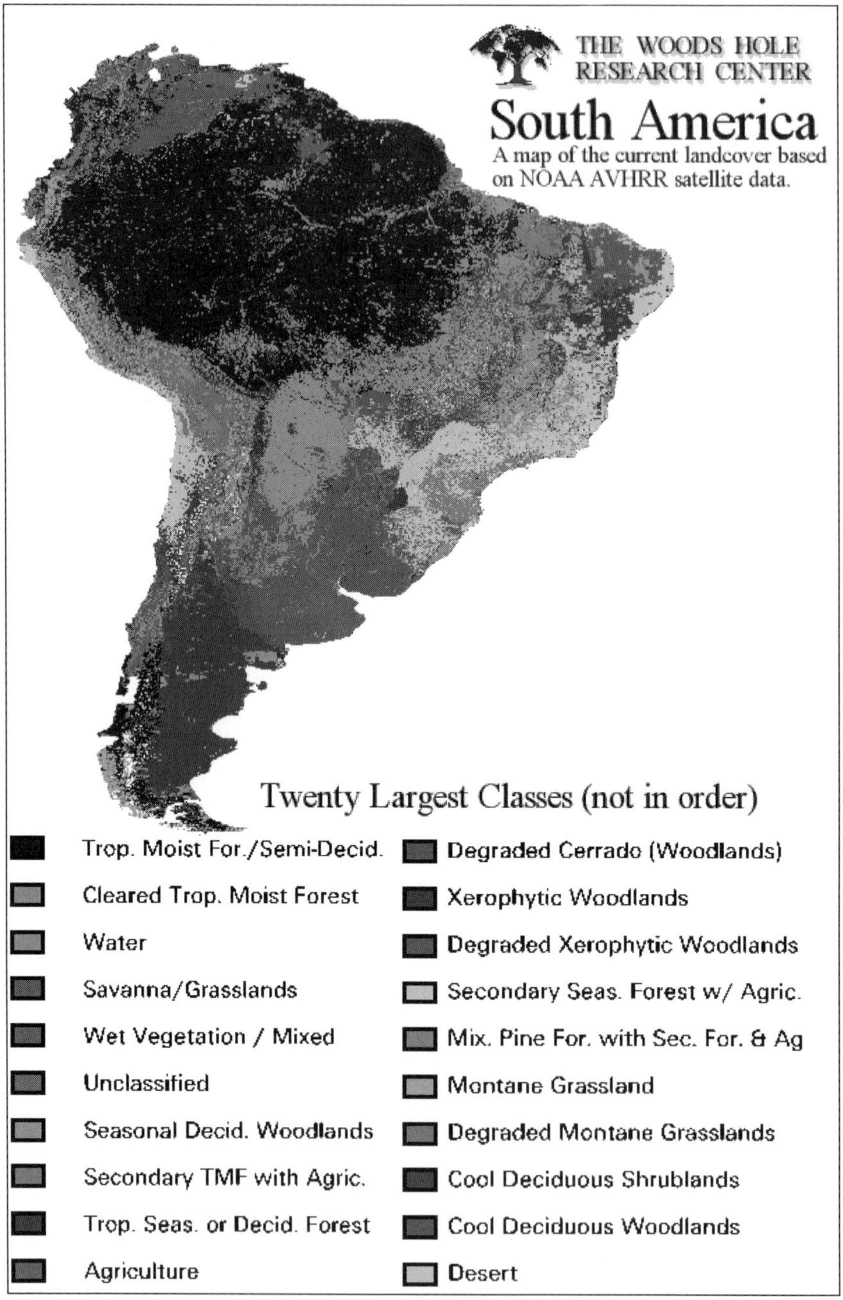

Fig. 7.10 Forest map of South America derived from meteorological satellite data from the AVHRR sensor (courtesy NASA, Woods Hole Research Centre).

- *Preventing ground water pollution* – GPS is used to perform exact location inventories of wells and potential contamination sources and map the migration of toxic plumes in ground water.
- *Monitoring health of the food chain* – If microscopic plankton die, marine life higher on the food chain also perish. GPS is used to collect scientific samples for studies on the effects of ultraviolet radiation and other hazards on bacterio-plankton.
- Air pollution measurement – GPS receivers can be coupled with gas sensors to map satellite data with ground-based samples.
- *Mapping of sub surface contamination* – GPS has been used to map pollutants in Arctic ice. Results showed that even in the remote arctic, industrial chemicals have penetrated into the ecosystem.
- *River management to avert natural disasters* – In Paraguay, GPS is used to detect natural canals and precisely determine the height of new channels to facilitate correct sediment deposit, and thereby controlling water flow and preventing floods.
- *Flood control facilities* – GPS is used to plan, design, construct and maintain needed flood control and drainage facilities and to protect and increase the quantity and quality of ground water in many parts of the western United States.
- *Oil spill tracking and cleanup* – The National Oceanic and Atmospheric Administration (NOAA) uses GPS affixed to buoys to track the movement of oil spills and to monitor how fast a spill is spreading. This information helps to manage the work of emergency crews more efficiently.
- *Management and maintenance of roads* – Using GPS, Italian companies have developed a system of computers and imaging equipment to gather information about road networks with inventories of features and attributes required for administration and maintenance.
- *Hazardous waste site investigation* – With the protective gear necessary at a waste site, it is difficult to use conventional surveying techniques. GPS is often used to perform real-time surveys using digital terrain modelling at sites heavily contaminated with asbestos, lead dust, PCBs and other hazardous materials.
- *Precise location of stored hazardous materials* – GPS is used in Department of Energy efforts to stabilise nuclear wastes from the Rocky Flats nuclear weapons plant, which operated from 1950 to 1980. Surveyors and engineers used GPS to precisely map the locations of hazardous waste sites.
- *Time-tagging of hazardous materials* – Spills or other incidents, for example, the remediation of contaminated sites such as the reclamation of a former Texaco refinery site in Illinois.
- *Time-tagging of dredge dumping operations* – GPS was used on drainage barges to tag the location and time of dumping to ensure adherence to Environmental Protection Agency (EPA) regulations and public safety concerns.
- *Monitoring of natural gas and oil pipelines* – A German firm has used GPS to

Figure 7.12 shows Landsat 7 images of the Euphrates River in south-east Turkey and north-west Syria. The large reservoir to the north-east is from the Ataturk dam, which was completed in early 1990. As of 18 May 2000 the waters of the Euphrates were rising by 30 centimetres a day and had already submerged the villages of Belkis and Apamee. The progress can be clearly monitored using geo-referenced multi-temporal imagery.

Another example of the use of EO for flood monitoring is the RAPIDS (Real-time Acquisition and Processing Integrated Data System) which covers the Ganges and the Brahmaputra–Jamuna floodplain (Downey *et al.* 1998). Over the long term, the radar image archive of Bangladesh in the monsoon season will facilitate mapping of hazards in areas where the population fishes and farms (growing rice in particular), when the floodplain is subject to unusual flooding or droughts. The radar archive also serves to identify river migration and coastal processes due to the yearly monsoon and flooding.

7.3.2 Volcanoes

Earth observation can be used for a number of applications in vulcanology, including:

- detection of new eruptions
- distinguishing between volcanoes and other hot spots such as forest fires
- use of SAR interferometry for mapping of topographic change
- thermal mapping of lava lakes and lava flows
- analysis of gas content (such as sulphur dioxide) and particle size distribution in volcanic clouds.

Satellite data can be used in complementary ways – for example, using high-resolution imagery to provide close-up coverage of active vents combined by daily regional coverage of the thermal distribution on a wider area. One specific example of the utility of satellite images for volcanic monitoring was the detection of snow-melt on the top of the Nevado Sabencaya mountain in Columbia (Abiodun 1992) which provided warning of the imminent volcanic eruption that took place on 14 July 1990. Other examples of applications of remote sensing for vulcanology include observing volcanic ash clouds using AVHRR data (Prata 1989). Figure 7.13 (colour section) shows an image of the Mount Vesuvius volcano acquired by the Terra Advanced Spaceborne Thermal Emission and Reflection Radiometer (ASTER) on 26 September 2000.

One forthcoming application of the MODIS sensor on Terra is to observe the progress of volcanic eruptions and to develop near real-time alarms for hot data flows. GIS-referenced data to be acquired includes the characteristics of specific eruptions, dispersal of eruption plumes and the geology of individual volcanoes (Mouginis-Mark *et al.* 2001). Another new technique being developed is the detection and quantification of the changes in area of new lava flows using repeat-pass radar interferometry. This technique has been demonstrated using data from the second SIR-C flight in October 1994 and can be used to build GIS databases against which future eruptions can be compared.

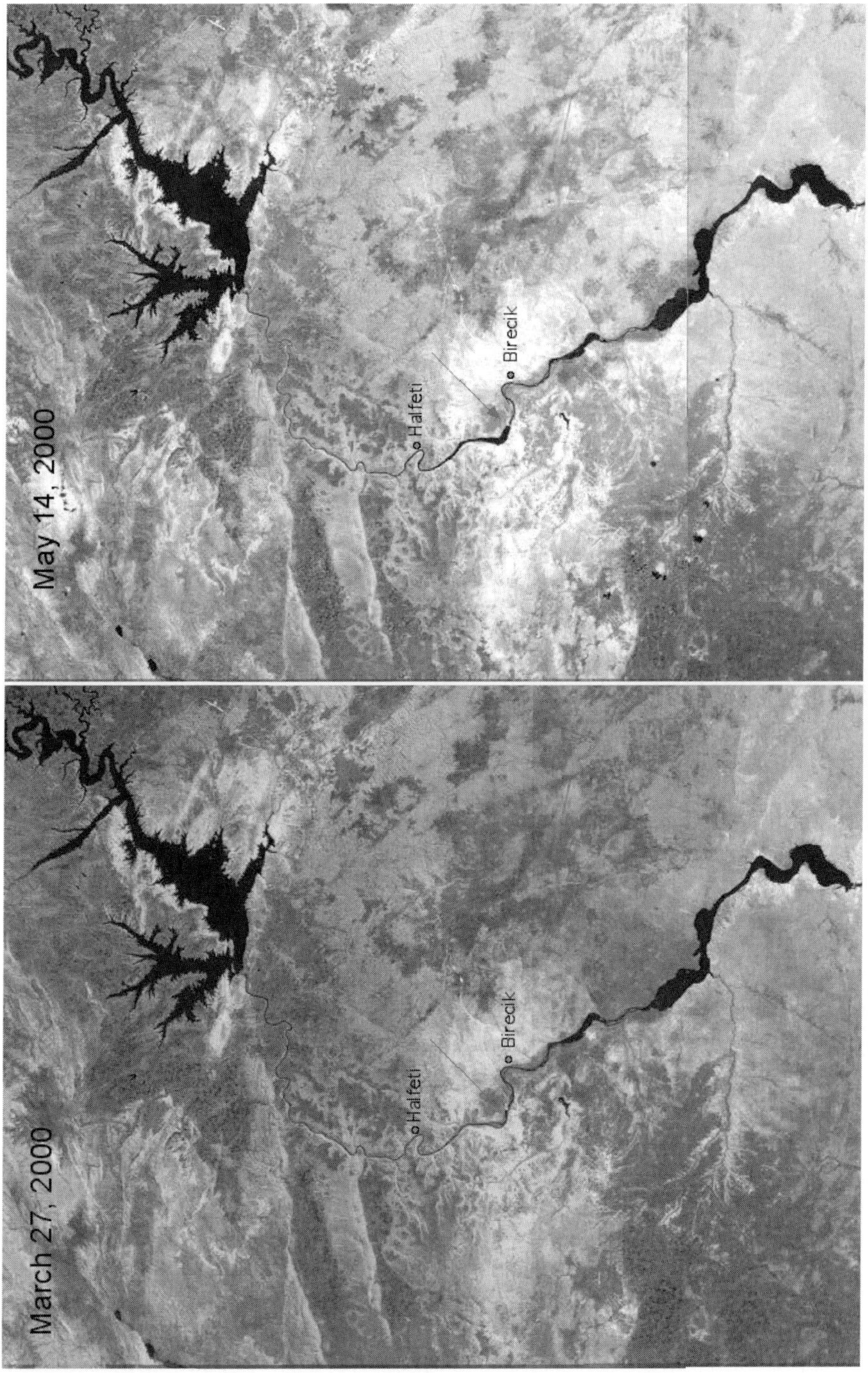

Fig. 7.12 Landsat images (27 March, 2000, 14 May, 2000) of Ataturk Dam. The arrow points to the reservoir of the new Birecik Dam, which began filling at the end of April 2000. (Copyright 2000 Dartmouth Flood Observatory.)

7.3.3 Earthquakes and landslides

One of the major problems with earthquakes is their extreme unpredictability and, because of this, EO in common with most technologies is of very limited use in the pre-earthquake phase. However, in terms of the mitigation of the effects of earthquakes, EO has a role to play, particularly in developing countries, for base-mapping for emergency relief logistics and estimation of settlement and structure vulnerability (CEOS 2000). Following an earthquake GIS-based damage mapping is of use not only to relief agencies that need to locate possible victims and structures at risk, but also the insurance industry (which needs to assess losses).

SAR interferometry (InSAR) is increasingly used for the mapping of seismic ground deformation and was widely used for the Izmit (Turkey) earthquake of 17 August 1999. InSAR is useful in the response phase as ground displacement can correlate with damage in built environments. The technique is, however, limited by data availability and processing constraints which make it difficult to use globally or routinely. One way around this is the development of a technique known as Permanent Scatterer InSAR, which makes use of natural and artificial corner reflectors. This technique could eventually provide a cost-effective supplement to expensive ground-based GPS and laser-ranging networks.

In addition to earthquakes, landslides pose serious threats to settlements and structures supporting transportation, natural resources management, and tourism. Landslides can be triggered by other natural hazards (earthquakes, floods, volcanic eruptions) or by man-made problems such as deforestation). Underwater landslides can also cause destructive waves (tsunami). Present Earth Observation satellite data can be used to assess damage, assist in warnings, or assess the risk of landslides and the use of SAR is again a tool with great potential (see Fig. 7.14).

7.3.4 Fire

Every year more than 50 million hectares of forest is burnt down, the majority of which has been directly ignited by humans rather than being attributable to natural causes such as lightning and volcanoes. Approximately 20 per cent of the forests burnt are boreal and 80 per cent tropical rainforest. Forest fires impact directly on human lives and indirectly through atmospheric pollution. Because it is important to be able to observe the evolution of fire through time, the ability to acquire and interpret multi-temporal image sets of a threatened region is of paramount importance. For example, a three-year time series of Landsat TM and SPOT XS images was used to analyse areas devastated by fire on the island of Sardinia (Marini and Mellis 1992).

Another example of the utility of EO data for fire monitoring (Lohi *et al.* 1999) is a project to detect forest fires operationally in Finland and neighbouring countries. Potential fires are detected using an algorithm based on NOAA AVHRR channel 3 (3.7 μm) data:

- All pixels with a 'brightness temperature' above a certain threshold are selected.
- GIS spatial analysis tools are used to group adjoining hot pixels into 'fire patches'.

Fig. 7.14 Subsidence mapping using satellite interferometry (courtesy NPA).

- Using a classification technique on channels 2 (near infrared) and 4 (thermal infrared) – which excludes scattered sunlight – each fire patch is tested to see if it is in a cloud-free zone.
- A screening algorithm is used to remove areas of Sun glint.
- Finally, digital maps of urban boundaries can be used to exclude heat sources in industrial sites.

A similar approach, using AVHRR images and GIS spatial analysis techniques, has been employed to develop algorithms which can be used to provide daily fire monitoring across Canada (Li 2001). Some of the geospatial datasets obtained are shown in Fig. 7.15. This shows the ability to detect hot spots by thermal emissions and smoke plumes. Also, for extended areas of fire the satellite-derived fire maps (below) show much more information than conventional ground-based monitoring which can only determine the boundaries of fires.

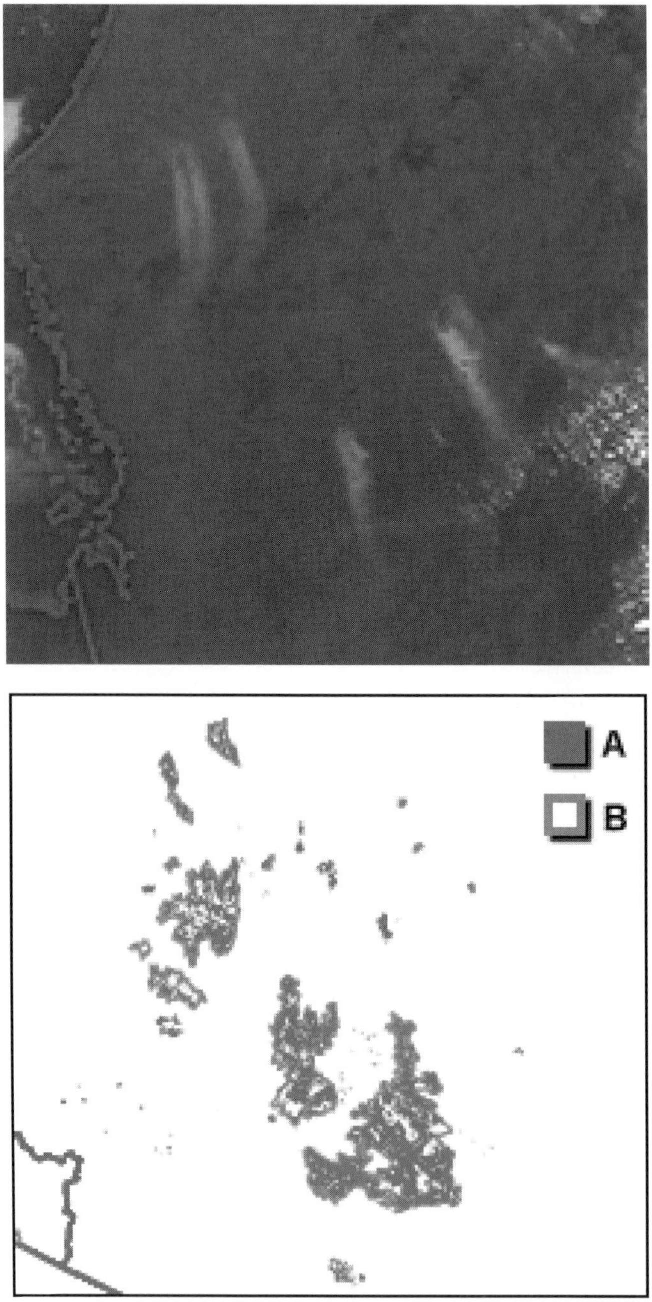

Fig. 7.15 (a) AVHRR image showing smoke plumes and computer-generated hotspots in red; (b) comparison of fires detected by satellite (A) and conventional means (B). (Both figures courtesy of Canada Centre for Remote Sensing.)

The above examples show once again how AVHRR data has proved of value for applications far outside the original mission objectives. The success of AVHRR has been used a blueprint for other missions such as the Terra component of EOS. The MODIS instrument aboard Terra has special 1-km resolution 'fire channels' at 4 and 11 μm, with high saturation at 500 and 400 K respectively, which allow four daily fire observations to be made (Kaufman *et al.* 1998). Such data can be used to monitor burn scars and smoke aerosols, and contributes to models of the fire process and its relationship with climate, atmosphere and ecosystems. By using GIS to overlay and perform spatial analysis on MODIS fire products, it should be possible to model the spatial and temporal distribution of fires regionally and globally, and to track the progress of fire frontiers and wild fires. Although reliable fire detection and monitoring using EO has been proved on a global basis with AVHRR sensors on the NOAA satellites, it is not operationally effective for real-time emergency response because the repeat period is too infrequent and the level of detailed information is insufficient. To help to overcome these limitations a dedicated satellite mission, known as FUEGO, is currently under development. The initial constellation of four FUEGO satellites will provide system service eight hours per day to 2005, and subject to approval of European Space Agency funding the system will be fully operational 24 hours per day with 12 satellites for seven years (FUEGO 2001).

7.3.5 Oil-slick detection

The three main causes of off-shore oil slicks are:

- tanker accidents, when large amounts of oil are spilled into the sea
- illegal oil discharges by ships during 'normal operations'
- natural oil seepage.

A range of conventional methods (Lankester 1998) are used to detect and monitor oil slicks. In ports and estuaries visual identification from surface vessels can be used but on the open seas aircraft surveys are needed. Typically a patrolling aircraft will fly at 1 km altitude using Sideways Looking Airborne Radar (SLAR) to detect the smoothing of the sea surface caused by an oil slick. Once a potential slick has been identified, the aircraft can fly at a lower altitude and use the ultraviolet (UV) and infrared (IR) scanners to positively identify the nature of the slick. If the vessel suspected of the discharge is in the area, then photographic and video evidence may also be gathered. However, such surveys over large areas (e.g. the Mediterranean Sea) to check for the presence of oil are limited to daylight hours in good weather conditions.

Satellite imagery can be used as a 'focusing tool', identifying probable spills over very large areas and then guiding aerial surveys for precise observation of specific locations.

SAR data has proved to be the most effective means of monitoring oil pollution, with oil slicks appearing as dark patches on SAR images because of the damping effect of the oil on the backscattered signals from the radar instrument. In addition, ships in the area will appear as bright 'corner reflectors' with characteristic wake patterns. The location of the oil spills is crucial to the success of the exercise, so GIS

Fig. 7.16 (a) ERS SAR image showing an oil slick, the ship responsible and the wake of the ship; (b) slick statistics in GIS form (Courtesy Earth Observation Sciences Limited.)

technology is of value to extract the information and present the results (Jensen *et al.* 1990). Figure 7.16 shows SAR detection of oil spills, firstly in image format and then as a 'vector' representation of the slick being studied.

7.4 POPULATION DYNAMICS AND SECURITY

All the previous examples in this chapter (environmental monitoring, hazard and disaster mitigation) show the use of satellite data in many areas that can have an acute or chronic impact on human society. In this section we examine in more depth other aspects of 'population dynamics and security' where GIS-enabled satellite data also has a role to play. The topics addressed are:

- urban development and dynamics
- disease transmission
- food security
- humanitarian operations and rehabilitation.

7.4.1 Urban development and dynamics

Earth observation, particularly coupled with GIS, can provide a major source of information for a wide range of urban development studies including monitoring dynamic patterns of growth, assessing trans-border development and describing infrastructures for transport, energy and communications. The resulting land use data can then be integrated with more traditional socio-economic datsets in a GIS to produce indicators for sustainable coverage. An idea of this approach is given by a study of the city of Bratislava made by the European Union (Ehrlich and Mehl 2001). Since the end of the Second World War, Bratislava has seen many changes due to political situations as well as its location as a major gateway between Eastern and Western Europe. Figure 7.17 (colour section) shows the evolution of the city using 'reconstructed' land use maps for the years 1949, 1969, 1985 and 1997; initially with aerial photography and subsequently with satellite data.

The use of urban GIS is also a keystone of the 'Murbandy' Project of the European Joint Research Centre (JRC) which aims to measure, analyse, understand and forecast the urban expansion of some European cities (Ehrlich *et al.* 1998). The European cities being studied are Helsinki (Finland), Tallinn (Estonia), Göteborg (Sweden), Copenhagen (Denmark), Sunderland (England), Dublin (Irish Republic), Brussels (Belgium), Essen, Dresden and Munich (Germany), Praha (Czech Republic), Bratislava (Slovakia), Vienna (Austria), Lyon, Grenoble and Marseille (France), Milan, Padova-Mestre and Palermo (Italy), Bilbao (Spain), Porto and Setubal (Portugal), Iraklion (Greece) and Nicosia (Cyprus). Murbandy is divided into three modules:

- measuring city dynamics by creating datasets of current and past land uses and transportation networks
- analysing the changes and relating them to socio-economic and physical models

Fig. 7.18 (a) IRS-1C image of Waterloo District of Brussels; (b) Derived GIS layers showing land use. (Institut de Gestion de l'Environnement et d'Aménagement du Territoire (IGEAT), Université Libre de Bruxelles (ULB).)

- developing spatial models of the urban expansion for the next few years.

The project makes use of existing GIS data and current and historical aerial photography as well as satellite images. Figure 7.18 shows the development of land use databases for the Waterloo district of Brussels with IRS-1C data and aerial photography (Fricke and Wolff 2001).

Satellite data is also a key part of a US project to model long-term urban growth using cellular automata and GIS (Clarke and Gaydos 1998). The model consists of a set of temporal GIS databases mapping human-derived land transformations in large metropolitan areas over the previous century. These datasets are based on historic maps and Landsat images (acquired since 1972) in conjunction with digital line graphs (DLG), digital elevation models and local land use maps. Computer animation is used effectively to illustrate the database and emphasise the impact of incremental urban growth, conveying the dramatic land use changes that result from urbanisation.

7.4.2 Disease transmission

Disease transmission remains a major problem in many areas of the world; for example, the World Health Organisation (WHO) estimates some 110 million clinical cases of malaria world wide per year, the vast majority of which are in sub-Saharan Africa (Connor *et al.* 1995). Vector-borne disease transmission is often correlated with spatial and temporal changes in the local environment, both of which can be assessed to a certain extent using satellite images. Because of the rapidly changing nature of diseases it is important to be able to obtain frequent imagery of the area of interest. This time series of images can then be geocoded, classified and incorporated within a GIS containing other spatially related information such as locations of known incidences and the spatial and temporal dynamics of the disease being tracked.

An example of the GIS approach is a study into the transmission of malaria in the Nadiad Taluka of Kheda district, Gujarat in western India (Malhotra and Srivastava 1995). The aims of this study were to:

- map the geomorphological, physical and climatic characteristics of the region using satellite images and existing paper maps
- identify relevant factors having a direct or indirect bearing on malaria transmission in Nadiad Taluka
- develop a GIS to charactertise the spread of malaria.

The study concluded that every village in the Nadiad region has at least one pond from which water is gathered. In an attempt to fight famine in the Nadiad area, these village ponds were deepened. Although this has created a year-round supply of pond water, it also creates conditions of high humidity where malaria disease-vectors develop rapidly. The information on pond location was extracted from topological sheets at a scale of 1 : 250,000 (showing topography and terrain of Nadiad) and combined with geographic information derived from satellite images of the region to provide the data layers to the GIS.

7.4.3 Food security

The security of food supplies continues to be of grave concern in many countries of the world. The Famine Early Warning System (FEWS) is an information system designed to help decision makers prevent famine in sub-Saharan Africa. FEWS specialists in the USA and Africa (McGuire 1997) look for early indicators of potential famine areas by assessing satellite imagery, weather data and ground-based crop information. The most important use of satellite data is to separate agricultural from non-agricultural land by stratification of NDVI imagery. Where more detailed local information is required, Landsat images are also used. In conjunction with this information other factors affecting local food availability and access are incorporated to identify the most vulnerable population groups that may require further assistance. These assessments are updated regularly to provide decision makers with timely and accurate information.

Satellite imagery such as AVHRR offers a cheap method of obtaining synoptic conditions on the habitat of the desert locust and hence of ensuing famines (Cherlet *et al.* 2000). Rainfall is the first indicator of a potential habitat, but estimates based on satellite data are often not adequate over desert areas. A certain amount of success has, however, been achieved by using AVHRR NDVI time series to measure the secondary effects of rainfall such as soil moisture and vegetation growth. More accurate data is potentially obtainable using the SPOT Vegetation instrument (see Section 7.1.5) and such data has been successfully introduced into the operational activities of the Desert Locust Control Group in the United Nations Food and Agricultural Organisation (FAO). The imagery is used to derive an NDVI which is overlaid in a GIS window representing the complete Desert Locust recession area.

7.4.4 Humanitarian operations and rehabilitation

The European Union (EU) is one of the largest participants in humanitarian aid operations around the world. As part of the ongoing post-conflict rehabiliation and reconstruction activities in Kosovo the EU has produced a prototype GIS (Ehrlich and Mehl 2001) to provide territorial and socio-economic information for field officials and also to provide a 'mobile GIS' that can be used for rapid field data collection of georeferenced data for humanitarian aid situations. The GIS includes a variety of spatially-referenced data (such as digital map, satellite images and digital elevation models) and attribute data (such as photographs, reports and links to internet sites). The system can be used for relief workers or reconstruction teams engaged in Kosovo – for example, the DEM data can be used to provide perspective views of the area where operations are taking place. The GIS can be installed on a laptop enabled with GPS and a digital camera for field data collection; in addition links to mobile telecommunication devices give rapid delivery of ground data to a central control site.

REFERENCES

Abiodun AA (1992) Natural disasters and their mitigation using space technology. *Proceedings of the Central Symposium of the 'International Space Year' Conference, Munich, Germany*, 30 March – 4 April 1992, ESA ISY-1, pp. 573–80.

Barrett EC and Beaumont MJ (1992) Alternative strategies for the use of satellite data analysis systems for hazard monitoring and disaster mitigation. *Proceedings of the*

Central Symposium of the 'International Space Year' Conference, Munich, Germany, 30 March – 4 April 1992, ESA ISY-1, pp. 581–6.

Belward AS and Valenzuela CR (eds) (1991) *Remote Sensing and Geographical Information Systems for Resource Management in Developing Countries*. Kluwer Academic Publishers, Dordrecht, The Netherlands.

CEOS – Disaster Management Support Group (2000) *Earthquake Hazard Team Report*.

Cherlet M, Mathoux P, Bartholomé E and Defourny P (2000) Vegetation contribution to the desert locust habitat monitoring, *Vegetation-2000 Conference, Lake Maggiore, Italy*, 3–6 April 2000.

Clarke KC and Gaydos L (1998) Long-term urban growth prediction using a cellular automaton model and GIS: applications in San Francisco and Washington/Baltimore. *International Journal of Geographical Information Science*.

Connor SJ, Thompson MC, Flasse S and Williams JB (1995) The use of low-cost remote sensing and GIS for identifying and monitoring the environmental factors associated with vector-borne disease transmission. In de Savigny D and Wijeyaratne P (eds) *GIS for Health and the Environment*. IDRC.

DLR – Deutches Zentrum fur Luft- und Raumfahrt (2000) Small satellite project BIRD web-site. http://www.ba.dlr.de/NE-WS/projects/bird/

Downey ID, Williams JB, Archer DJ, Stephenson JR, Stephenson R and Looyen R (1998) Enabling local user access to remote sensing data: RAPIDS a practical and affordable X-band ground station for developing countries. *Proceedings of 27th International Symposium on Remote Sensing of Environment, Tromso, Norway*, 8–12 June 1998.

Ehrlich D, Annoni A and Lavalle C (1998) Monitoring the sustainability of Europe's urban areas the MURBANDY project. *Proceedings of 27th International Symposium of Remote Sensing of Environment, Tromso, Norway*, 8–12 June 1998.

Ehrlich D and Mehl H (2001) Humanitarian relief information system. GMES web-site. http://gmes.jrc.it/download/Humanitarian/human_relief.pdf

Fisher T (1993) Use of 3D Geographic information systems in hazardous waste site investigations. In Goodchild M, Parks O P and Steyaert L T (eds), *Environmental Modelling with GIS*. Oxford University Press, New York, pp. 239–47.

Fricke R and Wolff E (2001) The development of land use databases for the Brussels area with IRS-1C data and aerial photography.
http://www.ulb.ac.be/igeat/telgis/murbandy/murbandy_eng.html

FUEGO (2001) Project web-site. www.insa.es/fuego

Haines-Young RH, Barr CJ, Black HIJ, Briggs DJ, Bunce RGH, Clarke RT, Cooper A, Dawson FH, Firbank LG, Fuller RM, Furse MT, Gillespie MK, Hill R, Hornung M, Howard DC, McCann T, Morecroft MD, Petit S, Sier ARJ, Smart S M, Smith GM, Stott A P, Stuart RC and Watkins JW (2000) *Accounting for Nature: Assessing Habitats in the UK Countryside*. DETR, London (ISBN 1 85112 460 8).

Harrison BA, Jupp DLB, Hutton PG and Mayo KK (1989) Accessing remote sensing technology, the microBRIAN example. *International Journal of Remote Sensing*, **10** (2), 301–09.

Holben BN (1986) Characteristics of maximum-value composite images from temporal AVHRR data. *International Journal of Remote Sensing*, **7** (11), 14–17.

Kaufman YJ, Justice C, Flynn L, Kendall J, Prins E, Giglio, Ward DE, Menzel P and Setzer A (1998) Monitoring Global Fires from EOS-MODIS. *Journal of Geophysical Research*, **103**, 32215–38.

King MD and Herring DD (2000) Monitoring Earth's vital signs. *Scientific American*, **282**, 72–7.

Kuchler DA, Jupp DLB, Claasen DBR and Bour W (1986) Coral reef remote sensing applications. *Geocarto International*, **4**, 3–15.

Lankester T (1998) OilWatch – Development of an International Oil Slick Detection System using Spaceborne SAR. http://oilwatch.eos.co.uk/TomDoc.htm

Li Z (2001) Satellite-based Forest Fire Monitoring. Canada Centre for Remote Sensing. http://www.ccrs.nrcan.gc.ca/ccrs/tekrd/rd/apps/em/fires/foreste.html

Lohi A, Ikola T, Rauste Y and Kelhä V (1999) Forest fire detection with satellites for fire control, IUFRO forest fire session. *IUFRO Conference on Remote Sensing and Forest Monitoring, Rogow, Poland*, 1–3 June, 1999.

Macdonald JS (1992) Delivery of information from Earth observation satellites, *Proceedings of Satellite Symposia 1 and 2: Navigation and Mobile Communications, and Image Processing, GIS and Space-assisted Mapping, from the 'International Space Year' Conference, Munich, Germany*, 30 March – 4 April 1992, ESA ISY-2, pp. 243–9.

Malhotra MS and Srivastava A (1995) Diagnostic features of malaria transmission in Nadiad using Remote Sensing and GIS. In de Savigny D and Wijeyaratne P (eds) *GIS for Health and the Environment*, IDRC.

Malingreau J-P (1986) Global vegetation dynamics: satellite observations over Asia. *International Journal of Remote Sensing*, **7** (9), 1121–46.

Marini A and Mellis MT (1992) TM-SPOT multi-temporal analysis on areas devastated by fire. *Proceedings of Satellite Symposia 1 and 2 : Navigation and Mobile Communications, and Image Processing, GIS and Space-assisted Mapping, from the 'International Space Year' Conference, Munich, Germany*, 30 March – 4 April 1992, ESA ISY-2, pp. 423–5.

McGuire ML (1997) *Status Report on Crop Use Intensity* (CUI*) Applications in the FEWS Project*. FEWS Working Paper.

Mehl H and Hiller K (1992) Application of multi-temporal Landsat-MSS and -TM data for the development of an inundation-risk-map. *Proceedings of the Central Symposium of the 'International Space Year' Conference, Munich, Germany*, 30 March – 4 April 1992, ESA ISY-1, pp. 591–4.

Montaperto AT and Rossi F (1992) Use of remote sensing spacecrafts in real time support to disaster management: interfaces with ground segment facilities and services. *Proceedings of the Central Symposium of the 'International Space Year' Conference, Munich, Ger*many, 30 March – 4 April 1992, ESA ISY-1, pp. 599–604.

Mouginis-Mark P, Flynn L, Garbeil H and Scott Rowland (2001) EOS volcanology, thermal and topographic studies, Santa María Volcano. http://eos.pgd.hawaii.edu/ppages/m-m.html

Prata AJ (1989) Observations of volcanic ash clouds in the 10-12 μm window using AVHRR/2 data. *International Journal of Remote Sensing*, **10** (4 and 5), 751–61.

Rao UR (1991) Space and agricultural management. In '*Space and Agricultural*

Management', Special Current Event Session, International Aeronautical Federation, 42nd IAF Congress, Montreal Canada, 1991, pp. 1–10.
Rast M and Readings CJ (1992) The first mission of the ESA Earth Observation Polar Platform. *Proceedings of the Central Symposium of the 'International Space Year' Conference, Munich, Germany*, 30 March – 4 April 1992, ESA ISY-1, pp. 1131–4.
Remote Sensing Applications Consultants (RSAC) (2000) *Synergistic uses of ASAR, MERIS and AATSR* data, UK Envisat Exploitation Programme.
Sahai B, Bhattacharya A and Hegde VS (1991) IRS-1A applications for groundwater targetting. *Current Science (India)*, **61** (3 and 4), 172–9.
Sahai B, Sood RK and Sharma SC (1985) Ground water exploration in the Suarashtra peninsula. *International Journal of Remote Sensing*, **6** (3 and 4), 433–41.
SOS (1999) Clean Seas website. www.satobsys.co.uk/CSeas
Stone TA, Schlesinger P, Woodwell GM and Houghton RA (1994) A map of the vegetation of South America based on satellite imagery. *Photogrammetric Engineering and Remote Sensing*, **60** (5), 541–51.
SSTL – Surrey Satellite Technology Limited (2001) Company home page. http://www.sstl.co.uk/
Townshend JRG and Justice CO (1986) Analysis of the dynamics of African vegetation using the Normalized Difference Vegetation Index. *International Journal of Remote Sensing*, **7** (11), 1435–45.
US Department of Commerce (1998) *GPS, Market Projections and Trends in the Newest Global Information Industry*. The International Trade Administration Office of Telecommunications, US Department of Commerce.
Vinogradov BV and Frolov DE (1992) Aerospace monitoring for ecological hazard and disaster mitigation. *Proceedings of the Central Symposium of the 'International Space Year' Conference, Munich, Germany*, 30 March – 4 April 1992, ESA ISY-1, pp. 615–18.
Wyatt BK and Fuller RM (1992) European applications of space-borne earth observation for land cover mapping. *Proceedings of the Central Symposium of the 'International Space Year' Conference, Munich, Germany*, 30 March – 4 April 1992, ESA ISY-1, pp. 655–9.
Yamagata Y and Akiyama T (1988) Flood damage analysis using multi-temporal Landsat Thematic Mapper data. *International Journal of Remote Sensing*, **9** (3), 503–14.

Further reading

Byrne GF, Crapper PF and Mayo KK (1980) Monitoring land cover change by principal component analysis of multi-temporal Landsat data. *Remote Sensing of Environment*, **10**, 175–84.
Churchill PN and Sieber AJ (1991) The current status of ERS-1 and the role of radar remote sensing for the management of natural resources in developing countries. In Belward AS and Valenzuela CR (eds), pp. 111–43.
Csaplovics E (1992) Analysis of human impact on degradation and desertification of the West African Sahel by means of multi-level remote sensing techniques. *Proceedings of Satellite Symposia 1 and 2 : Navigation and Mobile Communications,*

and Image Processing, GIS and Space-assisted Mapping, from the 'International Space Year' Conference, Munich, Germany, 30 March–4 April 1992, ESA ISY-2, pp. 413–16.

Glaze LS,Wilson L and Mouginis-Mark PJ (1999) Volcanic eruption plume top topography and heights as determined from photoclinometric analysis of satellite data. *Journal of Geophysical Research*, **104**, 2989–3001.

Hayes L (1985) The current use of TIROS-N series of meteorological satellites for land-cover studies. *International Journal of Remote Sensing*, **6** (1), 35–45.

Massonet D *et al.* (1993) The displacement field of the Landers earthquake mapped by radar interferometry. *Nature*, **364**, 138–42.

Sebastián-López A, San Miguel-Ayanz J and Libertà G (2000) *An Integrated Forest Fire Risk Index for Europe*.

8

Geomatics

In this final chapter we look at the emergence of 'geomatics', a generic term which covers the disciplines of surveying, GPS, Earth Observation, digital mapping and cartography, location-based services and GIS. This is a vast field but has become evermore intertwined and productive for the end user. There are many emerging themes which have been covered in the previous chapters of this book, and in this chapter a few of the most promising new developments in geomatics are addressed:

- very high resolution (VHR) satellite images
- developments in positioning and navigation
- on-line access to geographic databases
- geomatics case studies in satellite communication, precision farming and animal tracking.

8.1 VERY HIGH RESOLUTION SATELLITE IMAGES

The recent advent of Very High Resolution (VHR) imagery is seen by many in the EO business as the step function that is needed to provide significant growth in the market. The new generation of VHR provides one-metre imagery which enables the viewing of houses, cars and aircraft, and makes it possible to create highly precise digital maps and three-dimensional fly-through scenes. The 4-metre multi-spectral imagery available contemporaneously provides colour and infrared information to further characterise cities, rural areas and undeveloped land from space. This section looks at the history of VHR images, some current VHR missions and potential markets for the data.

8.1.1 History and perspective

One of the key factors in the utilisation of satellite imagery is the resolution available. Until very recently the main source of high-resolution imagery has been SPOT panchromatic with 10-metre pixels. The main constraint to achieving higher resolutions is not technical but political/security as military sources have had access to sub-metre data for years.

History of US reconnaissance

From 1960 the US reconnaissance satellite programme, designated CORONA (Richelson, 1999), focused on development of a satellite that would physically return its images in a canister. CORONA was to become a mainstay of the United States space reconnaissance programme for over a decade. The camera carried on the CORONA mission (KH for Keyhole) was capable of producing images with effective resolution of about 10 metres. The system also vastly enhanced the imaging capability of the USA, providing more images of the Soviet Union in its single day of operation than did the entire U-2 high-altitude aerial programme. As the programme developed, a succession of new camera systems – the KH-3, KH-4, KH-4A, and KH-4B – were flown, which by 1972 could provide images with approximately 2 metres resolution. Altogether there were 145 missions which yielded over 800,000 images of the Soviet Union and other areas of the world, such as coverage of the 1967 Six-Day War in the Middle East. Figure 8.1 shows an image of Moscow, with the Kremlin magnified, captured by CORONA at the height of the Cold War.

The CORONA programme was complemented by the GAMBIT programme which provided very high resolution images of specific targets rather than wide-area reconnaissance. The first GAMBIT camera , KH-7, had a resolution of 50 cm and the second and last model KH-8, flown up to 1984, provided an even higher resolution. An additional capability was provided by the HEXAGON system from 1971 and 1984, which provided coverage of much larger areas than CORONA and also, via the KH-9 system, a resolution of better than 1 metre. Another significant development was the deployment of KH-11 in 1976, the first reconnaissance satellite to send electronic signals via a relay satellite rather than using film canisters. As a result, the USA could obtain satellite images of a site or activity virtually simultaneously with a satellite passing overhead.

Declassification of imagery

In 1987, not too long after the entrance into service of the SPOT satellite producing imagery with a 10-m ground pixel size, came the Russian decision to allow space photography, taken using its KFA-1000 camera, to be sold on a world-wide basis (Petrie 1999). This photography had been taken originally for intelligence-gathering purposes and had a true ground resolution in the range 5 to 10 metres. This was followed by the even more unexpected decision by the Russian government in 1992 to allow the sale of still higher resolution space photography. This had been acquired with the KVR-1000 and KFA-3000 cameras producing photographs having a ground resolution of 2 to 3 metres. Prior to these developments (under Presidential Directive 37 of the Carter administration issued in 1978) the US government had restricted the ground resolution of American space imagery to 10 metres, an example of the practical implications of the constraints being placed on the geometry of the Large Format Camera (LFC) launched in 1983.

The result of the first Russian decision was the easing of this restriction by the Reagan administration. After the second Russian decision, came a further response by the Clinton administration. Under Presidential Directive 23 issued in March 1994, the development of commercial satellites that produced imagery to the 1-metre

Fig. 8.1 A KH-4B image of the Moscow, with an insert of the Kremlin. In the enlargement of the Kremlin, individual vehicles can be identified as trucks or cars, and Lenin's Tomb in Red Square can be seen. (Source: CIA, National Reconnaissance Office.)

ground pixel level was allowed. Somewhat inevitably, among the first beneficiaries of licences were the American-based companies that had previously been suppliers of hardware and software to the military space reconnaissance programmes. As a further gesture President Clinton took the decision in February 1995 to declassify and release into the public domain 860,000 high-resolution space photographs acquired between 1960 and 1972 through the CORONA satellite reconnaissance programme; thus providing an valuable 'prequel' to the Landsat programme which commenced in 1972.

8.1.2 Missions

This section provides a summary of the main commercial VHR systems in operation, or planned at the time of writing (January 2001).

Ikonos

Owned by Space Imaging, the Ikonos satellite launched in 1999 can collect 4-metre multi-spectral data and 1-metre panchromatic data. The two data streams can also be merged in a 'pan-sharpened' product which combines the spatial content of the 1-metre panchromatic data with the spectral content of the 4-metre multi-spectral data. Figure 8.2 shows a pan-sharpened image of the Year 2000 Olympic Site in Sydney, Australia.

Fig. 8.2 Ikonos image of Sydney from space showing the 2000 Olympic Park, 14 km west of downtown Sydney. Olympic Park venues seen in the image include the Main Stadium, Superdome, Athletics Centre and warm-up arena, Aquatic Centre, and the hockey, tennis and baseball venues. (Credit: 'spaceimaging.com.')

OrbView

OrbView 3 will also produce 1-metre resolution panchromatic and 4-metre resolution multi-spectral imagery with product specifications similar to those from Ikonos. The satellite will revisit each location on Earth in less than three days with a nadir swath width of 8 km and ability to turn from side-to-side up to 45°. The follow-up mission, OrbView 4, will also have the ability to produce hyper-spectral imagery using an instrument with 200 imaging channels.

Quickbird

Quickbird, owned by EarthWatch Inc. (Colorado, USA), is another mission based on 4-metre multi-spectral and 1-metre panchromatic data. Quickbird is a follow-up to EarlyBird, intended to be the first commercial VHR mission, but which failed a few days after launch in December 1997. Quickbird 1 also failed in November 2000 and is to be replaced by the higher specification Quickbird 2.

Radarsat 2

Radarsat 2 (currently planned for 2003) will provide the same kind of SAR data as Radarsat 1 with additional capabilities including new modes, a higher 3-metre resolution (compared to 8.5 metres for Radarsat 1), multi-polarisation and shorter imaging cycles, with a revisit frequency twice that of Radarsat 1.

8.1.3 Markets

The applications of VHR are widespread, competing head-on with aerial imagery and offering improved timeliness, spectral resolution and customisation. Selected markets are described in Table 8.1.

Table 8.1 Markets for VHR imagery (based on Petrie 1999)

Agribusiness	Infrared imaging capability will directly detect stressed crops within individual farm fields before the stress becomes visibly apparent. This early detection capability will enable farmers to take corrective action sooner and increase overall crop yields.
Mapping	With the increased geometric accuracy and rapid update rate, VHR imagery will be highly attractive to National Mapping Agencies (and their commercial rivals) especially small-scale topographic mapping and map revision. Ortho-rectified images can be produced at scales up to 1:25,000 or perhaps even larger and readily incorporated into GIS.
Utilities	Gas, electric and water utilities are heavy users of high-resolution geographic information. Detailed up-to-date maps that can be provided by VHR are critical to the planning, design, construction, operation, marketing and regulatory compliance of the utilities' widely dispersed networks. For example, Ikonos images show detailed features such as houses, urban growth boundaries, trees, utility towers, utility corridors, right-of ways, and building facilities. Customers can use the digital images for developing, maintaining or

	updating maps planning gas, water and electric transmission and distribution, identifying areas experiencing vegetation encroachment to prevent outages and emergencies, assessing land use and land cover or creating emergency plans for outages due to natural disasters.
Telecommunications	As the telecommunications industry continues to grow, so will the importance of detailed topography and land use data. By using high-resolution imagery to determine the topography and land cover classifications within new service regions, telecommunications companies can optimise the expansion of their wireless networks.
News media (TV networks, newspapers, etc.).	Images of areas experiencing natural disasters such as floods, tornados, forest fires, earthquakes, volcanic eruptions, major landslides and avalanches, etc., or man-made disasters such as oil spills, shipwrecks, explosions, pollution, etc., are expected to achieve a ready sale.
Insurance	Companies can use VHR multi-spectral imagery to map vegetation land cover around residential and industrial development to assess fire and flood risk.
Hazard, disasters	The timely acquisition of high-resolution imagery should also assist the assessment of the scale and impact of such disasters and the planning of relief and evacuation operations by the appropriate agencies.
Construction	Companies can use data to plan large-scale construction projects, including addressing regulatory compliance issues such as Environmental Impact Assessment.
Exploration	High-resolution 1-metre data will aid in producing up-to-date maps of remote regions around the world. These maps will be useful for such applications as planning equipment transport, seismic field testing and drilling operations for exploration activities and for mapping their production facilities to more effectively maintain and expand their production infrastructure.
Leisure and retail	There are many potential consumer-oriented applications for VHR imagery. These applications include assisting travel agents in vacation planning or real estate agents in showing homes to prospective buyers.
Military	By monitoring military activities in restricted areas VHR will help national security agencies manage international relations on a global scale. Near real-time high-resolution imagery, will help governments to monitor borders; gather intelligence on potential conflicts; identify and target enemy troops and assets; plan air, ground and naval missions; deploy resources and assess battle damage.

8.1.4 Geocoding issues

Although the principles of geocoding are the same for VHR imagery as for those images with larger pixel sizes, the production of images accurately located to the order of 1 metre requires a more rigorous approach to geometric model definition and collection of GCP data. The geometric accuracy specified for Ikonos products (Main 2000) is defined as one of two levels:

- Level 0 has 0.2-metre horizontal and 0.6-metre vertical accuracy
- Level 1 has 0.7-metre horizontal and 1.5-metre vertical accuracy.

To achieve this accuracy requires the use of GPS surveying, in the case of Level 1 products 'real time' GPS is sufficient but the more stringent requirements of Level 0 require differential post-processing. An average of one GCP is selected for each 10 km by 10 km of imagery. The rules of selection are similar to the traditional definition of 'good' GCPs with some extra considerations:

- Avoid points beneath trees as they may be obscured in the image.
- Avoid locations beneath power lines or near transmission antennae that may interfere with GPS reception.
- The best locations are high-contrast markings on asphalt roads, tennis court markings, etc., although care should be taken on areas that may be obscured by vehicles during image acquisition.

In addition, the generation of orthoimages requires a high-quality (10-metre resolution) digital elevation model. The above limitations mean that some development is necessary for the automated production of precision geocoded VHR imagery and mosaics.

8.2 DEVELOPMENTS IN POSITIONING AND NAVIGATION

This section describes some of the recent developments in positioning and navigation:

- High-precision GPS
- Galileo – the European counterpart to GPS
- 'Personal infomobility' applications.

8.2.1 High-precision GPS

Over the last few years GPS has become the ubiquitous positioning technology and a considerable expertise has developed in 'squeezing the last centimetre' from the technology. One technique to improve GPS performance is to use the Phase Variable which is less affected by multi-path (differing path lengths due to signals bouncing off objects) and has less than 1 mm of noise. This approach is usually called Kinematic DGPS (Clynch 2000) and has been used since the mid 1980s as a post-processing technique, whereby GPS data is acquired over a period of several hours at static sites and then collated into a common measurement of position. The accuracy obtained is a function of baseline length and, as originally developed, about one part per million (1 cm per 10 km) was attainable. Nowadays relative accuracies of between 5 and 10 times this are possible and the processing can be achieved in real time. The question of what absolute accuracy can be obtained by kinematic methods is directly related to the length of the baseline that can be successfully employed with the process. Consistent success has been achieved over short baselines (less than 20 km). Current techniques (Moore and Roberts 1997) have managed to extend on-the-fly kinematic GPS to baselines of up to 500 km between the base station and the mobile receiver.

8.2.2 The Galileo programme

The Galileo programme, planned to become operational in 2008, is intended as a European alternative and a complement to the United States GPS and Russian Glonass. It is hoped that Galileo will become an important part of the international transport infrastructure, as well as a useful scientific tool. Although GPS and, to a certain extent, Glonass are very well established as navigation systems, Galileo (Barboux 2001) will address several of the potential shortcomings of these systems:

- Despite safety and economic consequences of errors no guarantee or liability cover is provided by the data suppliers.
- Reliability of signal is uncertain, users are not informed immediately of errors.
- World-wide coverage can be patchy, for example in high latitudes or under cover of foliage.
- 'Out of the box' precision is not sufficient for some applications.
- GPS is first and foremost a US military system, available on a 'grace and favour' basis which could possibly be revoked depending on changing political situations.
- The future of Glonass is uncertain.

Galileo will consist of a series of 30 satellites, which will be placed in orbits at around 23,200 km (56° inclination) and monitored by a network of ground control stations, in order to provide world cover. Based on the space infrastructure shown in Fig. 8.3, Galileo will be integrated with various terrestrial systems and technologies to meet the needs of users, including those in towns and cities (where satellite transmissions without ground-based relays can be blocked by buildings), in high-risk areas (building sites, factories, depots), in isolated areas (where the cost of installing and maintaining terrestrial systems is prohibitive) and in high latitudes (where satellite signals are weak).

An initial survey (European Commission 2000a) identified four basic services that the users of Galileo are likely to require:

- *Position, Velocity and Time Service* (PVT) – a basic service aimed at mass market applications, intended as a free service analogous to GPS.
- *Accuracy and Integrity Service* (AI) – for subscribers only a high accuracy and availability service for professional markets.

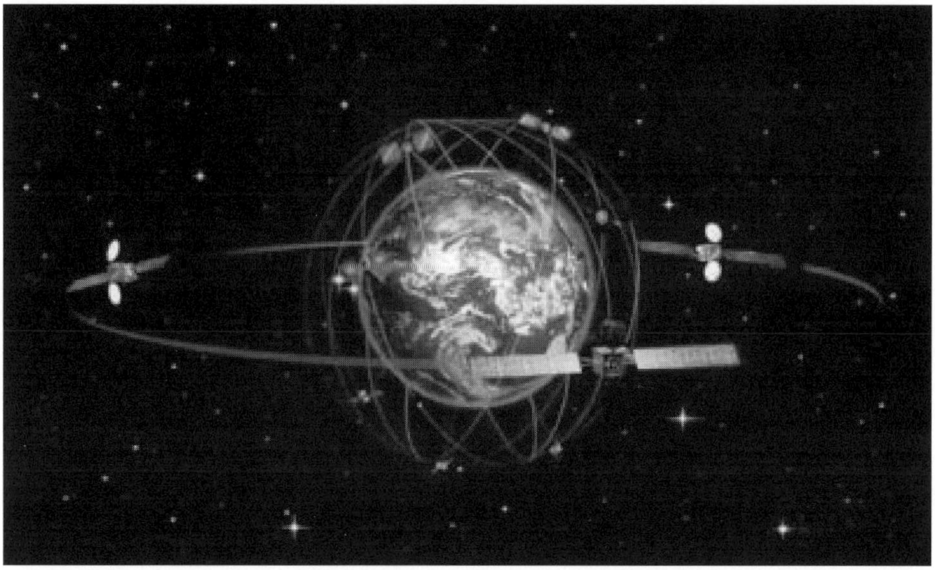

Fig. 8.3 Galileo system diagram (courtesy European Commission, European Space Agency).

- *Ranging and Timing Service* (RT) – a very precise ranging, positioning and timing service for more specialist professional subscribers.
- *High Integrity Service* (HI) – for 'trusted' subscribers only; a service with the highest integrity, availability, continuity and resistance to signal interference for very demanding safety of life markets.

Galileo is intended to be interoperable with existing systems, by making use of established international standards such as UTC (for timing) and WGS 84 (as the geodetic reference system) and by designing a complementary architecture to ground navigation systems (GPS, LORAN C, Glonass) or derived navigation systems such as UMTS. Some of the expected socio-economic benefits of Galileo are shown in Table 8.2.

8.2.3 Personal infomobility

Recreational GPS

GPS is increasingly being used for a wide range of outdoor pursuits such as recreational activities such as skiing, mountain climbing, orienteering (obviously not in competitions ...), yachting, camping, hunting, fishing or trekking. Hand-held GPS devices have been available for a number of years and those currently available (April 2001) are sufficiently small to be incorporated in mobile phones, digital cameras or even watch-size devices.

GPS can be used recreationally for a variety of innovative applications including:

Table 8.2 Anticipated socio-economic benefits of Galileo (European Commission 2000a)

Better public services	• More efficient transport services through increased traffic capacity • Greater access to on-line traffic information • Improved response from emergency and search and rescue services
New mass market services	• Advanced in-car navigation system, leading to shorter journey times • Seamless road-tolling systems • Personal tracking systems to assist supervision of vulnerable individuals
New professional services	• Aviation free-routing services, reducing air travel times by 5–10% • Fleet management services to reduce fuel consumption and consequential environmental impact • Monitoring of marine traffic use of Galileo linked to tachographs to improve safety of the road sector and transportation of dangerous goods

- recording of fishing sites
- mapping and following of bicycle trails
- hiking and cross-country navigation using devices with built-in digital maps
- competitions based on locating other players or hidden objects via GPS 'clues'
- location information for emergency services for people in difficulties.

Mobile GIS

This is a generic technology that has specific applications within different professions or commercial sectors. Mobile GIS showing current features of the user's location is already well established (for example, in the utilities sector) using expensive bespoke systems based on professional high-performance GPS. In the last few years, however, it has become very easy to assemble a modestly priced Mobile GIS using commercially available technology:

- Most GIS products can now be installed on a laptop PC or even more compact Personal Digital Assistant (PDA).
- Commercially available GPS receivers are easily interfaced to the above hardware.
- Standard mobile phones can be used to transmit location information or map updates ('red lines') for incorporation in corporate GIS repositories.

Increasingly, all the above features are becoming available in a single item of equipment, although until the advent of third-generation telecoms it is likely that large digital datasets (such as base-mapping) will either be pre-installed on the portable equipment, or loaded in the field via Compact Disc. An example of a mobile GIS can be seen in Fig. 8.4.

Fig. 8.4 ArcPad Mobile GIS (courtesy ESRI, Compaq).

8.3 ON-LINE ACCESS TO GEOGRAPHIC DATABASES

'Metadata' is the term used to describe the summary information or characteristics of a set of data and is the key to on-line access to geographic databases (Aalders 2001). To paraphrase Rudyard Kipling, metadata provides the 'What, Why, When, Who, Where and How' of the data. The main difference between geographic metadata and the many other metadata sets being collected for libraries, academia, professions and elsewhere is the emphasis on the 'Where' element.

8.3.1 Discovery metadata

There are a variety of different terms for defining types of metadata; these are all however broadly equivalent and constitute three levels of access. These levels (Fraser and Smith 1998) can be described as:

- *Resource discovery* – browsing over the web to find data resources which answer the question 'What datasets hold the sort of data I am interested in?'

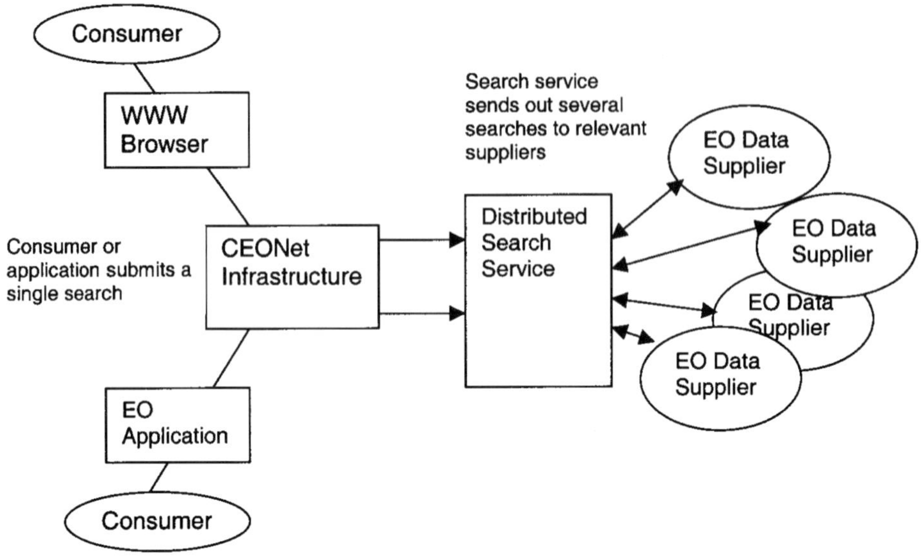

Fig. 8.5 Access to geographic databases (adapted from Fraser and Smith 1998).

- *Resource evaluation* – the detailed examination of the metadata that is necessary to select the datasets that are to be acquired.
- *Resource Access* – the process of obtaining and using the data that is required to enable the interface to load the geographical data into the application.

The typical process invoked by this three-level access to geographic data is shown in Fig. 8.5, in this case access to the Canadian CEONet geospatial network.

Discovery metadata is the minimum amount of information that needs to be provided to answer the six key questions about geospatial datasets:

- *What?* The title and description of the dataset.
- *Why?* An abstract detailing reasons for the data collection.
- *When?* When the dataset was created and the update cycles if any.
- *Who?* The originator and data supplier.
- *Where?* The geographical extent based on latitude and longitude, polygon boundaries, geographical names or administrative areas.
- *How?* Instructions on how to obtain more information or order the datasets, formats, media access and restrictions on data use.

The broad categories are only few in number to reduce the effort required to collect the information while still conforming to the requirement to convey to the enquirer the nature and content of the data resource. Discovery metadata is of most value when organised to conform to agreed standards to facilitate interchange of information. An example of this is the Directory Interchange Format (DIF), which is a widely used *de facto* standard. A DIF consists of a collection of fields which detail specific information about the data. Five fields are required in the DIF; the others expand upon and clarify the information. Some of the fields are text fields, others require the use of valid values. Other examples of metadata standards which may be encountered include:

- The United States Federal Geographic Data Committee's (FGDC) Content Standard on Digital Geospatial Metadata (CSDGM)
- Catalogue Interoperability Protocol (CIP) designed primarily for the exchange of information from Earth Observation catalogues based on the protocol of Z39.50 (Mills *et al.* 1998)
- Z39.50 itself – the International Standard, ISO 23950: '*Information Retrieval (Z39.50): Application Service Definition and Protocol Specification*' used widely for indexing of bibliographical data.

A typical DIF, in this case containing metadata for an ERS altimeter data archive (ESA 2000), is shown in Table 8.3.

Table 8.3 DIF description of altimeter data archive (ESA 2000)

Entry ID: ERS.ALT.WAP	
Temporal coverage:	
Start date:	1991-07-30
Stop date:	
Originating centre:	ESA
Distribution:	
Distribution Media:	Exabyte 8200
Distribution Size:	35 MB/orb, 43 orbit/exabyte
Distribution Format:	
Fees:	
Disciplines:	
Earth Science Land	
Earth Science Ocean	
Location keywords:	
Global	
Global Ocean	
Spatial coverage:	
Southernmost latitude:	82° S
Northernmost latitude:	82° N
Westernmost longitude:	180° W
Easternmost longitude:	180° E
Parameters:	
Atmospheric Winds > Surface Winds	
Ocean Waves > Significant Wave Height	
Ocean Waves > Wave Height	
Ocean Winds > Surface Winds	

Platform Characteristics > Orbital Characteristics

General Keywords:
EOSDIS
Oceanography

Sources:
European Remote Sensing Satellite 1 (ERS 1)
European Remote Sensing Satellite 2 (ERS 2)

Sensors:
ERS Radar Altimeter (RA)

IDN node:	ESA/ESRIN
DIF revision date:	1997–10–15
Dataset progress:	IN PROGRESS

8.3.2 Resource evaluation

The resource evaluation phase provides more information about what geospatial data is actually available. This may include the provision of sampled of geospatial datasets, browse products (for example, 'thumbnails' of images) and on-line technical and user guides. One useful resource for understanding which satellite images are available is an overpass map which shows when images have been acquired or when they are planned to be acquired. For satellites with fixed nadir-pointing sensors (such as Landsat) the dates when images are acquired are completely determined and predictable even years in advance; for steerable systems such as SPOT, the images to be acquired may change on a daily basis depending on programming requests. Figure 8.6 shows a typical image overpass map, in this case for Landsat 7 over Australasia in April 2001.

Figure 8.7 (colour section) shows a typical interface to a browse facility enabling information about the contents of an ATSR archive to be displayed. Since most geospatial archives contain thousands or even millions of individual objects, it is extremely important to be able to filter the data by means of a query which can specify categories of data, data ranges of interest and particularly regions of interest. Although the geographic location can be defined in terms of attribute data (e.g. country of coverage), or by typing in the coordinates of a lat–long 'bounding box', a more appropriate and intuitive approach is to use the power of GIS to specify a map-based interface.

8.3.3 Resource access

There are a number of current initiatives to provide improved access to Earth Observation archives and other geospatial resources. Many of these are now available on-line via the internet.

Global Change Master Directory (GCMD)

NASA's Global Change Master Directory (GCMD) is a comprehensive directory of descriptions of datasets of relevance to global change research. The GCMD

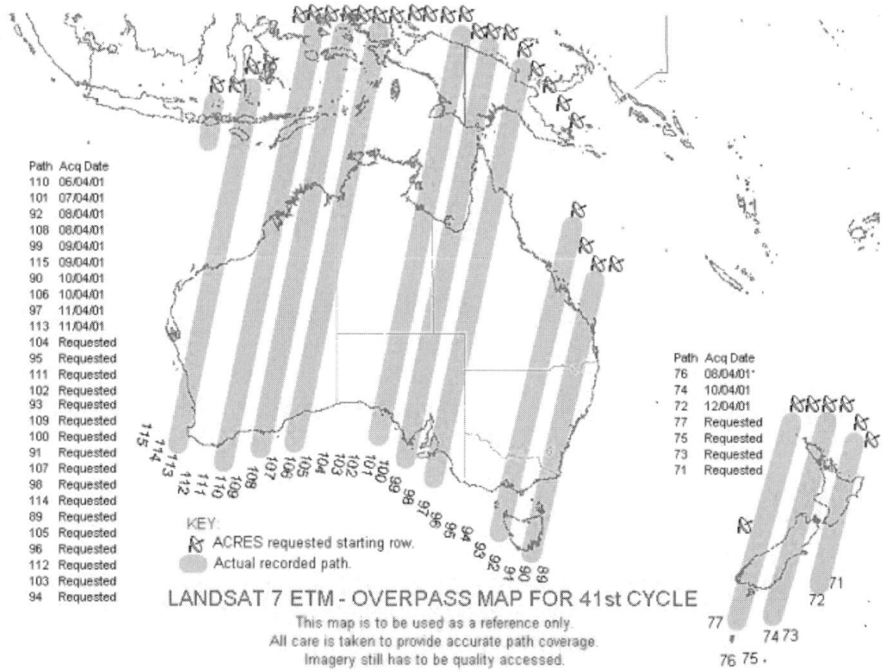

Fig. 8.6 This overpass map shows the path of the Landsat 7 satellite over Australasia in its 41st cycle since launch. The 41st cycle extends from 6 April to 22 April 2001. Paths are only shown where imagery was acquired. (Courtesy AUSLIG, Australia.)

database includes descriptions of datasets in DIF format) covering climate change, agriculture, the atmosphere, biosphere, hydrosphere and oceans, geology, geography and human dimensions of global change. The GCMD collaborates with several projects, like NASA's Earth Observing System Data and Information System (EOSDIS – see below) as well as with the US Global Change Research Program (USGCRP), other government agencies and university data projects.

CEOnet

The Canadian Earth Observation Network (CEOnet) is an initiative by the Canadian government to create a national infrastructure for providing improved access to Earth Observation archives and other complementary geospatial databases. This initiative is being driven by the requirements of Canadian users for better access to geomatics data, and by the opportunities for Canadian industry afforded by the rapid growth of the international market for geomatic data, services and network systems.

INFEO

Information on Earth Observation (INFEO) is a descendant of the European-Wide Service Exchange (EWSE), an initiative of the European Centre for Earth

Observation (CEO). INFEO is based on a 'virtual shopping mall' approach to presenting information, and includes consumer and supplier registration facilities. Many of the services provided INFEO have similarities to CEOnet. Both programmes are working closely together to provide a gateway that will pass organisation searches though to each other's advertising services. Similarly, the distributed search protocols used by the two systems are being aligned to enable interoperability.

EOSDIS

Earth Observing System Data and Information System (EOSDIS) is a comprehensive data and information system developed by NASA under the Earth Observing System (EOS) programme (see Section 7.1.1). EOSDIS manages data from NASA's past and current Earth science research satellites and field measurement programmes, and provides data archive, distribution and information management services. The objectives of EOSDIS (Shaffer 1992) are three-fold:

1. To improve, distribute and archive data products from pre-cursor missions to EOS (mainly extensions to current programmes such as Landsat and NOAA/AVHRR).
2. To process, distribute and archive data from EOS itself.
3. To control the EOS spacecraft.

Because of the size and complexity of EOSDIS it has been developed in a evolutionary manner with extensive use of prototyping and feedback from users to constrain the final implementation. The first activity for EOSDIS used existing global datasets as a platform for the geographic database for EOSDIS. From 1999 onward EOSDIS has been in operational use producing, archiving and distributing data from Landsat 7, Terra and ACRIMSat. The approximate volume of data archived in EOSDIS is truly phenomenal (Ramapriyan and Aitken 1999): in September 1999 the EOSDIS Distributed Active Archive Centres (DAACs) held over 250 terabytes (2.5×10^{14} bytes) of data and information about the Earth. By the end of 2000, the volume of data had more than doubled to over half a petabyte (5×10^{14} bytes) and was increasing at the rate of more that a terabyte (10^{12} bytes) per day.

8.4 GEOMATICS CASE STUDIES

This section looks at some geomatic case studies where complementary technologies such as EO, GIS, satellite communications and GPS have been integrated to provide an enhanced product. The 'synergistic' products described here are:

- use of satellite communications for EO data dissemination
- precision farming
- animal tracking
- location-based telecommunication services.

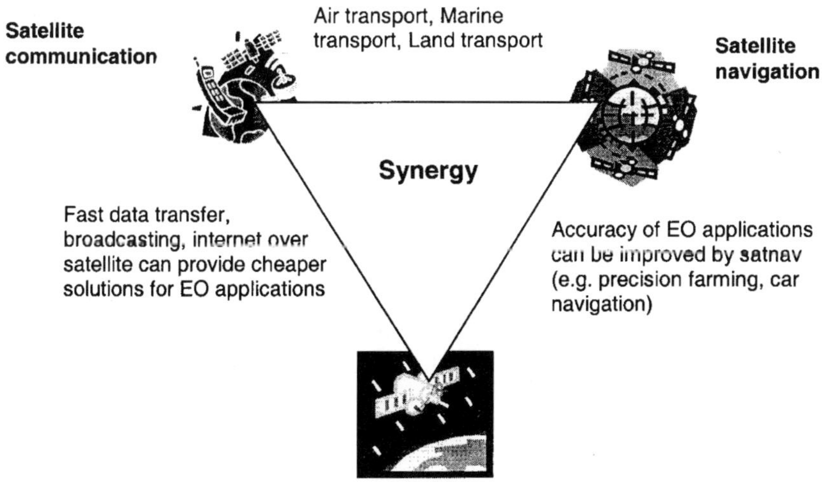

Fig. 8.8 The ASTRON synergy concept (courtesy European Commission).

8.4.1 Use of satellite communications for EO data dissemination

The European ASTRON (Applications on the Synergy of Satellite Telecommunications, Earth Observation and Navigation) programme confirmed that the synergy between Earth Observation, telecommunications and satellite navigation could be effective in a number of application areas, including personal infomobility, safety of life and resources and big events such as the Olympic Games. The basic concept of the ASTRON programme is shown in Fig. 8.8.

One of the areas investigated was the benefits attained by using satellite communications as a mechanism for distributing Earth Observation data. Examples noted (European Commission 1999, 2000b) include:

- The European Meteorological Satellite organisation (Eumetsat) uses a communications payload on the Meteosat spacecraft to disseminate imagery, products and meteorological information.
- The UK Defence and Evaluation Research Agency (DERA) took part in a two -month mission to provide Near Real Time (NRT) Radarsat images to the Navy ship HMS *Endurance* to assist with ice navigation in the Waddell Sea.
- Southampton University investigated the discrepancies between sea surface temperatures derived from AVHRR images and those obtained locally from floating buoys.

The third of these examples is of particular note as in this case it is the 'ground truth' from the buoys that was transmitted (using the ARGOS satellite communication system) rather than the satellite image data. ARGOS is a satellite-based system which collects, processes and disseminates environmental data from fixed and mobile platforms worldwide, developed jointly by Centre National d'Etudes Spatiales

(CNES, the French space agency), the National Aeronautics and Space Administration (NASA) and the National Oceanic and Atmospheric Administration (NOAA, USA). For over 20 years, ARGOS has provided data to environmental research and protection communities that, in many cases, was otherwise unobtainable (see Fig. 8.9 in the colour section).

8.4.2 Precision farming

Precision farming in agri-business
Nowadays every link in the agri-business chain is looking at ways to assimilate new information in a cost-effective manner to gain competitive advantage, improve productivity and plan for the future (Logica 1998). For example the European Commission is already the leading purchaser of civil satellite data in Europe for its agricultural monitoring programme. One major anticipated application area is precision farming, based on the combined use of various GIS technologies to allow farmers to produce detailed maps of the condition and productivity of their fields. As well as imagery and digital maps, precision farming systems typically link onboard GPS with yield-mapping and variable rate precision fertiliser applications As well as the potential economic advantages, precision-farming technologies may be required to comply with increasingly stringent regulations on pollution and use of fertilisers which could compel farmers to report in some detail the chemicals they are using and where they are putting them.

Precision-farming technology
Current precision farming makes use of a number of interrelated technologies including location-enabled combine harvesters fitted with yield meters. Using GPS data onboard the combine, the farmer can make use of specialised GIS software to produce digital maps of local variations in yield. The digital maps can in turn be used as input to variable rate crop sprayers (Fig. 8.10, colour section) which can be programmed with precisely located regimes for fertiliser application. As well as the use of GPS satellite technology for navigation, high-resolution satellite or aerial images are frequently used for crop identification and detection of in-field variations in crop vigour and identification of disease. Typical precision farming products are shown in Fig. 8.11 (colour section).

Farm Management Information Systems
The various digital map layers can be incorporated in a Farm Management Information System (FMIS) which consists of a simple land-parcel based GIS, digital map data and farm records from previous years (Steven and Millar 1997). This is used to determine the year-on-year improvements in a precision farming regime, for planning crop rotations and for adjusting temporary field boundaries, for example, to assist with preparation of claims for subsidies. A typical FMIS interface is shown in Fig. 8.12. Another example of an FMIS is the SABRES (Services to Agri-Business by Remote Sensing) project (Blakeman *et al.* 2000) which is developing a process that uses VHR satellite imagery to provide a cost-effective

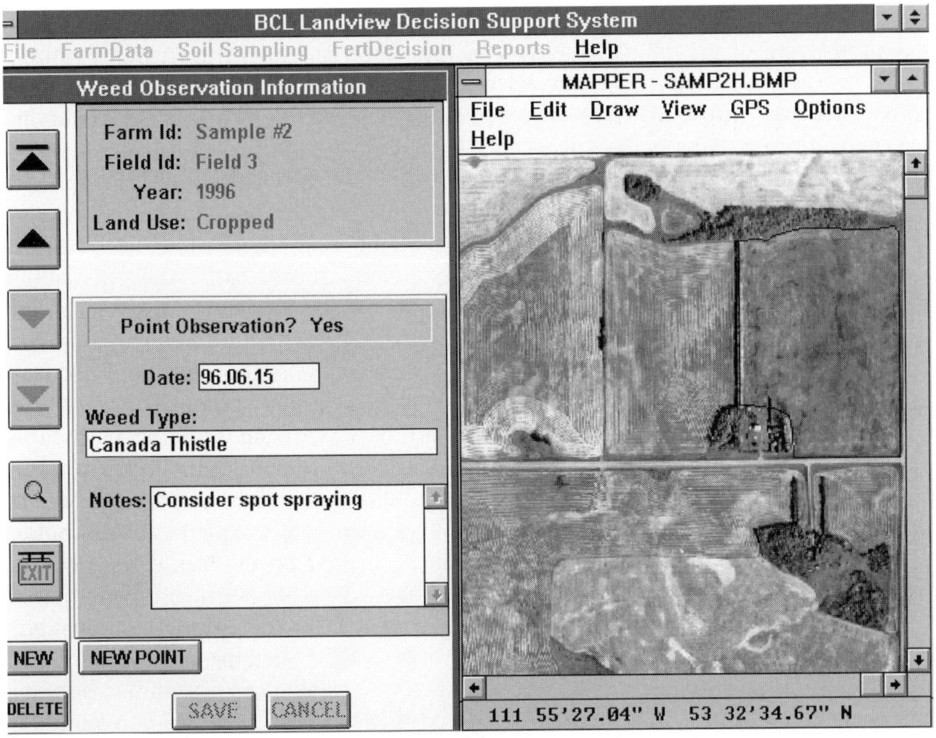

Fig. 8.12 Farm Management Infrmation System (copyright Ames Remote)

overview of crop condition during the growing season and prior to harvest. This additional data could be used to extract accurate and timely information on crop state and actual area grown. In addition to taking advantage of the high spatial resolution of these datasets SABRES will also deliver specialist image products, such as fertiliser and fungicide recommendation maps for the FMIS.

8.4.3 Animal tracking

Due to the increasing miniaturisation and mass-production of GPS devices there is a rapidly growing market in the digital tracking of animals. Many of these applications also make use of other geomatic technology such as GIS, satellite images and mobile communications. In this section we consider four types of application area:

- Monitoring of agricultural animals
- Location of domestic animals
- Tracking of large mammals
- Tracking of marine animals.

Monitoring of agricultural animals

The monitoring of agricultural animals such as cows and sheep is currently mainly performed as a research activity to determine the 'social' behaviour of animals. The GPS device may also be located with other equipment such as jaw-motion monitors to determine grazing patterns. It is likely that farm animal tracking will become more of an operational proposition as a combination of (1) adoption of precision farming technology for livestock as well as arable applications and (2) potential tightening of regulations on cattle movement and detailed record keeping. (It is conceivable, for example, that a government may at some future stage require automated 'cattle passports' in the wake of such problems as BSE in beef cattle and foot and mouth disease).

Location of domestic animals

Tracking of domestic pets, primarily cats and dogs, is a huge potential market; for example, the Dutch company Petfinder Technologies (www.petfinder.net) currently (April 2001) offers a service to locate lost cats in Amsterdam. The cat to be tracked wears a tiny micro-transmitter which is robust and transmits for one year on a single 3V lithium battery as used in watches. Every 2.3 seconds it transmits a radio pulse, which carries a 500-metre flat range and can be picked up by Petfinder vans and plotted using GIS as they move through the city.

Until recently GPS devices have only been used for tracking larger mammals but their use on domestic animals is becoming increasingly feasible, even on smaller animals such as pigeons. Understanding the pigeon's navigation system has been an intriguing challenge for many years. Tracking methods have included conventional radio tracking, following the pigeons in an aeroplane, and a route recorder containing a strong magnetic compass that measures the direction of flight. Researchers at the University of Frankfurt Zoology Institute have developed a GPS-based system which yields high-precision results (von Hünerbein and Rueter 2000). The work is based on an ultra-light GPS harness, capable of storing 90,000 positions and weighing less than 40 grams. This is still quite heavy for a pigeon, so the birds were subjected to a three-month weight-training programme starting with much lighter weights and gradually increasing the load. Initial trials successfully tracked pigeons for periods of up to three hours. Two challenges remain: reducing the device's weight even further and minimising the residual magnetic field that may affect the bird's navigation method.

Tracking of large mammals

GPS is widely used for tracking wild animals, primarily large mammals such as deer, elk, Grizzly Bear, moose and elephants. Animal tracking is often combined with habitat monitoring (see Section 7.2.1) to correlate changes in foraging patterns with changes in the habitat and establish the consequences for the species. GPS-based approaches are also used to log the location of food sources for specific species. Figure 8.13 shows a typical GPS harness for the larger animal.

One species that has been particularly well-tracked is the elephant. Elephants are motivated by three basic factors: finding food and water, social interaction with other elephants, and avoiding danger (Save the Elephants 2001). The GPS radio

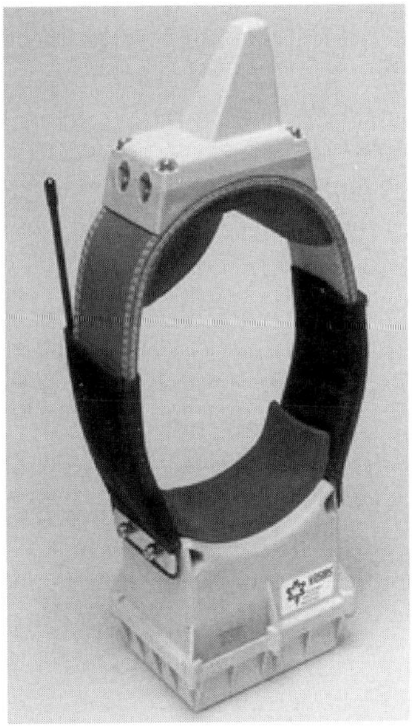

Fig. 8.13 GPS-1000 animal tracking collar suitable for large mammals (Lotek Wireless, Canada).

collars on the elephants are set to record data at set intervals and downloaded remotely every few months to an aircraft flying over the elephant's location. GIS software can then be used to plot the movements of each elephant since the last download and analyse the decisions that elephants make about where they go, when they travel, what areas they avoid, and the distances they travel. Figure 8.14 shows the track left by a matriach called Ngalatoni who, after her collar was fitted, spent three months in the protected region of Samburu National Reserve (Kenya) with her family, not moving far. One evening she reached the edge of the reserve, and travelling steadily through the night across an area that is favoured by bandits and poachers, she and her family reached the safety of the ranches of Il Ngwesi and Borana. They stayed there for a few months before repeating the 'streak' across the unprotected area back to Samburu National Reserve. This is a perfect example of how GPS tracking and GIS can highlight important elephant 'corridors' which can be used to establish relevant elephant management policies for that area.

Monitoring marine animals

As well as the direct use of GPS for tracking animals, indirect methods using satellite imagery are also useful, particularly for marine environments. For example, when tracking whales a good starting point is to know that they gather in locations that are rich in plankton and squid. Thus to find these whales it is desirable to find isolated locations in the ocean that are rich in their favourite foods. Using data from NASA's TOPEX/Poseidon satellite (Barth *et al.* 1999), scientists located places in the Gulf of Mexico that have all the right conditions to host plankton and squid and to attract whales. The most promising spots for finding plankton, squid, and whales are

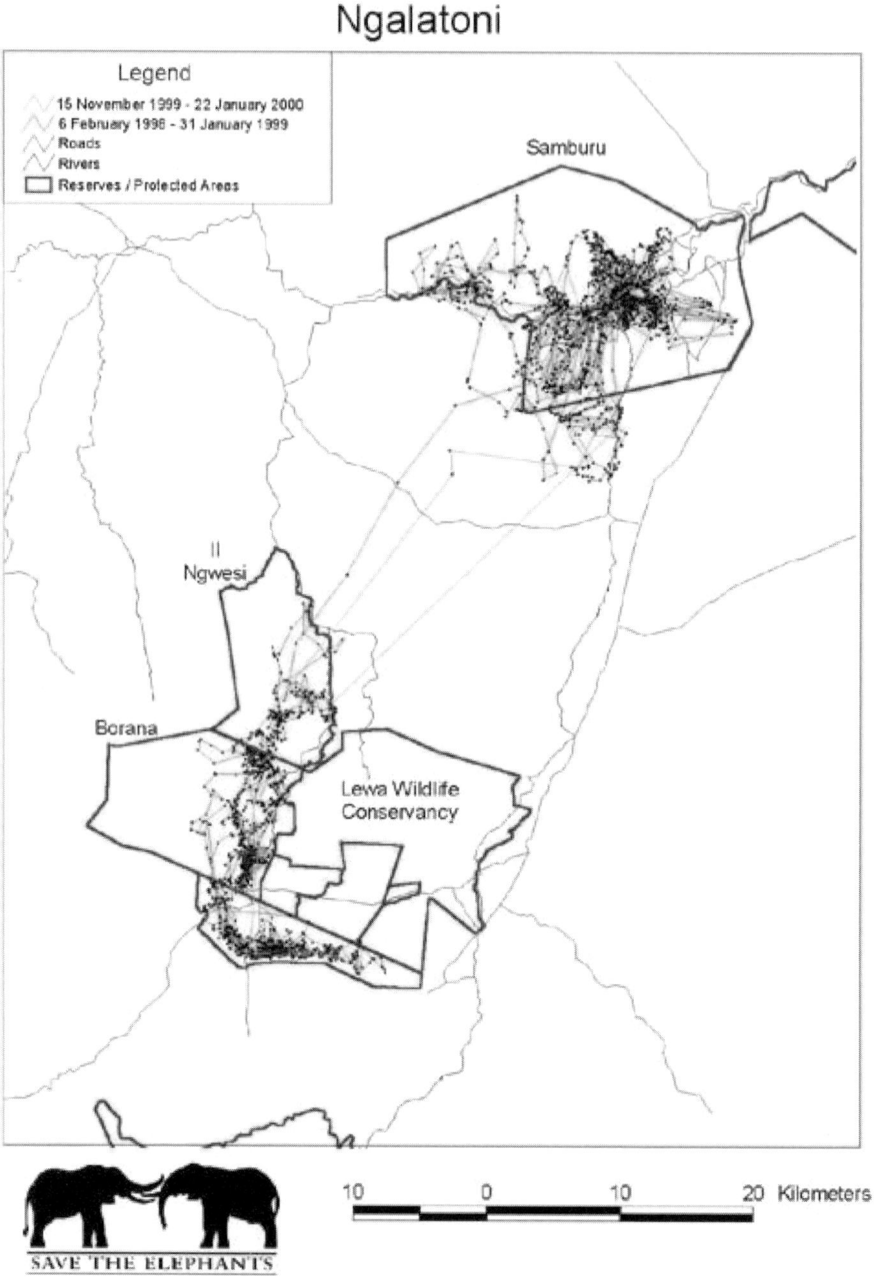

Fig. 8.14 Elephant tracking (courtesy, Save the Elephants).

called ocean 'cyclones' – circular systems, rotating counter-clockwise (in the Northern Hemisphere), with cooler water in the centre, with lower-than-average sea surface heights, and with an up-welling of water from the deep ocean to the surface. The cyclones pull up nutrients from the deep ocean, creating areas with an abundance of plankton, attracting squid and in turn the whales.

The TOPEX/Poseidon data is used to detect areas with lower-than-average sea surface heights correspond to ocean cyclones, the ideal places to watch for whales. Figure 8.15 shows a TOPEX image product of the Gulf of Mexico showing height anomalies together with the locations of whales subsequently spotted from local surveys. These satellite-guided surveys help to produce a better understanding of the population size, diversity, distribution and habitat of marine mammals in the Gulf and to provide a basis for future studies of marine life.

8.4.4 Location-based services

This final section provides a speculative look at the consumer segment which represents a completely new set of vertical markets for Earth Observation and GIS. With the advent of third generation (3G) telecommunications there is likely to be a huge development in location-based services (LBS, alternatively called Location Enabled Applications – LEA) for individual consumers using a combination of mobile phones equipped with GPS, GIS data layers and products derived from Earth Observation. A recent study (Ketselides 2000) identified new mass market applications exploiting synergies between EO and mobile positioning for business to consumer (B2C) opportunities. Four of the most promising were identified as:

- Personal weather forecaster
- Intelligent tour guide
- Sports assistant
- Virtual gaming.

Personal weather forecaster

This application would provide local area, short-term weather forecasts as well as related environmental and pollution information. The following functionality could be included:

- Users would be able to tailor the information they receive to suit themselves, e.g. skiing conditions, pollen count, pollution level.
- The user would be able to set an alarm to go off when the pollen count, for example, went beyond a specified level.
- The application would be linked automatically to public safety applications such as a hazardous weather warning.
- The application would be linked to the emergency services. For example, if the users were sailing and a storm blew up, the coast guard would be able to call the sailor to check on them, and if there was indeed a problem, the coast guard would be able to use positional information from the terminal to help in a rescue effort.

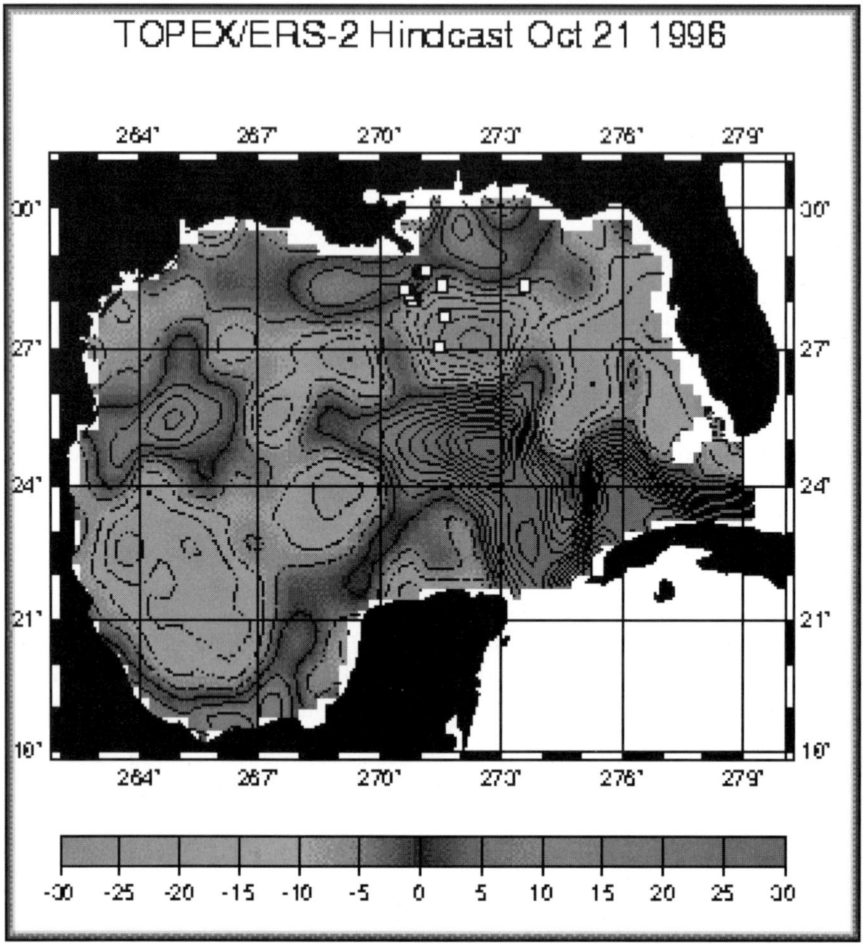

Fig. 8.15 White squares indicate whale sightings around an eddy shown by sea surface height anomaly (SSHA) in the Gulf of Mexico.

- The application would be interoperable with other applications such as in-car navigation or a personal 'travel assistant'.

Intelligent tour guide

This application would provide services to aid tourists or business travellers. Its functionality could include:

- Showing a map, satellite image or a '3D' view of the user's whereabouts with a 'you are here' spot marked.
- Showing the best route to get to a user-specified destination (e.g. hotel, beach).

- Performing the role of a tour guide by describing the surrounding buildings or landscape.
- Showing the location of nearby banks, post offices, hotels, 24-hour chemists, etc.
- Showing a map of sea temperature and water pollution levels for the relevant area of coastline.

Sports assistant
This application would enable outdoor sports enthusiasts to keep track of their geographical position on a satellite image and the geographical position of their team mates. It could also be connected to the emergency services to help them to mount a rescue effort in the event of an accident.

Virtual gaming
This application would enable real-world gaming and simulation. It would be able to track the gamer's position on a satellite image, possibly overlaid on a 3D view. 3G service providers are currently ranking gaming very highly as a potential 'killer application' for 3G.

The above examples give some idea of the potential consumer market for 3G services incorporating EO data. There are, however, a number of supporting developments needed in the geomatics market for this potential to be fulfilled (Marigold and Stuttard 2001):

- More operational satellite missions dedicated to specific information objectives. There are several initiatives that give cause for optimism such as FUEGO, EROS, Quickbird 2 and Ikonos 2.
- Linkage of satellite ground segments into operational information delivery channels beyond the established meteorological services. This is already happening for medium-resolution and high-resolution sources, but more needs to be done to open up access to satellite imagery as soon as possible after it is received.
- Availability of 3G mobile communications in less populated areas: as a minimum in intra-urban and rural areas. Ideally global coverage including wilderness areas, developing countries and oceans is required.
- Development of geospatial 'data brokers' providing on-line, mosaicked geo-information, including satellite and airborne imagery.

Together with Business to Business (B2B) and Business to Government (B2G) applications the future looks bright for the continued development of geomatics market-based GIS products incorporating satellite data.

REFERENCES

Aalders JGL (2001) *Data Searching by Metadata*, GIS, Ostrava.

Barboux J-P (2001) Galileo: is it worth the trouble? *GEOEurope*, April, pp. 40–1.

Barth S, Hansen E and Leben B (1999) Monitoring marine mammals from Colorado. http://www-ccar.Colorado.EDU/~altimetry/applications/whales/

Blakeman RH, Bryon RJ and Dampney P (2000) Assessing crop condition in real time using high resolution satellite imagery. In *Aspects of Applied Biology* **60**, *Remote Sensing in Agriculture*, pp. 163–71.

Clynch JR (2000) Stand alone GPS position errors. *United States Institute for Mathematics and its applications (IMA) 'Hot Topics' Workshop: Mathematical Challenges in Global Positioning Systems (GPS)*, 16–18 August 2000

ESA – European Space Agency (2000) ERS-1/2 Radar Altimeter Wave Form Product (ALT.WAP),
http://earth.esrin.esa.it:81/pgersaltwap

European Commission (1999) *Inventory of Projects with a European Dimension where Satellite Communications are used for Earth Observation Applications*, Report EUR 18675 EN.

European Commission (2000a) *Galileo Definiton Phase Initial Results*, European Commission, Directorate-General for Energy and Transport (DG-TREN).

European Commission (2000b) *Compendium of Space-Technology Applications Projects*. Report EUR 18971 EN.

Fraser D and Smith S (1998) Earth Observation Networks: the virtue of simplicity. In: Strobl J and Best C (eds) *Proceedings of the Earth Observation and Geo-Spatial Web and Internet Workshop, Salzburg*.

von Hünerbein K and Rueter E (2000)
http://www.gpsworld.com/0900/0900contest.html

Ketselides M (2000) *A market Perspective for ASTRON: Benefits for Europe from Integrating Space Technologies*. European Commision, Joint Research Centre, S.P.I. 98.155.

Logica (1998) *Earth Observation for Agribusiness*. European Commission, Centre for Earth Observation (CEO) Programme, Information Paper.

Main JD (2000) Precise ground control is essential for spatial accuracy. *Imaging Notes*, July/August, pp. 6–7.

Marigold G and Stuttard M (2001) *The Potential for Earth Observation in 3G Mobile Telecommunications*. British National Space Centre (BNSC) EO Sector Studies, Final Report.

Mills S, Kjeldsen A and Shipp J (1998) INFEO Search-A. *Proceedings of the Earth Observation and Geo-Spatial Web and Internet Workshop '98*.

Moore T and Roberts GW (1997) Centimetric GPS navigation to the North Pole. *Proceedings of ION-GPS-97, The 10th International Technical Meeting of the Satellite Division of the Institute of Navigation, Kansas City, USA*, September 1997, pp. 1189–96.

Petrie G (1999) Characteristics and applications of high-resolution space imagery. *Mapping Awareness*, November 1999.

Ramapriyan H K and Aitken S (1999) EOSDIS Earth Observing System Data and Information System, EOSDIS website.
http://spsosun.gsfc.nasa.gov/New_EOSDIS.html

Richelson JT (1999) US *Satellite Imagery*. National Security Archive Electronic Briefing Book No. 13.

Save the Elephants (2001) Project web-site. www.savetheelephants.com

Shaffer LR (1992) The Earth Observing System. *Proceedings of the Central Symposium of the 'International Space Year' Conference, Munich, Germany*, 30 March – 4 April 1992, ESA ISY-1 pp. 945–50.

Steven MD and Millar C (1997) Satellite monitoring for precision farm decision support. In Stafford JV (ed.) *First European Conference on Precision Agriculture 1997*. BIOS Scientific Publishers, pp.191–200.

9

Summary and Conclusions

In this final chapter we present a summary of the major themes that have been presented throughout the book, and note a few of the particular success stories where GIS-processed satellite data has played a major role. This is followed by a description of some of the areas where further work may still be needed and concludes with the author's personal viewpoint of future developments.

Summary

The major themes presented throughout this book may be summarised as follows:

- There is a wide range of satellite data available in image format. An image acquired by an optical or radar (SAR) sensor will need to be geocoded into a standard map projection before it can usefully be combined with other data layers. These can include other suitably processed images or location information from the Global Positioning System (GPS).
- Before scientifically useful information can be extracted form satellite images they must be pre-processed to remove distortions introduced by the Earth and its atmosphere. The basic technique for obtaining landcover information from optical images is multispectral classification. Other specialist algorithms are used for extracting information from SAR images, monitoring vegetation and analysing time series of data.
- Satellite data can provide a major contribution to the production of Digital Elevation Models (DEMs) for a wide variety of applications. Digital stereo-matching can be applied to optical or SAR stereo-pairs. Interferometric techniques are also used for SAR 'tandem pairs'.
- The most effective way to process and analyse satellite data is by using a Geographic Information System (GIS). The variety of geospatial data structures and tools in a GIS provides an effective platform for knowledge-based image classification, spatial analysis and visualisation of results.
- Scientific research and the use of GIS techniques are helping to develop a market for land applications of satellite data. These include agricultural monitoring and yield prediction; forestry; geological and civil engineering

applications and telecommunications and media. GPS is also playing a major role in these applications.
- GIS, GPS and satellite images are also being combined for a wide variety of oceanographic, atmospheric and 'frozen world' applications. These make use of dedicated meteorological platforms and specialist instruments such as coastal zone scanners and satellite altimeters.
- Earth observation has a vital role to play in understanding the many environmental challenges facing the planet. These include mapping of vegetation dynamics, modelling global climate change, disaster monitoring and tracking population dynamics. A large number of satellite missions dedicated to environmental monitoring are scheduled for the next decade.
- The use of satellite data is becoming increasingly interwoven with the emerging field of geomatics. Areas of rapid development include Very High Resolution (VHR) images, high-accuracy GPS location, on-line access to geographic databases and the mobile multi-media marketplace.

The reader is encouraged to re-visit any of the preceding chapters to enhance their appreciation of the diversity of applications of 'geographic information from space'.

Success Stories

In this section we take the opportunity to highlight a few of the particular 'success stories' of the last few years, that demonstrate the diversity of global applications of geographic information from space:

- GPS – originally developed for the US military, GPS is now the basis of a mass-market for low-cost, personal location products.
- Meteorological Satellites – as a key component of TV weather forecasting now a part of many people's daily lives. In particular AVHRR, is used for a huge range of global agricultural, forestry and climate modelling applications far beyond its original remit.
- Landsat – for nearly thirty years this programme has been in the vanguard of land cover and environmental applications of satellite images.
- SPOT Image – this organisation has provided the template for the commercial development of remote sensing applications.
- IRS – this programme has demonstrated how a country can define its own satellite missions and successfully apply them to national development issues.
- SAR interferometry – an innovative and highly complex piece of theoretical scientific research which has led to many practical applications.
- Production of DEMs – the use of a variety of satellite data sources and processing techniques is providing an accurate source of digital elevation models around the world.

Underpinning all these success stories is the ubiquitous availability of GIS – ranging from handheld mobile devices to the massive distributed geospatial databases of the EOS Mission.

Work to be Done

It has been over forty years since the launch of the first weather satellite by the United States gave the first hint of the potential of data obtained from Space. Since then there have been enormous strides made in the collection, processing and dissemination of geographic information derived from satellite data. There do however remain a number of areas where ongoing research and development effort continues to be needed, including:

- Information extraction – research into maximising the amount of useful information that can be derived from satellite data, in particular 'decoding' the complicated and interwoven components of SAR images.
- Data fusion – developing techniques for the optimal combination of data from different sources such as SAR and optical images.
- Image understanding – definition of algorithms to automatically extract features, such as crop types, from images with the minimum amount of human intervention .
- Commercial evolution – development of a true mass market for value-added products that is not dependent on government subsidy and intervention.
- Continued improvements in global positioning services to provide higher reliability, more accurate location and better performance in situations with limited satellite visibility.

Conclusions

Finally I would like to conclude with my personal opinion of where the state of the art might be in a few years time. Old Williams' Almanack predictions for 2010 include:

- Earth observation platforms will be characterised by the continuation of 'big science' EOS missions. These will be successfully complemented by quicker, cheaper, more focussed Smallsat missions.
- VHR imagery will be well-established and accepted as a key source of geospatial information however its further development will be constrained by international concerns about 'shutter control' and privacy issues surrounding the use of sub-metre data.
- Constellations of very high resolution radar sensors will lead to the development of accurate, rapid revisit, all weather products for time critical applications such as agriculture and disaster monitoring.
- There will be an established and growing commercial market for image products, but still with some government and international bodies as 'anchor tenants'. The vast majority of these products will be browsed for, ordered and delivered via the internet with a turnaround time of less than a day.
- The use of personal location services, from wristwatch to backpack to car, will be sustained by a combination of precision products from GPS and the newly-available Galileo system. Personal location services will also be a standard component of 'fifth generation' telecoms products.
- Mobile multi-media products will also feature selected information derived

from satellite images, such as weather forecasts and schematic maps; however the delivery of large images to mobile handsets will still be constrained by the bandwidth available.

Most importantly of all, there will be huge leaps in the effective use of information from satellites for global long-term modelling, which will be ever more necessary to determine the destiny of our planet.

Glossary of terms

3G third generation mobile telecommunications
AATSR Advanced Along-track Scanning Radiometer (Envisat)
ACRIM Active Cavity Radiometer Irradiance Monitors (EOS/ACRIMSat)
across-track stereo stereo imagery obtained by sensors pointing to the right and left of the satellite track (in contrast to along-track stereo)
active sensor a sensor (such as SAR) that generates its own energy source
ADEOS-II Advanced Earth Observing Satellite II (Japan)
affine transform a simple linear function used to model the geometric transformation of an image (e.g. for bulk geocoding)
agro-meteorological models mathematical models for forecasting crop growth based on agricultural (e.g. soil type, planting date) and meteorological parameters
AIRS Atmospheric InfraRed Sounder (EOS)
albedo the proportion of light or radiation reflected by the Earth's surface
aliasing phenomenon in SAR interferometery where different signal path lengths (and hence elevations) are associated with the same shift in phase.
along-track stereo stereo imagery obtained by sensors pointing along the satellite track, for example fore and aft (in contrast to across-track stereo)
AMI Active Microwave Instrument (ERS)
AMM Antarctic Mapping Mission
AMSR Advanced Microwave Scanning Radiometer (ADEOS-II)
animation a powerful tool to add the third dimension of time to a two-dimensional map
Aqua an EOS satellite which will provide a multidisciplinary study of the Earth's interrelated processes
arc-node a model of GIS topology where an arc is a sequence of connected vectors which begins and ends in a node (which in turn can be the point of intersection between two or more arcs)
ASAR Advanced Synthetic Aperture Radar (Envisat)
ascending node the point on a spacecraft's orbit at which it crosses the equator from south to north
aspect direction of greatest slope at a location in a digital elevation model

ASTER Advanced Spaceborne Thermal Emission and Reflection (Terra)

ASTRON Applications on the Synergy of Satellite Telecommunications, Earth Observation and Navigation

atmospheric window a spectral band in which atmospheric attenuation of the electromagnetic signal is small thus making passive remote sensing possible

ATSR Along-Track Scanning Radiometer (ERS)

attitude the three-dimensional vector which describes the direction in which a spacecraft platform is pointing

attribute data additional information linking properties to a geographic object, for example the ownership of a field

Aura an EOS satellite which will focus on measurements of atmospheric trace gases and their transformations

autolinear automatically derived linear contrast stretch designed to map the original dynamic range (or a designated percentage thereof) onto the maximum available dynamic range

automated cartography *see* digital cartography

AVHRR Advanced Very High Resolution Radiometer

AVNIR Advanced Visible and Near Infrared Radiometer (ADEOS-II)

azimuth angle of pointing of a satellite or instrument measured from a fixed point in the plane of the horizon

backscatter the proportion of the energy incident on a target re-radiated towards an active microwave instrument

band *see* spectral band

band fabrication the creation of a set of artificial image bands from pre-existing ones

band ratio the division of pixels in one spectral band by those in a second to try to eliminate gross radiometric errors, e.g. Sun-angle effects, or as the basis of a vegetation index

Bezier curve a curved line determined by mathematical parameters which define its endpoints and its curvature

Bhaskara an experimental Indian remote-sensing programme

binary image an image having only two possible pixel values, often used as an overlay for masking

biosphere the part of the planet in which we all live

blending function a function comprising a linear combination of two component functions – typically used to provide a smooth join between two adjacent sections of an image mosaic

blunder detection method of validation based on the trapping of gross errors or physically impossible results

box classifier a simple algorithm used for supervised classification with class values being determined by the range within each spectral band of a training sample (also known as parallelepiped classifier)

brightness the overall intensity (i.e. total reflected energy) of an image pixel, often contained in the first component of an image transform such as the IHS transform or tasseled cap transform

browse image *see* quick-look image
buffer zone in a GIS a region of interest defined as all points within a fixed distance of a given feature, e.g. 10 km of a particular city. Often used as the basis for spatial analysis
bulk geocoding the use of simple models of the satellite imaging process to produce roughly accurate geocoded images
business rules a series of well-defined methods which govern every aspect of GIS operation and ensure that the geographic information is accurate, current, consistent, and reliable

C-band radar frequencies between 4 and 8 GHz
calibration conversion of data values from arbitrary engineering units (such as pixel values) to meaningful geophysical quantities such as temperature
CCD Charge Coupled Device
CEC Commission of the European Communities
centroid the mathematically-defined centre of a shape such as a polygonal vector boundary or a cluster in feature space
CEONet (Canadian Earth Observation Network) an initiative by the Canadian government to a national infrastructure for providing improved access to Earth Observation archives and other complementary geospatial databases
CEOS Committee on Earth Observation Satellites
cerebral processing visual photo-interpretation of image (specifically SAR) products using the 'mark one eyeball'
CERES Clouds and the Earth's Radiant Energy System (Terra)
CFCs chlorofluorocarbons
change image an image representing the change between two members of a multi-temporal image set obtained, for example, by subtracting the oldest from the youngest
channel alternative name for (spectral) band, often used for meteorological images
chip a small extract taken from a reference image used for the automatic relocation of ground control points for subsequent use in geocoding
CIP Catalogue Interoperability Protocol
class sorting the introduction of additional geographic information after an image classification
classification the use of statistical techniques to segment images into different feature classes, usually based on multi-spectral characteristics
classifier modification the introduction of additional geographic information during an image classification
clinometry an experimental technique used to derive height information from a single image based on local grey-scale variation (also known as shape from shading)
clipping calculating where map vectors cross an image boundary
CloudSat an EOS satellite whose primary goal is to provide data for global cloud prediction models and to understand the role of clouds in climate change
cluster the grouping together of points on a scatter diagram (or in a feature space) characteristic of the response from a single feature class

cluster analysis *see* unsupervised classification
clutter man-made and natural features that may affect radio signal propagation
CoastWatch a NOAA programme which makes satellite and associated readings from environmental buoys available to scientists and other users
coherence a measure of the spatial variance of the phase difference between two SAR image acquisitions
colour gamut the range of colours available in a particular medium, e.g. printing or visual display
compositing *see* temporal compositing
confusion matrix a method used to validate a multi-spectral classification by comparison of the class values of a sample of points against their actual classes as determined from ground data
contrast stretch the process of modifying the histogram of an image, usually to enhance visual appearance
convolution filtering of an image by multiplying the neighbours of each pixel by the weights in a kernel and summing the results
CORINE CoORdination of INformation on the Environment
CORONA a US reconnaissance satellite program
cross-polarisation *see* polarisation
cryosphere the 'frozen world'
CTH Cloud Top Height
cubic convolution a resampling algorithm based on a cubic approximation of the ideal interpolator (sinc function)
CZCS Coastal Zone Colour Scanner

DAAC Distributed Active Archive Centre (EOSDIS)
data compression any algorithm for encoding large datasets (typically images) so that less computer storage is required. If the original dataset can be exactly reconstituted (decompressed) the algorithm is known as 'lossless' compression
data model a logical representation of the objects in a database and the relationships between them
datum a (geodetic) datum defines the parameters used in a map projection such as size and shape of the earth, origin and orientation
DBMS Database Management System
DCS Data Collection System (ADEOS)
descending node the point on a spacecraft's orbit at which it crosses the equator from north to south
density slicing an image enhancement algorithm which operates by dividing the image histogram into a number of non-overlapping ranges and allocating a colour to each; can also be used as a very simple method of image classification
depression angle the angle at a radar sensor between horizontal and its range direction
destriping removal of regular stripe patterns in an image caused by miscalibrations within arrays of sensors; such patterns may be aligned vertically or horizontally (or occasionally both)

DFT Discrete Fourier Transform

dielectric constant the property of a material that determines the relative speed that an electrical signal will travel in that material

DIF Directory Interchange Format

difference of gaussian (DoG) a non-linear filter though to be a good approximation of the 'edge detector' employed in human visual system

differential GPS use of correction data generated by a base station, whose position is accurately known, to remove the errors in the position measured by a GPS receiver

digital cartography the drawing of detailed hardcopy maps by computer (also referred to as automated cartography)

digital elevation model (DEM) representation of the topography of the Earth's surface by an array of numbers representing heights above a reference datum

digital map any representation of a map in numerical format, for example storing line-segments as digital vectors

digital parallax model (DPM) an intermediate model used in the production of a digital elevation model. The DPM represents the apparent shift (parallax) in position at a location viewed from two different angles. This can then be converted to the actual height value at the location.

digital terrain model (DTM) often used as a synonym for digital elevation model, a DTM may have additional thematic information, such as land cover classes, as well as height information

digitisation the process of converting information to digital format, specifically the conversion of paper map sheets to vector or raster format

direct image referencing *see* georeferencing

disaster a serious life-threatening situation which has arisen when a hazard has exceeded some critical limit

discovery metadata the minimum amount of information that needs to be provided to answer the key questions about geospatial datasets

doppler shift the change in frequency of a signal (typically radar) due to the motion of the transmitter or the target

DORIS Doppler Orbitography and Radiopositioning Integrated by Satellite (Envisat)

dynamic model a model which represents how an object's state changes with time

dynamic range the range of digital (signal) values present in an image

EAI Environmental Impact Assessment

easting the horizontal direction in a rectangular map coordinate system

elevation the height of a point on the Earth's surface above a fixed datum (as in a digital elevation model; angle of pointing of a satellite above (or below) the plane of the horizon

empirical model a mathematical model based largely on observed parameters rather than rigorous scientific theory

Envisat a satellite mission, developed by the European Space Agency for Environmental monitoring

EOS Earth Observing System

EOSDIS EOS Data and Information System

ephemeris the continuous set of orbital data describing the position and attitude of the satellite platform at any given time

epipolar image geometry meaning that, given a point in the first image of a stereo pair, the match point must lie on a known line in the second image

ERD Entity–Relation Diagram

ERTS-1 Earth Resources Technology Satellite 1, later renamed Landsat 1 (USA)

ESA European Space Agency

ESE Earth Science Enterprise

ESI Environmental Sensitivity Index

ETM+ Enhanced Thematic Mapper Plus (Landsat)

Eumetsat European Meteorological Satellite organisation

EVI Environmental Vegetation Index

false colour the selection of set of image bands whereby terrestrial features are not portrayed in their natural colours

feature code a means of identifying 'layers' of different features in a GIS by giving each a unique code, e.g. feature code 94 = inland waterways

feature dependent symbology GIS display based on feature types or attributes, e.g. using a different size symbol for towns depending on their population

feature space the *n*-dimensional equivalent of a scatter diagram where each dimension represents the pixel value in a particular spectral band

filler data data used to pad-out non-rectangular images so that they may be stored more conveniently

foreshortening the effect (in SAR images) of points on slopes facing towards the transmitter appearing closer together

Fourier transform a mathematical algorithm based on the analysis of a signal into components of different frequencies

frost filter non-linear filter used to reduce speckle noise in a SAR image

Fuego a proposed smallsat constellation which will provide early fire outbreak detection and high resolution fire-line monitoring (Spain)

functional model a model based on the use of algorithms which transform sets of inputs to a physical system to a set of outputs

GAC Global Area Coverage (AVHRR)

GAI Green Area Index

Galileo a proposed European satellite system intended as an alternative and a complement to the United States' GPS and Russian Glonass

GAMBIT a US reconnaissance satellite programme

gamut *see* colour gamut

gaussian The statistical 'bell curve' distribution, also known as the normal distribution

GCM General Circulation Model

GCMD NASA's Global Change Master Directory (GCMD) is a comprehensive directory of descriptions of datasets of relevance to global change research.

generalisation the errors and uncertainties introduced as the result of deriving maps at one scale from those at a different scale

geocoding production of an image where pixel values have been resampled onto a regular grid in a standard map projection the process of attaching spatial-references, such as addresses to database objects

geodatabase, geographic database the component of a GIS where the geographic information is actually stored, also known as a GIS repository

geodetic datum *see* datum

geographic information system (GIS) a system for capturing, storing, checking, integrating, manipulating, analysing and displaying data (such as geocoded images) which are spatially referenced to the Earth

geoid the equipotential surface of the Earth's gravity field that most closely approximates the mean sea surface

geolocation obtaining the position of a device (e.g. mobile phone, combine harvester) in a known coordinate system

geomatics a generic term which covers *inter alia* the disciplines of surveying, GPS, Earth Observation, digital mapping and cartography, location-based services and GIS

geometric correction obsolete term for geometric transformation/geocoding

geometric transformation the process of creating a geocoded image (synonym for geocoding); the matrix used to define the geocoding function

geoprocessing the processing of geographically referenced data

georeferencing producing an image where the transformation to map coordinates is known but may not actually be applied to the image data itself (also known as direct image referencing)

geostationary a satellite whose orbit allows it to remain above a fixed location on the Earth's surface (strictly such a satellite is geosynchronous as its orbit will generally deviate slightly from a fixed point)

GIS repository *see* geodatabase

GLAS Geoscience Laser Altimeter System (EOS)

GLI Global Imager (ADEOS)

Glonass The Russian equivalent to the Global Positioning System (GPS)

GOMOS Global Ozone Monitoring by Occultation of Stars (Envisat)

GPS Global Positioning System

GRACE Gravity Recovery and Climate Experiment (EOS)

ground control point (GCP) a point whose position is accurately known on an image and in map coordinates

ground segment the part of an Earth Observation mission comprising date reception, processing, archiving and distribution facilities, as opposed to the space segment

Gruen's algorithm an adaptive least squares correlation algorithm used extensively for feature matching

hazard a condition capable of exerting adverse effects on a human life, property or activity

Hexagon a US reconnaissance satellite programme

histogram equalisation a form of contrast stretch where the histogram of the original image is adjusted (equalised) to have a uniform distribution

histogram matching a form of contrast stretch where the histogram of one image is modified to correspond to that of a reference image (used extensively for mosaic production)

homogeneous referring to an image area having little variation in value between neighbouring pixels (i.e. lacking in texture)

Hough transform an image transform used to detect features whose shape can be expressed as parameterised equations such as straight lines and circles

HRV High Resolution Visible (SPOT)

hyperspectral an instrument having a much larger number (typically hundreds) of very narrow spectral bands, thus providing a finer granularity of spectral discrimination

ICESat an EOS 'smallsat' mission flying the Geoscience Laser Altimeter System (GLAS)

IfSAR *see* SAR Interferometry

IHS (Intensity-Hue-Saturation) Transform An algorithm used to convert image bands from multi-spectral (RGB) space to the more intuitive HIS space. The intensity (I) component is related to the overall brightness of a pixel, hue (H) to its colour and saturation (S) to its purity.

IIPF Interactive Image Processing Facility – the 1990's precursor to the current generation of desktop GIS

Ikonos a VHR satellite capable of acquiring high-resolution multi-spectral and panchromatic images

ILAS-II Improved Limb Atmospheric Spectrometer II (ADEOS)

image a data structure storing intensity or other values as a regular array of pixels (also known as a raster, or raster image)

image map a product generated from a geocoded satellite image together with additional cartographic information

image processing any processing (such as contrast stretch, edge enhancement) performed on data in image formats

image registration an 'unmodelled' alternative to geocoding where no map information is available. The two images are directly overlaid by the use of pairs of tie-points

image repair substitution of new data values for missing or inaccurate data in an image, one example being destriping

image stratification *see* stratification

image understanding the use of knowledge-based and other advanced tools to obtain detailed semantic information about the real-world scene portrayed in an image

incidence angle in SAR images the angle on the ground between vertical and the range (the complement of the depression angle)

INFEO Information on Earth Observation, a descendant of the European-Wide Service Exchange providing a catalogue service for Earth Observation datasets
infomobility Any technology capable of providing information to 'people on the move'
InSAR *see* SAR Interferometry
instantaneous field-of-view (IFOV) the portion of the Earth's surface that can be seen at any one instant by a satellite instrument
interferogram an image representing the phase difference between two SAR images of the same area
interpolation any one of a number of algorithms used for image resampling, e.g. nearest neighbour, cubic convolution; the estimation of intermediate values from known data points
intervisibility analysis the determination of which areas can be seen from a given point
ionosphere the part of the upper atmosphere where free electrons occur in sufficient density to affect the propagation of radio signals
IR Infrared
IRS Indian Remote-sensing Satellite
ISODATA Iterative Self-Organising Data Analysis (algorithm for unsupervised classification)
iso-doppler lines points of the Earth's surface with constant doppler frequency for a particular SAR image

Jason a joint USA–France mission designed to monitor global ocean circulation
JERS Japanese Earth Resources Satellite
jitter high-frequency attitude variation in the orbit of a satellite

kernel The matrix of numbers used to define a filtering (convolution) operation on an image
KH Keyhole (US reconnaissance satellites)
kinematic GPS a post-processing technique, whereby GPS data is collated over a period of several hours at static sites to provide a highly accurate measurement of location

L-band Radar frequencies between 1 and 2 GHz
LAC Local Area Coverage (AVHRR)
LACIE Large Area Crop Inventory Experiment
LAI Leaf-Area Index
Lambertian reflector An optical source modelled by Lambert's cosine law, i.e. having an intensity directly proportional to the cosine of the angle from which it is viewed
land cover a description of the surface type at a location on Earth, e.g. the type of vegetation growing in an area
Landsat a series of Earth Observation satellites starting with 'Earth Resources Technology Satellite' (ERTS-1) in 1972 and continuing to the present day (USA)

land use the purpose to which an area of land is put. This is distinct from land cover, e.g. a land cover of grass could be used for cattle grazing or as a golf course.
Laplacian filter a non-directional linear filter used to enhance edges in an image
layer stack a data structure consisting of a large number of images and GIS data layers of the same region to which a spatial algorithm, e.g. a multi-spectral classification is applied simultaneously
layover the extreme case of foreshortening in SAR images where the terrain slope is greater than that of the incident signal
LEA location enabled applications
Lee filter a non-linear filter used to reduce speckle noise in a SAR image
LFC Large Format Camera (LFC)
lidar Light Detection and Ranging
lineament mapping the use of image-processing algorithms to detect linear features in images, particularly for geological applications
linear interpolation the estimation of values intermediate to two data points by using a straight-line approximation to the function being interpolated
line spread function (LSF) one-dimensional analogue of a point spread function
LISS Linear Imaging Self-scanned Sensor (IRS)
location-based services (LBS) provision of consumer services based on a user's position e.g. 'Where is my nearest restaurant?'. Also known as location enabled applications (LEA)
look-up table (LUT) a table of numbers specifying a transfer function to convert input data values to required output values, e.g. as defined in a contrast stretch.

MARS Monitoring of Agriculture by Remote Sensing (EC project)
masking use of a binary image to define which parts of an image are to be included in an image map or mosaic (for example, to remove areas of sea or cloud)
maximum likelihood classifier an algorithm used for supervised classification with class values being determined by the probability that a point lies within the classes defined by a training sample
MERIS MEdium Resolution Imaging Spectrometer (Envisat)
metadata a term used to describe the summary information or characteristics of a set of data ('data about data').
Meteosat a geostationary weather satellite (Eumetsat)
mid-infrared (MIR) the portion of the electomagnetic spectrum between 3.5 and 5.0 μm available for remote sensing
mie scattering atmospheric scattering which occurs when the wavelengths of visible light are approximately equally scattered
minimum distance classifier algorithm used for supervised classification with class values being determined by the nearest class centroid defined by the training sample
MIPAS Michelson Interferometer for Passive Atmospheric Sounding (Envisat)
MISR Multi-angle Imaging Spectro-Radiometer (Terra)
Misregistration The misalignment of supposedly coincident points in a multi-temporal image set caused by inaccuracies in the geocoding process

mixel a pixel corresponding to an area on the ground with a mixture of land cover type
mobile GIS a device that can be used for collection and dissemination of GIS data 'in the field'
MODIS MODerate resolution Image Spectroradiometer (Terra)
MOPITT Measurements of Pollution in The Troposphere (Terra)
mosaic a combination of two or more geocoded images to form a larger image product
MSS Multi-spectral Scanner (Landsat)
multipath the corruption of the direct GPS signal by one or more signals reflected from the local surroundings
multi-temporal analysis analysis of an area using collocated images acquired at different times. Multitemporal analysis can be subdivided into (1) change detection (2) multi-temporal classification
multi-temporal classification the inclusion of images from more than one date in a classification set in order to achieve more accurate results, for example by taking into account differences in spectral response during a crop's growing cycle
MURBANDY Monitoring Urban Dynamics
MWR MicroWave Radiometer (Envisat)

nadir referring to an instrument pointing directly downwards from a satellite platform
NASA National Aeronautics and Space Administration (USA)
NASDA National Space Development Agency (Japan)
navigation (1) sometimes used as a synonym for geocoding, (2) the process of route-finding using GPS locations
NDVI Normalised Difference Vegetation Index
near-infrared (NIR) The portion of the electomagnetic spectrum between 1.0 and 2.4 μm available for remote sensing
neatline a border line drawn around the extent of a map.
neural network a non-algorithmic computational paradigm consisting of a number of interconnected 'neurons'. The output of a neuron is determined by a function which combines the values of all the inputs connected to that neuron
NIMA United States National Imagery and Mapping Agency
El Niño a disruption of the ocean atmosphere system in the tropical Pacific having important consequences for weather around the globe
NOAA National Oceanographic and Atmospheric Administration (USA)
northing The vertical 'y' direction in a rectangular map coordinate system
NSCAT NASA Scatterometer

object model a static model of an object's structure and relationships
OCR Optical Character Reader
orthoimage a precison geocoded image fully corrected for terrain effects
orthoradar an orthoimage generated from SAR data

PA Panchromatic (SPOT) also referred to as SPOT-PAN
PAF Processing and Archive Facility (ERS)
panchromatic an image having a single spectral band, also referred to as monochrome
parallelepiped classifier *see* box classifier
passive sensor a sensor that uses an external energy source such as the Sun (contrast active sensor)
PDA Personal Digital Assistant
PDO Pacific Decadal Oscillation
percentile in a (cumulative) image histogram the value below which a certain percentage of the image values lie e.g. The 90th percentile is the value below which 90 per cent of the image values lie. The 50th percentile is also known as the median and the 25th and 75th percentiles the lower-quartile and upper-quartile respectively
perigee the point nearest the Earth in a satellite's orbit
petabyte 1024 terabytes or roughly 10^{15} bytes
phase two waves are considered in phase if they are perfectly superimposable. Otherwise they are out of phase
phase aliasing *see* aliasing
photogrammetry the use of photography for surveying, specifically the measurement of elevations on the earth's surface
PICASSO-CENA Pathfinder Instruments for Cloud and Aerosol Spaceborne Observations – Climatologie Etendue des Nuages et des Aerosols (EOS)
Piecewise linear a function composed of a number of straight-line segments (i.e. function is linear between pairs of defined points) often used to provide a customised contrast stretch
pixel a picture element the smallest component of an image. Also referred to (in the case of SAR images) as a resolution cell
pixel size the terrestrial area represented by a single pixel in a satellite image
POES Polar Orbiting Environmental Satellites (Eumetsat)
point spread function (PSF) a two-dimensional function characterising the response of a sensor to the area surrounding the point on which it is centred
polarisation when a radar transmits microwaves, they are polarised into a vertical (V) or horizontal (H) plane. The antenna also detects the scattered signal in a particular polarisation this may differ from that transmitted. This gives four possible polarisations referred to as HH, VV, HV and VH. The last two are referred to as cross-polarisations
POLDER POLarisation and Directionality of the Earth's Reflectance (ADEOS-II)
precision farming the application of advanced spatial technology such as GIS, Earth Observation, GPS and yield sensors to improve agricultural productivity
precision geocoding geocoding using detailed models of satellite geometry and orbit, digital elevation models, etc.; to produce an accurate transformation of an image to a map projection
pre-processing any preliminary processing such as contrast stretch, image repair performed on the radiometric values of an image prior to geocoding and subsequent product generation

Presidential Directive 23 issued in March 1994, authorised the development of commercial satellites producing imagery to the 1 metre ground pixel level
Principal Components Transform (PCT) an algorithm which produces a decorrelated linear combination of the original image bands
process validation a method of validation using the mathematical model of a particular process to also generate quality control parameters as an intrinsic part of the process
product validation a method of validation based on quality control of the final output product against expected accuracy standards
proximity map an image where the values represent the distance of each grid point from features of interest such as roads or urban areas
pseudo-colour the mapping of the grey-level values of a panchromatic image onto specific colours for presentation purposes. Often used in conjunction with density slicing
pushbroom sensor a linear array of charge-coupled devices (CCDs). The two-dimensional image is acquired by the motion of the satellite relative to the ground, like a broom sweeping across a floor
PVI Perpendicular Vegetation Index

quadtree an alternative method for storage of raster images, based on the successive division or the image into quarters until a hierarchy of homogeneous blocks is created
query language a form of interactive computer language used to ask questions of a computer database. An example is Sequential Query Language (SQL)
quick-look image a simplified image used to indicate the overall appearance and quality of an image product. Also referred to as a browse image or thumbnail
QuikScat Quick Scatterometer (USA)

RA Radar Altimeter
radargrammetry photogrammetric processed carried out using stereo pairs of SAR images
radarmaps image maps or mosaics derived from SAR images
Radarsat Earth Observation satellite flying SAR instruments (Canada)
radiometric correction the alteration of image values to account for Sun-angle, atmospheric attenuation and other illumination effects
range the distance between a remote sensing instrument (such as SAR) and its terrestrial target
raster, raster image *see* image
ratio *see* band ratio
Rayleigh scattering the selective scattering of shorter wavelengths (blue and violet light) by atmospheric gases
RBV Return Beam Vidicon (Landsat)
red lines markings on a map to indicate areas where the map is out-of-date and needs changing. Now used to refer to annotations on a GIS layer where updates are required to the geographic database

repeat cycle the time taken by an orbiting satellite between revisits to a particular location

resampling the process of deriving suitable values for a geocoded image pixel from the original image data values

residual vector a measure of the accuracy of the geocoding process defined as the difference between a GCP's true position and that given by the calculated transformation

resolution the size of a target that a particular sensor is able to discriminate (which may be variable depending on target characteristics, constrast with background etc.). Often used loosely as a synonym for pixel size or intermediate field-of-view

resolution cell *see* pixel

RGB Red Green Blue

RIS Retro-reflector In Space (ADEOS-II)

RMS Root Mean Square (error)

ROI Region Of Interest

S-band Radar frequencies between 1.55 and 3.9 GHz

SA Selective Availability (GPS)

SAGE Stratospheric and Atmospheric Gas Experiment (Russia)

SAR Synthetic Aperture Radar

SAR Interferometry a method of obtaining height information, based on the difference in phase of two SAR signals originating from two slightly different positions. Also referred to as InSAR or IfSAR

saturation (1) when the actual intensity associated with a pixel value exceeds the dynamic range of the sensor or processing system; (2) the 'purity' component of an IHS transform

ScaRaB Scanner for the earth's Radiation Budget (Envisat)

scatter diagram a two-dimensional plot of values in an image where the x-axis represents all the values in band 1 and the y-axis represents the corresponding values in band 2

scatterometer a high-frequency microwave radar used to measure ocean near-surface wind speed and direction.

SCIAMACHY Scanning Imaging Absorption Spectrometer for Atmospheric Cartography (Envisat)

Seasat pioneering oceanographic satellite (USA)

SeaWiFS Sea-viewing Wide Field-of-view Sensor

SeaWinds scatterometer (ADEOS-II)

seed point location used as the start for various search algorithms, such as the position of a ground control point

segmentation division of an image into discrete 'objects' on the basis of their characteristics in an image (e.g. colour, texture, edges) or by using external information (e.g. map-guided segmentation)

shadowing effect apparent in SAR images when the local slope of the terrain prevents certain areas from being imaged

shape from shading *see* clinometry
simulated annealing mathematical algorithm for finding a global minimum in a system based on an analogy with the way in which a metal cools and freezes into a minimum energy crystalline structure
sinc function the ideal mathematical interpolation function
SIR Shuttle Imaging Radar
slant range SAR image format in which one axis represents the distance from the satellite position to the target, whilst the orthogonal axis represents the doppler shift
SLAR Sideways Looking Airborne Radar
sliver polygon a spurious polygon obtained when the common boundary between two areas on a map is digitised twice
smallsat 'smaller, faster, cheaper' missions
sobel filter a directional linear-filter kernel for detecting edges in an image
SORCE Solar Radiation and Climate Experiment (EOS)
Space Oblique Mercator (SOM) a map projection defined in terms of the satellite position and direction when an image was acquired. The SOM is a conformal mapping along the ground track of the satellite and uses as axes the along-track and across-track directions
space segment the part of an Earth Observation mission comprising the satellite platform and instruments, as opposed to the ground segment
spatial analysis deriving information based on the topology of a GIS dataset
speckle coherent noise caused by phase variation in a SAR image
spectral band a specific range of wavelengths within the electromagnetic spectrum
spectral signature the characteristic amount of incident energy reflected by a particular surface type as a function of wavelength used as the basis of multi-spectral classification
specular reflection perfect 'mirror-like' reflection of radar signals from smooth surfaces such as lakes. Because little of the incident energy is scattered back towards the receiver such features will appear dark in a SAR image
SPOT Satellite Pour Observation de la Terre
stereo matching any algorithm or technique used to obtain height information from the parallax information contained within two images acquired at different angles (a stereo pair)
stereo pair *see* stereo matching
stereoplotter optical device for the plotting of contours from a photographic stereo pair
SRTM Shuttle Radar Topography Mission
SST Sea Surface Temperature
Stratification the introduction of additional geographic information prior to an image classification, e.g. to divide the area on the basis of height ranges
Sun-synchronous a satellite orbit which crosses the equator at the same local time each day
swiping image comparison technique where two image are shown either side of a line on a GIS display moved by the user

SWIR Short Wave Infrared
sigma-nought (σ_0) another term for the amount of backscatter in a SAR image
spatial index a method for storing objects in a geographic database that provides efficient computer access using their spatial location
symbology the combination of colours, line-styles, icons and so on used to provide a consistent interpretation of a map

tabular data *see* attribute data
tandem pairs SAR images of an area acquired by two satellites following the same orbit. Because the second image is acquired very shortly after the first they are particularly suited for SAR interferometry
target a terrestrial feature present in an image, usually in the context of radar
tasseled cap an algorithm similar to the principal components transform. Instead of deriving the coefficients of the transformation statistically typical values which correspond to attributes of land cover ('brightness', 'greenness' and 'wetness') are used
temporal compositing the averaging of values in adjacent members of a multi-temporal image set (typically AVHRR) in order to reduce the effects of cloud cover and atmospheric variation
terabyte 2^{40} (1,099,511,627,776) bytes. This is approximately 1 trillion bytes, hence the name
Terra the Terra satellite is the 'flagship' of EOS. It provides global data on the state of the atmosphere, land, and oceans and their interactions with solar radiation
texture generic term applied to a number of statistical measures of the variability between a pixel and its neighbours
thermal infrared (TIR) the portion of the electromagnetic spectrum between 8.0 and 22.0 μm available for remote sensing
thumbnail *see* quick-look image
tie-point a point which can be identified in two different images used for image-to-image registration. Similar to a ground control point except that the true map coordinate need not be known
TIN Triangulated Irregular Network
TLA Three Letter Acronym
TM Thematic Mapper (Landsat)
TOMS Total Ozone Mapping Spectrometer (ADEOS-II)
topology the spatial relationship between geographic objects, for example 'is contained within', 'is adjacent to'
training sample a representative sample of known feature types to be used as the basis of multi-spectral classification
Triana a mission to investigate the relationship between solar radiation and climate (EOS)
TRMM Tropical Rainfall Measuring Mission (EOS)
troposphere the portion of the atmosphere that lies between the Earth's surface and an altitude of approximately 15 km
TVI Transformed Vegetation Index

unsupervised classification multi-spectral classification performed without the use of training samples and based solely on the multi-variate statistical distribution of the image pixels (also known as cluster analysis)
unwrapping in SAR interferometry the algorithm used to convert phase differences in the signal into height values
UoSAT a series of smallsat missions (UK)
USGS United States Geological Survey
UTM Universal Transverse Mercator
UV ultraviolet

validation any method used to confirm the accuracy and reliability of a processing chain. Validation may be subdivided into blunder detection, process validation and product valdation.
VCL Vegetation Canopy Lidar (EOS)
vector format digital representation of map features as a series of straight-line segments together with associated attribute data
Vegetation instrument/mission providing global daily monitoring of terrestrial vegetation (SPOT)
vegetation index any algorithm used to transform image values into a more direct measure of the vegetation content of a pixel. Often based on a comparison of the response in the red portion of the spectrum to that in the near infrared
VHR Very High Resolution
VIS Visible band
visualisation depicting three-dimensional spatial information as well as temporal information
VSI Vegatative Sponge Index

windowing selection of map vectors that lie within a given boundary
world reference system (WRS) the path/row convention used for the designation of image location in terms of the repetitive orbit of a satellite (the path) and the distance along a particular orbit (the row)
WV Water Vapour

X-band Radar frequencies between 8 and 12 GHz
XS eXtended Spectrum (SPOT)

zero doppler the point of no frequency shift in SAR processing

Index